Studies in Computational Intelligence

Volume 546

Series editor

Janusz Kacprzyk, Polish Academy of Sciences, Warsaw, Poland
e-mail: kacprzyk@ibspan.waw.pl

For further volumes:
http://www.springer.com/series/7092

About this Series

The series "Studies in Computational Intelligence" (SCI) publishes new developments and advances in the various areas of computational intelligence—quickly and with a high quality. The intent is to cover the theory, applications, and design methods of computational intelligence, as embedded in the fields of engineering, computer science, physics and life sciences, as well as the methodologies behind them. The series contains monographs, lecture notes and edited volumes in computational intelligence spanning the areas of neural networks, connectionist systems, genetic algorithms, evolutionary computation, artificial intelligence, cellular automata, self-organizing systems, soft computing, fuzzy systems, and hybrid intelligent systems. Of particular value to both the contributors and the readership are the short publication timeframe and the world-wide distribution, which enable both wide and rapid dissemination of research output.

Nik Bessis · Ciprian Dobre

Editors

Big Data and Internet of Things: A Roadmap for Smart Environments

Editors
Nik Bessis
School of Computer Science
    and Mathematics
University of Derby
Derby
UK

Ciprian Dobre
Department of Computer Science
University Politehnica of Bucharest
Bucharest
Romania

ISSN 1860-949X   ISSN 1860-9503   (electronic)
ISBN 978-3-319-34481-2   ISBN 978-3-319-05029-4   (eBook)
DOI 10.1007/978-3-319-05029-4
Springer Cham Heidelberg New York Dordrecht London

Printed on acid-free paper

Springer is part of Springer Science+Business Media (www.springer.com)

Foreword

The amount of data around us is growing exponentially and the leaps in storage and computational power within the last decade underpin the unique selling proposition of Big Data of being able to provide better insights into various business processes or everyday life in a way that was not possible in the past. By analyzing large data sets and discovering relationships across structured and unstructured datasets, it is the driving force behind business analytics and marketed as a significant opportunity to boost innovation, production, and competition. As such, it is no surprise that Big Data is a booming topic of interest not only in the scientific community but also in the enterprise world.

The increasing volume, variety, and velocity of data produced by the Internet of Things (IoT) will continue to fuel the explosion of data for the foreseeable future. With estimates ranging from 16 to 50 billion Internet connected devices by 2020, the hardest challenge for large-scale, context-aware applications and smart environments is to tap into disparate and ever growing data streams originating from everyday devices and to extract hidden but relevant and meaningful information and hard-to-detect behavioral patterns out of it. To reap the full benefits, any successful solution to build context-aware data-intensive applications and services must to be able to make this valuable or important information transparent and available at a much higher frequency to substantially improve the decision making and prediction capabilities of the applications and services.

Unlike most big data books, this book *Big Data and Internet of Things: A Roadmap for Smart Environments* focuses on the foundations and principles of emerging Big Data technologies, and on creating an integrated collective intelligence in the areas of data-intensive, context-aware applications for smart environments. It touches three themes:

1. *Foundations and principles*, discussing big data platforms, management systems, and architectures for the IoT and analytics.
2. *Modelling*, covering various aspects on big data models, context discovery, and metadata management for a.o. the intelligent transportation and healthcare vertical domains.
3. *Applications*, providing a variety of data analytics use cases in the area of cloud-based M2M systems, social networks and media, logistics, disaster evacuation, and virtual reality.

This book provides timely and up-to-date information about the state-of-the-art and best practices in this area, and will help both specialists and non-specialists understand the various technical and scientific advances and challenges, and apply proposed solutions to their own problems. I am sure you will enjoy reading this book.

December 12, 2013   Davy Preuveneers

Preface

Introduction

Data-intensive computing is now starting to be considered as the basis for a new, fourth paradigm for science. Two factors are encouraging this trend. First, vast amounts of data are becoming available in more and more application areas. Second, the infrastructures allowing to persistently store these data for sharing and processing are becoming a reality. This allows unifying knowledge acquired through the three past paradigms for scientific research (theory, experiments, and simulations) with vast amounts of multidisciplinary data. The technical and scientific issues related to this context have been designated as "Big Data" challenges and have been identified as highly strategic by major research agencies.

On the other hand, the combination of the Internet and emerging technologies such as near-field communications, real-time localization, and embedded sensors, transform everyday objects into smart objects capable to understand and react to their environment. Such objects enable new computing applications, and are at the base of the vision of a global infrastructure of networked physical objects known today as the Internet of Things (IoT). The vision of an Internet of Things built from smart objects raises several important research questions in terms of system architecture, design and development, and human involvement. For example, what is the right balance for the distribution of functionality between smart objects and the supporting infrastructure? Can Cloud Computing and Big Data become true infrastructure-enablers for IoT? During the past few years, IT has seen several major paradigm shifts in technology. These shifts have been primarily in the areas of cloud computing, Big Data, mobility, and the Internet of Things. The convergence of these four areas already creates a new "platform" for enterprises to develop new business and mission capabilities. These capabilities will enable a more integrated view of IT architecture which we explore in this book.

The book presents current progress on challenges related to Big Data management by focusing on the particular challenges associated with this convergence, presenting current progresses and challenges associated with context-aware, data-intensive applications and services. A representative application category is that of Smart Environments such as Smart Cities, which cover a wide range of needs related to public safety, water and energy management, smart buildings, government and agency administration, transportation, health, education, etc. This application class is the subject of many R&D activities and a priority for several government strategic plans for the forthcoming years. Context-aware, data-intensive applications have specific requirements including but not limited to real-time data processing, data access patterns (frequent, periodic, ad-hoc, inter-related, location, device etc.), QoS, intelligent queries, etc.

During the last years, large storage infrastructures were continuously improved in response to the increasing demands of the data intensive applications. Many works address issues related to data intensive scientific applications and Web applications in Cloud environments. In scientific applications, big data volumes collected from high throughput laboratory instruments, observation devices, LHC, and others are submitted to complex software packages for Monte Carlo simulations, extensive analyses, data mining, etc. They are also used for generating graphical outputs and visualization, and are archived in digital repositories for further use. A central topic of research is the adoption of parallel processing solutions able to respond to scalability and performance requirements. One example is the application of the MapReduce model, used initially for the processing of large volumes of textual data, for High Energy Physics data analyses and Kmeans clustering. Web applications are interactive and based on unstructured data that are processes in a write-once read-many-times manner. Research issues include the scalability, fault tolerance, and adaptability to simple data models used by Web applications. For example, the distributed storage system Bigtable is designed to scale to a very large data size. It responds to a variety of workload profiles, from throughput-oriented batch-processing jobs to latency-sensitive data delivery to end-users. Dynamo is a highly available key-value storage system that supports retrieving data by primary key. It has a clearly defined consistency window and efficiently use its resources.

Other applications with high demands for Big Data services are context-aware. Some of their requirements are similar to those of scientific and Web applications. Other needs are shared with mobile computing applications (such as battery life and limited bandwidth) and are due to the presence of mobile devices (sensors, mobile phones) in context-aware applications. Context-aware paradigm is at the base of developing many applications that include people, economy, mobility, and governance. They enrich the urban environment with situational information, which can be rapidly exploited by citizens in their professional, social, and individual activities to increase city competitiveness. In addition, user-centric approaches to services can improve the core urban systems (such as public safety, transportation, government, agency administration and social services, education, healthcare) with impact on quality of life. They allow users interact with others (people), services, and devices in a natural way yet maintaining their normal flow of activities in the real world. Context-aware applications require and use large amounts of input data, in various formats, collected from sensors or mobile users, from public open data sources, or from other applications. While several applications might share the same business and context data, the access and processing profiles can be different from one application to another. For example, public safety applications should react rapidly to context changes to deliver real-time alerts for potential dangerous situations. Collaborative diagnosis allows physicians work together to combine their expertise for complicated diseases and lets them access more information to make a convenient assessment.

Thus, an aim of this book is to become a state-of-the-art reference discussing progress made, as well as prompting future directions on the theories, practices, standards, and strategies that are related to the emerging computational technologies and their association with supporting the Internet of Things advanced functioning for organizational settings including both business and e-science. Apart from interoperable and intercooperative aspects, the book deals with a notable opportunity namely, the current trend in which a collectively shared and generated content is emerged from Internet end-users. Specifically, the book presents advances on managing and exploiting the vast size of data generated from within the smart environment (i.e. smart cities) toward an integrated, collective intelligence approach. We believe interfunctionality between e-infrastructures will enable the store and generation of a vast amount of data, which if combined and analyzed through an agile, synergetic, collaborative, and collective intelligence manner will make a difference in the organizational settings and their user communities.

The book also presents methods and practices to improve large storage infrastructures in response to increasing demands of the data-intensive applications. We address here the category of context-aware applications, which have specific requirements and a special profile. Context data can be used not only to accurately understand the semantics of business data in the benefit of applications, but it can be exploited for improving the performance and facilitate the management of Big Data store service. Of particular interest are the features of a Big Data store Cloud service able to support context-aware applications and their applicability in bringing the Internet of Things to the functionality of an integrated collective intelligence approach on the future Internet.

Who Should Read the Book?

The content of the book offers state-of-the-art information and references for work undertaken in the challenging areas of IoT and Big Data. The book is intended for those interested in joining interdisciplinary and transdisciplinary works in the areas of Smart Environments, Internet of Things and various computational technologies for the purpose of an integrated collective computational intelligence approach into the Big Data era. These tools include but not limited to the use of distributed data capturing, crowd sourcing, context awareness and event the data-driven paradigm, integration and their analytics and visualizations.

Thus, the book should be of particular interest for:

Researchers and doctoral students working in the area of IoT sensors, distributed technologies, collective and computational intelligence, primarily as a reference publication. The book should be also a very useful reference for all researchers and doctoral students working in the broader fields of computational and collective intelligence, emerging and distributed technologies.

Academics and students engaging in research informed teaching and/or learning in the above fields. The view here is that the book can serve as a good reference offering a solid understanding of the subject areas.

Professionals including computing specialists, practitioners, managers, and consultants who want to understand and realize the aspects (opportunities and challenges) of using computational technologies for the Big Data coupled with Internet of Things advanced functioning and its integrated collective intelligence approach in various organizational settings such as in smart environments.

Book Organization and Overview

The book contains 19 self-contained chapters that were very carefully selected based on peer review by at least two expert and independent reviewers. The book is organized into three parts according to the thematic topic of each chapter. The following three parts reflect the general themes of interest to the IoT and Big Data communities.

Part I: Foundations and Principles

The part focuses on presenting the foundations and principles of on-going investigations, and presents an analysis of current challenges and advances related to Big Data management. The part consists of seven chapters. In particular:

"Big Data Platforms for the Internet of Things" discusses the challenges, state of the art, and future trends in context-aware environments (infrastructure and services) for the Internet of Things. It makes a critical analysis of current opportunistic approaches using the elements of a newly defined taxonomy. The chapter reviews state-of-the-art Big Data platforms, from an Internet of Things perspective.

"Improving Data and Service Interoperability with Structure, Compliance, Conformance and Context Awareness" revisits the interoperability problem in the IoT context, focusing on structure, compliance, conformance, and context-awareness particular challenges. The chapter proposes an architectural style, structural services, which combines the service modelling capabilities of SOA with the flexibility of structured resources in REST.

"Big Data Management Systems for the Exploitation of Pervasive Environments" focuses on Big Data management systems for the exploitation of pervasive environments. The chapter presents an original solution that uses Big Data technologies for redesigning an IoT context-aware application for the exploitation of pervasive environment addressing problems and discussing the important aspects of this solution.

"On RFID False Authentications" presents security challenges in the Internet of Things. Within this context, the chapter offers a state-of-the-art review of existing authentication protocols. The authors give a necessary and sufficient condition for false authentications prevention, and propose a naive semaphore-based solution which revises the pattern by adding semaphore operations so as to avoid false authentications.

"Adaptive Pipelined Neural Network Structure in Self-Aware Internet of Things" is concerned with self-healing systems that can auto-detect, analyze, and fixe or reconfigure issues associated with their self-behavior and performance. Self-healing processes should occur in real-time to restore the desired functionality as soon as possible. Adaptive neural networks are proposed as a solution to some of these challenges; monitoring the system and environment, mapping a suitable solution and adapting the system accordingly.

"Spatial Dimensions of Big Data: Application of Geographical Concepts and Spatial Technology to the Internet of Things" highlights the Spatial Dimensions of Big Data, and presents a case study on the application of geographical concepts and spatial technology to the Internet of Things and Smart Cities. By applying spatial relationships, functions, and models to the spatial characteristics of smart objects and the sensor data, the flows and behavior of objects and people in Smart Cities can be efficiently monitored and orchestrated.

"Fog Computing: A Platform for Internet of Things and Analytics" discusses another potential foundation of IoT platforms: fog computing. The chapter examines disruptions due to the explosive proliferation of endpoints in IoT, and proposes a hierarchical distributed architecture that extends from the edge of the network to the core called Fog Computing. In particular, the chapter discusses a new dimension that IoT adds to Big Data and Analytics: a massively distributed number of sources at the edge.

Part II: Advanced Models and Architectures

This part focuses on presenting theoretical and state-of-the-art models, architectures, e-infrastructures, and algorithms that enable the inter-cooperative and inter-operable nature of the Internet of Things for the purpose of collective and computational intelligence, in the Big Data context, and consists of six chapters. In particular:

"Big Data Metadata Management in Smart Grids" focuses on the Big Data management in Smart Grids. In such environments, data are collected from various sources, and then processed by different intelligent systems with the purpose of providing efficient system planning, power delivery, and customer operations. Three important issues in managing and solutions to overcome them are discussed. Concrete examples from the offshore wind energy are used to demonstrate the solutions.

"Context-Aware Dynamic Discovery and Configuration of *Things* in Smart Environments" proposes the Context-Aware Dynamic Discovery of Things (CADDOT) model. A tool is presented, SmartLink, used to establish direct communication between sensor hardware and cloud-based IoT middleware platforms. The prototype tool is developed on an Android platform. Global Sensor Network (GSN) is used as the IoT middleware for a proof-of-concept validation.

"Simultaneous Analysis of Multiple Big Data Networks: Mapping Graphs into a Data Model" proposes a new model to map web multinetwork graphs in a data model. The result is a multidimensional database that offers numerous analytical measures of several networks concurrently. The proposed model also supports real-time analysis and online analytical processing (OLAP) operations, including data mining and business intelligence analysis.

"Toward Web Enhanced Building Automation Systems" reviews strategies to provide a loosely coupled Web protocol stack for interoperation between devices in a building. Making the assumption of seamless access to sensor data through IoT paradigms, the chapter provides an overview of some of the most exciting enabling applications that rely on intelligent data analysis and machine learning for energy saving in buildings.

"Intelligent Transportation Systems and Wireless Access in Vehicular Environment Technology for Developing Smart Cities" focuses on Intelligent Transport Systems (ITS) and wireless communications as enabling technologies for Smart Cities. After reviewing main advances and achievements, the chapter highlights major research projects developed in Europe and the USA. The chapter presents the main contributions that ITS can provide in the development of Smart Cities as well as the future challenges.

"Emerging Technologies in Health Information Systems: Genomics Driven Wellness Tracking and Management System (GO-WELL)" presents GO-WELL, an example of future personal health record (PHR) concept. GO-WELL is based on clinical envirogenomic knowledge base (CENG-KB) to engage patients for predictive care. The chapter describes concepts related to the inclusion of personalized medicine, omics revolution, incorporation of genomic data into medical decision processes, and the utilization of enviro-behavioral parameters for disease risk assessment.

Part III: Advanced Applications and Future Trends

The part focuses on presenting cutting-edge Internet of Things related applications to enable collective and computational intelligence, as well as prompting future developments in the area. The part consists of six chapters. In particular:

"Sustainability Data and Analytics in Cloud-based M2M Systems" analyzes data and analytics applications in M2M systems for sustainability governance. The chapter presents techniques for M2M data and process integration, including linking and managing monitored objects, sustainability monitoring data and analytics applications, for different stakeholders who are interested in dealing with large-scale monitoring data in M2M environments. It presents a cloud-based data analytics system for sustainability governance that includes a Platform-as-a-Service and an analytics framework. The authors also illustrate their prototype on a real-world cloud system for facility monitoring.

"Social Networking Analysis" makes an analysis of the role of social network analysis on applications dealing with studies of kinship structure, social mobility, science citations, use of contacts among members of nonstandard groups, corporate power, international trade exploitation, class structure, and many other areas. Building a useful understanding of a social network is to sketch a pattern of social relationships, kinships, community structure, interlocking dictatorships, and so forth for analysis.

"Leveraging Social Media and IOT to Bootstrap Smart Environments" deals with the realization of Smart Environments, and presents lightweight Cyber Physical Social Systems to include building occupants within the control loop to allow them some control over their environment. The authors define the concept of citizen actuation, and present an experiment where they obtain a reduction in average energy usage of 26 %. The chapter proposes the Cyber-Physical Social System architecture, and discusses future research in this domain.

"Four-Layer Architecture for Product Traceability in Logistic Applications" describes the design of an auto-managed system for the tracking and location of products in transportation routes, called Transportation Monitoring System (TMSystem). A four-layer system is proposed to provide an efficient solution for the Real-Time Monitoring (RTM) of goods. Two Web Services are proposed, Location Web Service and Check Order Web Service, so that customers can easily access information about the shipment of their orders. Finally, a Web Application is developed to access those Web Services.

"Disaster Evacuation Guidance Using Opportunistic Communication: Potential for Opportunity-Based Service" proposes a disaster evacuation guidance using opportunistic communication where evacuees gather location information of impassable and congested roads by disaster into their smartphones by themselves, and also share the information with each other by short-range wireless communication between nearby smartphones. The proposed guidance is designed not only to navigate evacuating crowds to refuges, but to rapidly aggregate the disaster information.

"iPromotion: A Cloud-Based Platform for Virtual Reality Internet Advertising" presents a large-scale platform for distributing Virtual Reality advertisements over the World Wide Web. The platform aims at receiving and transmitting large amounts of data over mobile and desktop devices in Smart City contexts, is based on a modular and distributed architecture to allow for scalability, and incorporates content-based search capabilities for VR scenes to allow for content management.

Nik Bessis
Ciprian Dobre

Acknowledgments

It is with our great pleasure to comment on the hard work and support of many people who have been involved in the development of this book. It is always a major undertaking but most importantly, a great encouragement and definitely a reward and an honor when experiencing the enthusiasm and eagerness of so many people willing to participate and contribute towards the publication of a book. Without their support the book could not have been satisfactory completed.

First and foremost, we wish to thank all the authors who despite busy schedules, they have devoted so much of their time preparing and writing their chapters, and responding to numerous comments made from the reviewers and editors. Special gratitude goes to Dr. Davy Preuveneers from DistriNet research group, the Embedded and Ubiquitous Systems task force, at Katholieke Universiteit Leuven, Belgium for writing the foreword to this collection and also to the reviewers for their constructive commentaries and insightful suggestions.

Last but not least, we wish to thank Professor Janusz Kacprzyk, Editor-in-Chief of 'Studies in Computational Intelligence' Springer Series, Dr. Thomas Ditzinger, Ms. Heather King and the whole Springer's editorial team for their strong and continuous support through the development of this book.

Finally, we are deeply indebted to our families for their love, patience and support throughout this rewarding experience.

Nik Bessis
Ciprian Dobre

Contents

Part III   Advanced Applications and Future Trends

Part I
Foundations and Principles

Big Data Platforms for the Internet of Things

Radu-Ioan Ciobanu, Valentin Cristea, Ciprian Dobre and Florin Pop

Abstract This chapter discusses the challenges, state of the art, and future trends in context-aware environments (infrastructure and services) for the Internet of Things, an open and dynamic environment where new Things can join in at any time, and offer new services or improvements of old services in terms of performance and quality of service. The dynamic behavior is supported by mechanisms for Things publishing, notification, search, and/or retrieval. Self-adaptation is important in this respect. For example, when things are unable to establish direct communication, or when communication should be offloaded to cope with large throughputs, mobile collaboration can be used to facilitate communication through opportunistic networks. These types of networks, formed when mobile devices communicate only using short-range transmission protocols, usually when users are close, can help applications still exchange data. Routes are built dynamically, since each mobile device is acting according to the store-carry-and-forward paradigm. Thus, contacts are seen as opportunities to move data towards the destination. In such networks data dissemination is usually based on a publish/subscribe model. We make a critical analysis of current opportunistic approaches using the elements of a newly defined taxonomy. We review current state-of-the-art work in this area, from an IoT perspective.

R.-I. Ciobanu (✉) · V. Cristea · C. Dobre · F. Pop
University Politehnica of Bucharest, Spl. Independentei 313, Bucharest, Romania
e-mail: radu.ciobanu@cti.pub.ro

V. Cristea (✉)
e-mail: valentin.cristea@cs.pub.ro

C. Dobre
e-mail: ciprian.dobre@cs.pub.ro

F. Pop
e-mail: florin.pop@cs.pub.ro

N. Bessis and C. Dobre (eds.), *Big Data and Internet of Things:*
A Roadmap for Smart Environments, Studies in Computational Intelligence 546,
DOI: 10.1007/978-3-319-05029-4_1, © Springer International Publishing Switzerland 2014

1 Introduction

Every day we create 2.5 quintillion bytes of data; so much that 90 % of the data in the world today has been created in the last 2 years alone. This data comes from sensors used to gather climate information, from posts to social media sites, digital pictures and videos, purchase transaction records, or cell phone GPS signals, to name only a few. This data is *Big Data*. Analyzing large data sets already underpins new waves of productivity growth, innovation, and consumer surplus. Big data is more than simply a matter of size; it is an opportunity to find insights in new and emerging types of data and content, to make businesses more agile, and to answer questions that were previously considered beyond our reach. Until now, there was no practical way to harvest this opportunity. But today we are witnessing an exponential growth in the volume and detail of data captured by enterprises, the rise of multimedia, social media and Online Social Networks (OSN), and the Internet of Things (IoT).

Many of Big Data challenges are generated by future applications where users and machines will need to collaborate in intelligent ways together. In the near future information will be available all around us, and will be served in the most convenient way—we will be notified automatically when a congestion occurs and the car will be able to decide how to optimize our driving route, the fridge will notify us when we the milk supply is out, etc. Technology becomes more and more part of our daily life. New technologies have finally reached a stage of development in which they can significantly improve our lives. For example, our cities are fast transforming into artificial ecosystems of interconnected, interdependent intelligent digital "organisms". They are transforming into *smart cities*, as they benefit more and more from intelligent applications designed to drive a sustainable economic development and an incubator of innovation and transformation that merges the virtual world of Mobile Services, IoT and OSN with the physical infrastructures of Smart Building, Smart Utilities (i.e., electricity, heating, water, waste, transportation, and unified communication and collaboration infrastructure). The transformation of the metropolitan landscape is driven by the opportunity to embed intelligence into any component of our towns and connect them in real-time, merging together physical world of objects, humans and virtual conversation and transactions. Town planners and administration bodies just need the right tools at their fingertips to consume all the data points that a town or city generate and then be able to turn that into actions that improve people's lives. Smart Cities of tomorrow will rely not only on sensors within the city infrastructure, but also on a large number of devices that will willingly sense and integrate their data into technological platforms used for introspection into the habits and situations of individuals and city-large communities. Predictions say that cities will generate over 4.1 TB/day/km^2 of urbanized land area by 2016. Handling efficiently such amounts of data is already a challenge.

We have barely begun to get a sense of the dimensions of this kind of data, of the privacy implications, of ways in which we can code it with respect to meaningful attributes in space and time. As we move into an era of unprecedented volumes of data and computing power, the benefits are not for business alone. Data can help citizens' access to government, hold it accountable and build new services to help themselves. In one sense, all this is part of a world that is fast becoming digital in all its dimensions. People will develop more easily their understanding and design ideas using digital representations and data. This will support the development of the new ideas for the future of urban and social life, weaved together under the umbrella of what is now called the future Internet of Things (IoT).

As part of the Future Internet, IoT aims to integrate, collect information from-, and offer services to a very diverse spectrum of physical things used in different domains. "Things" are everyday objects for which IoT offers a virtual presence on the Internet, allocates a specific identity and virtual address, and adds capabilities to self-organize and communicate with other things without human intervention. To ensure a high quality of services, additional capabilities can be included such as context awareness, autonomy, and reactivity. Things are very diverse. Very simple things, like books, can have Radio Frequency Identification—RFID tags that help tracking them without human intervention. For example, in an electronic commerce system, a RFID sensor network can detect when a thing left the warehouse and can trigger specific actions like inventory update or customer rewarding for buying a high end product [14]. In this simple case, RFIDs enable the automatic identification or things, the capture of their context (for example the location) and the execution of corresponding actions if necessary. Sensors and actuators are used to transform real things into *virtual objects* [43] with digital identities. In this way, things may communicate, interfere and collaborate with each other over the Internet [5]. Adding part of application logic to things transforms them into *smart objects*, which have additional capabilities to sense, log and understand the events occurring in the physical environment, autonomously react to context changes, and intercommunicate with other things and people. A tool endowed with such capabilities could register when and how the workers used it and produce a financial cost figure. Similarly, smart objects used in the e-health domain could continuously monitor the status of a patient and adapt the therapy according to the needs. Smart objects can also be general purpose portable devices like smart phones and tablets, that have processing and storage capabilities, and are endowed with different types of sensors for time, position, temperature, etc. Both specialized and general purpose smart objects have the capability to interact with people.

The IoT includes a hardware, software and services infrastructure for things networking. IoT infrastructure is event-driven and real-time, supporting the context sensing, processing, and exchange with other things and the environment. The infrastructure is very complex due to the huge number (50–100 trillion) of heterogeneous, (possibly) mobile things that dynamically join and leave the IoT, generate and consume billions of parallel and simultaneous events geographically distributed all over the world. The complexity is augmented by the difficulty to represent, interpret, process, and predict the diversity of possible contexts. The infrastructure must have important characteristics such as reliability, safety, survivability, security and fault tolerance. Also, it must manage the communication, storage and compute resources.

The main function of the IoT infrastructure is to *support communication* among things (and other entities such as people, applications, etc.). This function must be flexible and adapted to the large variety of things, from simple sensors to sophisticated smart objects. More specific, things need a communication infrastructure that is low-data-rate, low-power, and low-complexity. Actual solutions are based on short-range radio frequency (RF) transmissions in ad-hoc wireless personal area networks (WPANs). A main concern of the IoT infrastructure developers is supporting heterogeneous things by adopting appropriate standards for the physical and media access control (MAC) layers, and for communication protocols. The protocol and compatible interconnection for the simple wireless connectivity with relaxed throughput (2–250 kb/s), low range (up to 100 m), moderate latency (10–50 ms) requirements and low cost, adapted to devices previously not connected to the Internet were defined in IEEE 802.15.4. Other similar efforts refer to industrial and vehicular applications—ZigBee, open standards for process control in industrial automation and related applications—ISA100.11a and WirelessHART, and encapsulating IPv6 datagrams in 802.15.4 frames, neighbor discovery and routing that allow sensors to communicate with Internet hosts—6LoWPAN [6]. The main scope of IoT specialists is the world-wide network of interconnected virtual objects uniquely addressable and communicating through standard protocols. The challenge here is coping with a huge number of (heterogeneous) virtual objects.

The IoT architecture supports physical things' integration in Internet and the complex interaction flow of services triggered by events occurrence. The main concepts involved are the object-identification, sensing and connecting capabilities as the basis for the development of independent cooperative services and applications that address several key features for IoT architecture: Service Orientation, Web-base, distributed processing, easy integration via native XML and SOAP messaging, component-base, open access, N-tiered architecture, support for vertical and horizontal scalability [37]. These features allow "physical objects become active participants in business processes" [39]. As a consequence, the Web Services must be available to interact with the corresponding virtual objects over the Internet, query and change their state and any information associated with them. The new key features for the IoT architecture include persistent messaging for the highest availability, complete security and reliability for total control and compliance, platform independence and interoperability (more specific for middleware).

An IoT infrastructure considers *extended process functionality*, pathways and layered networks as main components. The IoT infrastructure should support object-connected technologies for both "Human-to-Objects" and "Objects-to-Objects" communications. Because of the wide heterogeneity of wireless devices involves, the communication platforms are heterogeneous, ad-hoc, and opportunistic. IoT is a large heterogeneous collection of things, which differ from each other. Even things that have the same nature, construction, and properties can differ from one another by their situation or context. Context means the conditions in which things exist in other words their surrounding world. Since virtual things in IoT are interconnected, the meaning of the data they exchange with other things and people is clearer only if they are interpreted in the thing's context. This is why the IoT infrastructure runs reliably and permanently to provide the context as a "public utility" to IoT services [13]. For human users, the context is the information that characterizes user's interaction with Internet applications plus the location where this interaction occurs, so that

Fig. 1 Context-aware
services

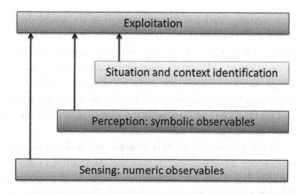

the service can be adapted easily to users' preferences, For things, we need another
approach. A very suggestive example is given in [24]. The authors explain the case
of a plant that is the target of an automatic watering service. In order to control the
watering dosages and frequency, the service has to sense the dryness status of the
plant, to use the domain knowledge of plants and find their watering "preferences",
and to ask the weather prediction service about the chances of rain in the next days.
So, the context of a thing includes information about thing's environment and about
the thing itself.

Several context modeling and reasoning techniques are known today, some of them
being based on knowledge representation and description logics. Ontology-based
models can describe complex context data, allow context integration among several
sources, and can use reasoning tools to recognize more abstract contexts. Ontologies
provide a formal specification of the semantics of context data that stay at the base
of knowledge sharing among different things in IoT. In addition, new context can be
derived by ontological reasoning. Ontology-based models can be used to organize
IoT infrastructure context-aware services as a fabric structured into multiple levels
of abstraction (see Fig. 1) starting with collecting information from physical sensors
(called low level context), which could be meaningless and consequently not useful
to applications. Next, higher level context is derived by reasoning and interpretation.
Finally, context is exploited by triggering specific actions [13].

The IoT infrastructure combines the context model with event-based organization
of services that support the collection, transmission, processing, storage and delivery
of context information. In the event-driven architecture vocabulary, events are generated by different sources, event producers, when for example a context change is
significant and must be propagated to target applications, event consumers. Producers
and consumers are loosely coupled by the asynchronous transmission and reception
of events. They don't have to know and explicitly refer each other. In addition, the producers don't know if the transmitted events are consumed ever. A publish/subscribe
mechanism is used to offer the events to the interested consumers. Other components
are used such as event channels for communication, and event processing engines
for complex event detection. To them, components for the event specification, event

management, and for the integration of the event-driven system with the application must be added. It is worth noting that events are associated with changes of things' context [14]. There are also non-functional requirements associated with IoT infrastructure [14]: large scale integration, interoperability between heterogeneous things, fault tolerance and network disconnections, mobility, energy saving, reliability, safety, survivability, protection of users' personal information (e.g., location and preferences) against security attacks, QoS and overall performance, scalability, self-* properties and transparency.

In the remainder of this chapter we make a thorough review of methods and techniques designed to support the adaptation of modern context-aware platform towards Big Data infrastructures designed to support the IoT vision. The rest of this chapter is organized as follows. We first make an analysis of context aware infrastructures for IoT. As previously mentioned, the communication support is an important component of any IoT platform, and in Sect. 3 we make a comprehensive study of opportunistic data dissemination support for the Internet of Things. In Sect. 4 we present future trends and research directions in Big Data platforms for the Internet of Things. Section 5 presents the conclusions and final remarks.

2 Context-Aware Infrastructures for the Internet of Things

This section presents up to date solutions and research results regarding the structure, characteristics, and services for context aware Internet of Things infrastructure. A ubiquitous computing environment is characterized by a diverse range of hardware (sensors, user devices, computing infrastructure, etc.) and equally diverse set of applications which anticipate the need of users and act on their behalf in a proactive manner [35]. The vision of ubiquitous computing is only recently becoming a reality in a scale that can be practically distributed to end users.

A context-aware system is generally characterized by several functions. It generally gathers context information available from user interface, pre-specified data or sensors and add it to a repository (*Context Acquisition and Sensing*). Furthermore, the system converts the gathered raw context information into a meaningful context which can be used (*Context Filtering and Modeling*). Finally, the system uses the context to react and make the appropriate context available to the user (*Context Reasoning, Storage and Retrieval*).

Several cross-device context-aware application middleware systems have been developed previously. In their majority these were Web service-based context-aware systems, especially the most recent ones. However, there has been a big variety of middleware systems, developed mainly in the early 2000, that do not rely on Web service technologies and are not designed to work on Web service-based environments [40]. In our analyzed we began by studying several popular context-aware platforms, considering their provided functions and particular characteristics. From the category of non-based on web service context-aware platforms we mention the following. **RCSM** [45] is a middleware supporting context sensitive applications based on an object model: context-sensitive applications are modeled as objects. RCSM supports situation awareness by providing a special language for specifying situation awareness requirements. Based on these requirements, application-specific object containers for runtime situation analysis will be generated. RCSM runtime system obtains context data from different sources and provides the data to object containers which conduct the situation analysis. The **JCAF** (Java Context Awareness Framework) [2] supports both the infrastructure and the programming framework for developing context-aware applications in Java. Contextual information is handled by separate services to which clients can publish and from which they can retrieve contextual. The communication is based on Java RMI (Remote Method Invocation). An example of application that use Java RMI is MultiCaR: Remote Invocation for Large Scale, Context-Aware Applications [15]. This application also address the issue of Big Data analytics.

The **PACE** middleware [25] provides context and preference management together with a programming toolkit and tools for assisting context-aware applications to store, access, and utilize contextual information managed by the middleware. PACE supports context-aware applications to make decisions based on user preferences. **CAMUS** is an infrastructure for context-aware network-based intelligent robots [30]. It supports various types of context information, such as user, place and environment, and context reasoning. However, this system is not based on Web services and it works in a close environment. **SOCAM** is a middleware for building context-aware services [22]. It supports context modeling and reasoning based on OWL. However, its implementation is based on RMI.

Web service-based context-aware platforms include the following. **CoWSAMI** is a middleware supporting context-awareness in pervasive environments [1]. The **ESCAPE** framework [40] is aWeb services-based context management system for teamwork and disaster management. ESCAPE services are designed for a front-end of mobile devices and the back-end of high end systems. The front-end part includes components support for context sensing and sharing that are based on Web services and are executed in an ad hoc network of mobile devices. The back-end includes a Web service for storing and sharing context information among different front-ends. The **inContext** project [40] provides various techniques for supporting context-awareness in emerging team collaboration. It is designed for Web services-based collaborative working environments. inContext provides techniques for modeling, storing, reasoning, and exchanging context information among Web services.

Being context-aware allows software not only to be able to deal with changes in the environment the software operates in, but also being able to improve the response to the use of the software. That means context-awareness techniques aim at supporting both functional and non-functional software requirements. Authors of [18] identified three important context-awareness behaviors:

1. The representation of available information and services to an end user.
2. The automatic execution of a service.
3. The tagging and storing of context information for later retrieval.

Such platforms are considered by many precursors of the Internet of Things vision, defined as a dynamic network infrastructure with self-configuring capabilities based on standard and interoperable communication protocols [43]. In this vision, physical and virtual "things" have identities, physical attributes, and virtual personalities and use intelligent interfaces, and are seamlessly integrated into the information network. In the IoT, *things* are expected to become active participants in business, information and social processes where they are enabled to interact and communicate among themselves and with the environment by exchanging data and information *sensed* about the environment, while reacting autonomously to the *real/physical world* events and influencing it by running processes that trigger actions and create services with or without direct human intervention. Interfaces in the form of services facilitate interactions with these *smart things* over the Internet, query and change their state and any information associated with them, taking into account security and privacy issues.

Sensors in IoT can run anywhere and on any objects. They are used to collect data such as biomedical information, environment temperature, humidity, ambient noise level. This is also a research issue close to Big Data science. The data provided by such sensors can be used by customized context-aware applications and services, capable to adapting their behavior to their running environment. However, sensor data exhibits high complexity (e.g., because of huge volumes and inter-dependency relationships between sources), dynamism (e.g., updates performed in real-time and data that can critical age until it becomes useless), accuracy, precision and timeliness. An IoT system should not concern itself with the individual pieces of sensor data: rather, the information should be interpreted into a higher, domain-relevant concept. For example, sensors might monitor temperature, humidity, while the information needed by a watering actuator might be that the environment is dry. This higher-level concept is called a situation, which is an abstract state of affairs interesting to applications [12].

Situations are generally representations (simple, human understandable) of sensor data. They shield the applications from the complexities of sensor readings, sensor data noise and inferences activities. However, in large-scale systems there may be tens or hundreds of situations that applications need to recognize and respond to. Underlying these situations will be an even greater number of sensors that are used in situation identification. A system has a significant task of defining and managing these situations. This includes capturing what and how situations are to be recognized from which pieces of contexts, and how different situations are related to each other. The system should know, for example, which situations can or cannot occur: a room hosting a "scientific event" and an "academic class" at the same time; otherwise, inappropriate adaptive behavior may occur. Temporal order between situations is also important, such as the inability of a car to go directly from a situation of 'parked' to 'driving on a highway'. Given the inherent inaccuracy of sensor data and the limitations of inference rules, the detection of situations is imperfect.

The research topics on situation identification for IoT involve several issues [47]. First, *representation* deals with how to define logic primitives used to construct a situation's logical specification. In representation, logical primitives should capture features in complex sensor data (e.g., acceleration data), domain knowledge (e.g., a spatial map or social network), and different relationships between situations. Also, an IoT system is assumed to be highly dynamic. New sensors can be introduced, that introduce new types of context. Therefore, the logical primitives should be flexibly extensive, such as new primitives to not cause modifications or produce ambiguous meanings to existing ones. *Specification* deals with defining the logic behind a particular situation. This can be acquired by experts or learned from training data. It typically relies on a situation model with apriori expert knowledge, on which reasoning is applied based on the input sensor data. For example, in logic programming [8] the key underlying assumption is that knowledge about situations can be modularized or digitized.

Other logic theories, such as situation calculus [46], have also been used to infer situations in IoT systems. Kalyan et al. [28] introduce a multi-level situation theory, where an intermediate level micro situation is introduced between infons and situations. An infon embodies a discrete unit of information for a single entity (e.g., a customer or a product), while a situation makes certain infons factual and thus support facts. Micro situations are composed of these entity-specific infons which can be explicitly obtained from queries or implicitly derived from sensors and reasons. Situations are considered as a hierarchical aggregation of micro situations and situations. This work aims to assist information reuse and support ease of retrieving the right kind of information by providing appropriate abstraction of information.

Ontologies have also increasingly gained attention as a generic, formal and explicit way to capture and specify the domain knowledge with its intrinsic semantics through consensual terminology and formal axioms and constraints. They provide a formal way to represent sensor data, context, and situations into well-structured terminology. Based on the modeled concepts, developers can define logical specifications of situations in rules. An exemplar rule on an activity 'sleeping' is given in [22]:

```
(?user rdf:type socam:Person),
(?user, socam:locatedIn, socam:Bedroom),
(?user, socam:hasPosture, 'LIEDOWN'),
(socam:Bedroom, socam:lightLevel, 'LOW'),
(socam:Bedroom, socam:doorStatus, 'CLOSED')
-> (?user socam:status 'SLEEPING')
```

Ontologies, together with their support for representation formalisms, can support reasoning, including detecting inconsistency, or deriving new knowledge. An ontological reasoner can be used to check consistency in a class hierarchy and consistency between instances, e.g. whether a class is being a subclass of two classes that are declared as disjoint or whether two instances are contradictory to each other (such as a person being detected in two spatially disjoint locations at the same time). Given the current sensor data, the reasoner will derive a new set of statements. In the above 'sleeping' example, if the reasoner is based on a forward-chaining rule engine, it can match the conditions of this rule against the sensor input. If all the conditions are satisfied, the reasoner will infer the conclusion of the rule. The reasoning will terminate if the status of the user is inferred, when the status of the user is set to be the default inference goal in this reasoner.

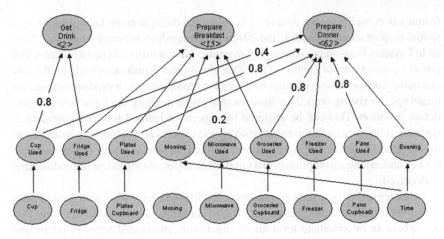

Fig. 2 An example of situation inferring using Dempster-Shafer theory (from [34])

Other solutions are based on the Dempster-Shafer theory (DST) [16], a mathematical theory of evidence, which propagates uncertainty values and consequently provides an indication of the certainty of inferences. The process of using DST is described as follows. First, developers apply expert knowledge to construct an evidential network that describes how sensors lead to activities. The left-hand side of Fig. 2 describes that the sensors on the cup and fridge are connected to context information (e.g., 'cup used'). Such context information can be further inferred or composed to higher-level context. The composition of context information points to an activity (e.g., 'Get drink') at the top. Developers can use such an approach to determine the evidence space and degree of belief in an evidence. For example, in the figure the values on the arrows represent the belief in particular sensor (also called the uncertainty of sensor observations). Generally, in reasoning situations are inferred from a large amount of imperfect sensor data. In reasoning, one of the main processes is called situation identification—deriving a situation by interpreting or fusing several pieces of context in some way. Specifying and identifying situations can have a large variability depending on factors such as time, location, individual users, and working environments. This makes specification-based approaches relying on models of a priori knowledge impractical to use. Machine learning techniques have been widely applied to learning complex associations between situations and sensor data. However, the performance of reasoning is usually undermined by the complexity of the underlying sensor data.

Bayesian networks and Hidden Markov Models (HMMs) have been applied in many context-aware systems. In HMMs statistical models a system being modeled is assumed to be a Markov chain that is a sequence of events [41]. A HMM is composed of a finite set of hidden states and observations that are generated from states. For example, a HMM where each state represents a single activity (e.g., 'prepare dinner', 'go to bed', 'take shower', and 'leave house') is presented in [41]. They represent observations in three types of characterized sensor data that are generated in each activity, which are raw sensor data, the change of sensor data, the last observed sensor data, and the combination of them. The HMM is trained to obtain the three probability parameters, where the prior probability of an activity represents the likelihood of the user starting from this activity; the state transition probabilities represent the likelihood of the user changing from one activity to another; and the observation emission probabilities represent the likelihood of the occurrence of a sensor observation when the user is conducting a certain activity.

Finally, Support Vector Machines (SVM) [20] is a relatively new method for classifying both linear and nonlinear data. An SVM uses a nonlinear mapping to transform the original training data into a higher dimension. Within this new dimension, it searches for the linear optimal separating hyper-plane that separates the training data of one class from another. With an appropriate nonlinear mapping to a sufficiently high dimension, data from two classes can always be separated. SVMs are good at handling large feature spaces since they employ over fitting protection, which does not necessarily depend on the number of features. Kanda et al. [29] use SVMs to categorise motion trajectories (such as 'fast', 'idle', and 'stop') based on the velocity, direction, and shape features extracted from various sensors (within a car for example). Different types of sensor data lead to different techniques to analyze them. Numerical data, for example, can be used to infer motions like 'walking' or 'running' from acceleration data. Situation identification at this level is usually performed in learning-based approaches, which uncover complicated associations (e.g., nonlinear) between continuous numerical data and situations by carving up ranges of numerical data (e.g., decision tree) or finding an appropriate algebraic function to satisfy or 'explain' data (e.g., neural networks or SVMs). Specification-based approaches can apply if the association between sensor data and situations are rather explicit and representable in logic rules. Situations can also be recognized from categorical features; for example, inferring a room's situation—'meeting' or 'presentation'—from the number of persons co-located in the room and the applications running in the computer installed in the room. This higher-level of situation identification can be performed using either specification- or learning-based approaches.

Uncertainty can also exist in the use of oversimplified rules that are defined in an ad-hoc way. In representing uncertainty of rules, Web Ontology Language (OWL), a family of knowledge representation languages for authoring ontologies endorsed by W3C, can be extended with a conditional probabilistic class to encode the probability that an instance belongs to a class respectively given that it belongs to another class. Although good at expressing uncertainty, these qualitative approaches need to be combined with other machine-learning techniques to quantify the uncertainty to be used in situation identification. Learning-based approaches have a stronger capability to resolve uncertainty by training with the real-world data that involves noise. These approaches not only learn associations between sensor data and situations, but also the effect that the uncertainty of sensor data has on the associations. For example, the conditional probabilities learned in Bayes networks include the reliability of sensor data as well as the contribution of the characterized sensor data in identifying a situation. A popular architectural model for IoT is composed of autonomous physical/digital objects augmented with sensing, processing, and network capabilities. Unlike RFID tags, smart objects carry an application logic that let them sense their local situation and interact with the environment through actuators. They sense, log, and interpret what's occurring within themselves and the work, act on their own, intercommunicate with each other, and exchange data [31].

According to the scenarios illustrated in [31], the architectural differences in the way smart objects understand (sense, interpret or react to) events, and interact with their environment in terms of input, output, control and feedback, classify them as either activity-aware objects, policy-aware objects or process-aware objects. A process-aware object represents the most accomplished type, and characterizes: awareness (a process-aware object understands the operational process that is part of and can relate the occurrence of real-work activities and events to these processes), representation (its model consists of a context-aware workflow model that defines timing and ordering or work activities), and interaction (a process-aware object providers workers with context-aware guidance about tasks, deadlines, and decisions).

With an understanding of what context is and the different ways in which it can be used, application builders can now more easily determine what behaviors or features they want their applications to support and what context is required to achieve these behaviors. However, something is still missing. Application builders may still need help moving from the design to an actual implementation. This help can come from a combination of architectural services or features that designers can use to build their applications from. For example, the Webinos EU funded project aims to deliver a platform for web applications across mobile, PC, home media and in-car devices [42]. The webinos project has over twenty partners from across Europe spanning academic institutions, industry research firms, software firms, handset manufacturers and automotive manufacturers. Its vision is to build a multiplatform, applications platform based on web technology that is fit for purpose, across a wide range of connected devices, taking computing to its next logical step, that of ubiquity. In order to do so, knowing the state of the device and the user at any given time, and making decisions based on that context is crucial. Context-awareness and true cross-platform and cross-device applications cannot be achieved without the developer having access to an environment able to synchronize content and application state between devices, adapt to changes in user context, and provide standard JavaScript APIs to let web applications access device features. Therefore webinos is a good illustration of a cross platform level of abstraction for procedural calls, but at the same time, incorporate an additional data abstraction layer for use in third party context-aware and context-sharing applications that are webinos-enabled [42].

3 A Study on Opportunistic Data Dissemination Support for the Internet of Things

Today, the way that people access information and communicate is radically changing, right before our eyes, in many ways that are not yet readily apparent. Wireless devices, such as mobile phones, connected devices and consumer electronics, are infiltrating all aspects of our lives. As the cost to mobilize these devices continues to drop, and as wireless networks become faster, ubiquitous and cheaper, it is easy to see a near future where almost everything and everyone are wirelessly online, 24×7. It is evident that the wireless universe of things is rapidly accelerating. This raises many questions as well as opportunities, especially for businesses that offer communications equipment and services, consumer electronics and other connected devices. People will increasingly expect to access a wide range of data and rich content on many devices. Up until recently, the most common mobile data types include voice, contacts (address books) and SMS (text messages). Today, people expect to wirelessly transparently access their email, social network messaging, calendars, photos, videos, files, music, games, apps and web pages. Behind we have a Big Data platforms and environments. Wireless applications need to easily manage and filter an avalanche of mobile data. This includes functions such as syncing, sharing, searching, transferring, archiving, deleting and caching. It means making it seamless to share mobile data and rich content with other devices, people, groups, businesses, schools, public agencies, systems, etc. But when things are unable to establish direct communication, or when communication should be offloaded to cope with large throughputs, mobile collaboration can be used to facilitate communication through opportunistic networks. These types of networks, formed when mobile devices communicate only using short-range transmission protocols, usually when users are close, can help applications still exchange data. Routes are built dynamically, since each mobile device is acting according to the store-carry-and-forward paradigm. Thus, contacts are seen as opportunities to move data towards the destination. In such networks data dissemination is usually based on a publish/subscribe model.

One type of such mobile networks that has been deeply researched in recent years is represented by opportunistic networks (ONs). They are dynamically built when mobile devices collaborate to form communication paths while users are in close proximity. Opportunistic networks are based on a store-carry-and-forward paradigm [36], which means that a node that wants to relay a message begins by storing it, then carries it around the network until the carrier encounters the destination or a node that is more likely to bring the data close to the destination, and then finally forwards it.

One of the main challenges of opportunistic networks is deciding which nodes should the data be relayed to in order for it to reach its destination, and do it as quickly and efficiently as possible. Various types of solutions have been proposed, ranging from disseminating the information to every encountered node in an epidemic fashion, to selecting the nodes with the highest social coefficient or centrality [26]. Prediction methods have also been employed [9], based on the knowledge that the mobile nodes from an opportunistic network are devices belonging to humans, which generally have the same movement and interaction patterns that they follow every day. The analysis of contact time (duration of an encounter between two nodes) and inter-contact time (duration between consecutive contacts of the same two nodes) has also been used in deciding a suitable relay node. Aside from selecting the node that the data will be forwarded to, research has also focused on congestion control, privacy, security, or incentive methods for convincing users to altruistically participate in the network.

An important topic in opportunistic networks is represented by *data dissemination*. In such networks, given the dynamic nature of the wireless devices in IoT, topologies are unstable. Various authors proposed different data-centric approaches for data dissemination, usually based on different publish/subscribe models, where data is pro-actively and cooperatively disseminated from sources towards possibly interested receivers, as sources and receivers might not be aware of each other and never get in touch directly. We analyze in the following lines existing work in the area of data dissemination in opportunistic networks. We analyze different collaboration-based communication solutions, emphasizing their capabilities to opportunistically disseminate data. We present the advantages and disadvantages of the analyzed solutions. Furthermore, we propose the categories of a taxonomy that captures the capabilities of data dissemination techniques used in opportunistic networks. Using the categories of the proposed taxonomy, we also present a critical analysis of four opportunistic data dissemination solutions. To our knowledge, a classification of data dissemination techniques for IoT platforms has never been previously proposed.

3.1 Opportunistic Data Dissemination and the Internet of Things

In recent years more and more IoT scenarios and use-cases based on delay-tolerant networks (DTNs) and opportunistic networks (ONs) have been proposed. These scenarios range from commercial (e.g. targeted advertising [23]) to information (e.g. floating content [17], context-aware mobile platforms [19]), smart cities [7] or even emergency situations (e.g. crowd location, disaster management [33], etc.). Such networks differ from traditional ones because they do not require a fixed infrastructure to function. Traditional networks generally assume that the nodes are fixed and the topology of the network is well known in advance. By analyzing the topology, a node that wishes to send a message will embed that message with the path and thus it will be sure it can reach the intended destination. With DTNs and ONs, the situation is different. Firstly, the nodes are mobile, so a node doesn't have the same neighbors at two different moments in time. Secondly, contact opportunities between two nodes only arise when they are in contact, so the contact duration is important in deciding whether data should be exchanged or not. Thirdly, nodes in ONs and DTNs are mobile devices, so they are restricted by buffer space. This means that, when two nodes encounter, simply exchanging all data between them is not feasible. ONs are based on a store-carry-and-forward paradigm, where nodes store messages and carry them until a suitable destination (in terms of probability of delivering a message to its intended target) is encountered, when they are forwarded. Such delay-tolerant and opportunistic networks have a real use in ensuring communication in IoT, when traditional methods are not available, are too costly, or alternatives are required.

Mobile online advertising nowadays has become ubiquitous, due to companies such as Google (on Android devices) or Facebook making their revenue by targeting advertisements to fit a user's preferences and tastes. However, users fear that, in order to show suitable ads, big companies collect too much precious information and privacy is lost. However, ONs can help correct this and ensure privacy, and such a proposed solution is MobiAd [23]. It is an application that presents the user with local targeted advertisements, while still preserving the privacy. The ads are selected from an ad pool received from local hotspots or even broadcast on the local mobile base-station, but statistical information about ad views and user clicks is encrypted and sent to the advertisement channel opportunistically, via the other ON devices in the network (or using static WiFi hotspots). Thus, other nodes and even the ad provider cannot know which ads a given node has viewed or clicked on.

Since ONs and DTNs are used to carry information opportunistically, a suitable scenario is represented by information-sharing. An example of this is floating content in open city squares [17], where mobile nodes may enter a geographically-fixed area that specifies the physical boundaries of the network (an anchor zone), spend a certain amount of time there, and then leave. While located in the anchor zone, devices (or a static access point) produce content and replicate it opportunistically to other nodes, which may use the data for themselves, or carry it in order to forward it to interested nodes. When a node exits the anchor zone, the zone-specific data is deleted, so the floating content's availability is probabilistic and strictly connected to the anchor zone. A real-life use of floating content in open city squares is a touristic information centre, where users receive information about a certain building, place or statue when they are located in its vicinity.

Information can be spread using opportunistic networks in other environments, such as academics, where ONs can be employed as the backbone of a framework that facilitates interaction between the members of a faculty (students, professors, etc.) possessing smartphones or other mobile devices. Thus, communication inside the faculty (in a limited physical area, like in open city squares) may be performed opportunistically, instead of using a fixed infrastructure such as WiFi or 3G. The benefits are limiting the used bandwidth and the battery consumption, since Bluetooth (mainly used for opportunistic communication with smartphones) consumes less power than both 3G and WiFi. A platform for supporting generic context-aware mobile applications is CAPIM [19], and it can fully benefit from ON integration. CAPIM (Context-Aware Platform using Integrated Mobile services) is a solution designed to support the construction of next-generation applications. It integrates various services that collect context data (location, user's profile and characteristics, environment information, etc.). The smart services are dynamically loaded by mobile clients, which take advantage of the sensing capabilities provided by modern smartphones, possibly augmented with external sensors.

Moving towards the future, DTNs and ONs are a very good fit for smart cities [7]. These cities monitor and integrate the conditions of all their critical infrastructures (such as roads, bridges, tunnels, rails, subways, airports, seaports, communications, water, power, etc.) in order to better optimize their resources and plan their preventive maintenance activities. They react to changes in any of these infrastructures by analyzing the environment and deciding what the suitable decision is. Smart cities connect the physical, IT, social and business infrastructures, for the purpose of leveraging the collective intelligence of the cities. Delay-tolerant and opportunistic networks can be employed to perform communication between various parts of a smart city. For example, the traffic lights system can be opportunistically connected to a service that offers information about traffic jams, crowded roads, accidents, etc., so it can adapt to the conditions of the environment. All data processing for smart cities applications require efficient Big Data platforms.

Another situation where ONs and DTNs may be used is in emergency situations, e.g. at crowded locations where groups of individuals get separated and need to locate each other. In such scenarios (e.g. an amusement park where a child gets separated from his parents, a concert where a group of friends split and they need to find each other, a football match, etc.), contacts between mobile devices happen often and for longer periods of time, but (due to the high number of mobile devices in a very small space), classic communication means such as 3G and mobile telephony may not work properly. In these cases, the mobile devices may be used to opportunistically broadcast the location of mobile device owners to interested receivers. For example, a child at an amusement park might carry a smartphone that opportunistically broadcasts the child's encrypted location, by leveraging the neighboring devices. The child's parents have a smartphone of their own which receives the broadcast from encountered nodes, thus being aware of the child's location at any point in time. Similarly, non-emergency scenarios may also benefit from this solution, where events regarding the bands that will be playing or any other announcements can be disseminated to participants of a music concert or any other type of social event with high participation.

Finally, opportunistic communication may also be used in disaster management scenarios [33], where regular communication might be down. Thus, mobile devices have the potential of forming a new opportunistic infrastructure to be used for communication, either by the rescue teams when searching for survivors, or even by the survivors to signal their presence and their location among the debris (since smartphones are generally equipped with a GPS). Rescue teams may even use the opportunistic infrastructure to communicate between each others, by placing temporary access points at key places in the disaster area.

These scenarios have shown that ONs and DTNs have the possibility of replacing traditional network communication in some situations, or even of working alongside it, in order to bring benefits for offering advance communication capabilities for future IoT platforms. Thus, we believe that research effort should be put into this area, since there are many issues that have yet to be explored.

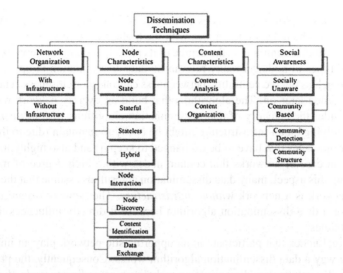

Fig. 3 A taxonomy for data dissemination techniques

3.2 A Taxonomy for Dissemination Techniques

We introduce the categories of a proposed taxonomy for data dissemination techniques in opportunistic networks (presented in Fig. 3). This taxonomy was created by analyzing the existing methods in the literature and highlighting the common characteristics between various methods. Previous attempts to categorize different opportunistic dissemination techniques are focused on specific aspects. The taxonomy proposed in [10] separates the forwarding methods according to mostly their knowledge about context, proposing a separation into three distinct classes: context-oblivious, partially context-aware and fully context-aware). Similarly, the taxonomy in [36] separates techniques based on their knowledge of the infrastructure, making a distinction between algorithms without infrastructure and algorithms where the ad-hoc networks exploit some form of infrastructure to forward messages. Algorithms without infrastructure can be further divided into algorithms based on dissemination (like Epidemic, Meeting and Visits and Networking Coding), and algorithms based on context (like CAR and MobySpace). Algorithms that exploit a form of infrastructure can also be divided into fixed infrastructure and mobile infrastructure algorithms. In case of fixed infrastructure algorithms (like Infostations and SWIM), special nodes are located at specific geographical points, whereas special nodes proposed in mobile infrastructure algorithms (like Ferries and DataMULEs) move around in the network randomly or follow predefined paths.

According to our proposed taxonomy, data dissemination algorithms can be categorized by the *organization of the network* on which they apply. In general, in an opportunistic network no assumption is made on the existence of a direct path between two nodes that wish to communicate. Nonetheless, some dissemination algorithms may exploit certain nodes called "hubs" and build an overlay network between them. The hubs are the nodes with the highest centrality in each community, where a node's centrality is the degree of its participation in the network. There are several types of node centrality relevant to data dissemination in opportunistic networks (such as degree centrality, betweenness centrality or closeness centrality), that are detailed later on. The algorithms that build an overlay network based on hubs fall under the category of data dissemination algorithms *with infrastructure*. However, relying on an infrastructure might be costly to maintain (due to the large number of messages that have to be exchanged to keep it) and also highly unstable, especially in case of networks that contain nodes with a high degree of mobility. Considering this aspect, many data dissemination algorithms assume that the opportunistic network is a network *without infrastructure*. The *network organization* is relevant for a data dissemination algorithm because it directly influences the data transfer policies.

The actual nodes that participate in an opportunistic network play an important part in the way a data dissemination algorithm works. Consequently, the proposed taxonomy also categorizes dissemination techniques according to *node characteristics*. A first characteristic of a node in an opportunistic network is the *node state*. Depending on its implementation, a dissemination technique can either follow a *stateful*, a *stateless* or a *hybrid* approach. An approach that maintains the state of a node requires control traffic (e.g. unsubscription messages in a publish/subscribe-based algorithm) that can prove to be expensive. Moreover, it suffers if frequent topology changes occur. On the other hand, a *stateless* approach does not require control traffic, but has unsatisfactory results if event flooding is used. The *hybrid* approach takes advantage of both the *stateful* and the *stateless* approaches.

Another important characteristic of a node in an opportunistic network is *node interaction*. As stated before, nodes in an opportunistic network generally have a high degree of mobility, so the interaction between them must be as fast and as efficient as possible. The reason for this is that contact duration (the time interval when two network devices are in range of each other) may be extremely low. According to the proposed taxonomy, there are three basic aspects of *node interaction*, the first one being *node discovery*. Depending on the type of mobile device being used, the discovery of nodes that are in the wireless communication range can be done in several ways, but it is usually accomplished by sending periodical broadcast messages to inform neighboring nodes about the device's presence. When two nodes come into wireless communication range and make contact, they each have to inform the other node of the data they store. Therefore the second aspect of *node interaction* is *content identification*, meaning the way in which nodes represent the data internally and how they "declare" it (usually using some form of meta-data descriptions). Nodes may also advertise the channels they have data from or they may present a hash of the data they store. The final subcategory of *node interaction* is *data exchange*, which is the way two nodes transfer data to and from each other. This refers not only to the actual data transferring method, but also to the way data is organized or split into units. The three *node interaction* steps presented here may also be done asynchronously for several neighboring nodes, and the way they are implemented affects the performance of a data dissemination algorithm.

As stated in [44], an interesting use case for opportunistic networks is the sharing of content available on mobile users' devices. In such a network, users themselves may generate content (e.g. photos, clips) on their mobile devices, which might be of interest to other users. However, content producers and consumers might not be connected to each other, so an opportunistic data dissemination method is necessary. Because there can be many types of content, each having different characteristics, the proposed taxonomy also classifies data dissemination algorithms according to the *content characteristics*.

An important aspect of the actual content is its *organization*. Most often, content is organized into channels, an approach used for publish/subscribe-based data dissemination. The publish/subscribe pattern is used mainly because communication is based on messages and can be anonymous, whilst participants are decoupled from time, space and flow. Time decoupling takes place because publishers and subscribers do not need to run at the same time, while space decoupling happens because a direct connection between nodes does not have to exist. Furthermore, no synchronized operations are needed for publishing and subscribing, so nodes are also decoupled from the communication flow. The approach allows the users to subscribe to a channel and automatically receive updates for the content they are interested in. Such an organization is taken further by structuring channels into episodes and enclosures.

Aside from the way content is organized at a node, the proposed taxonomy categorizes data dissemination techniques by *content analysis*. *Content analysis* represents the way in which the algorithm analyzes a certain content object and decides if it will fetch it or not. There are two reasons a node might download a content object from another encountered node: it is subscribed to a channel that the object belongs to, or the node has a higher probability of being or arriving in the proximity of another node that is subscribed to that channel than the node that originally had the information. Not all dissemination algorithms analyze the data from other nodes: some simply fetch as much data as they can, until their cache is full, like Epidemic routing, while others only verify if they do not already contain the data or if they have not contained it recently. More advanced *content analysis* can be accomplished by assigning priorities (or utilities) to each content object from a neighboring node. In this way, considering the amount of free cache memory, a node can decide what and how many content objects it can fetch from another node. A node can also calculate the priority for its own content objects, and advertise only the priorities. Thus, a neighboring node can choose the data that maximizes the local priority of its cache. One method of computing priorities is based on heuristics that compare two content objects. Heuristics can compare content objects by their age, by their hop count or by the number of subscribers to the channel the object belongs to. A more complex approach to computing the value of priorities is to use a mathematical formula that assigns weights to various parameters. This method is used especially in socially-aware dissemination algorithms, where users are split in communities, and each community is assigned an individual weight (more about socially-aware algorithms will be presented in the next paragraph).

The final category of the proposed taxonomy is the *social awareness*. Recently, the social aspect of opportunistic networking has been studied, because the nodes in an opportunistic network are represented by humans carrying wireless devices. The human factor can be an important dimension that is already considered by several data dissemination algorithms. When designing such an algorithm, it is important to know that user movements are conditioned by social relationships. The first subcategory of social awareness is represented by *socially-unaware* algorithms, which do not assume the existence of a social structure that governs the movement or interaction of the nodes in an opportunistic network. Data dissemination techniques of this type may be as simple as spreading the content to all encountered nodes, but they can also take advantage of non-social context information such as geographical location. Most of the recent data dissemination techniques that are aware of the social aspect of an opportunistic network are *community-based*. Such dissemination algorithms assume that users can be grouped into communities, based on strong social relationships between users. Even though there are several proposed representations of social behavior, the caveman model [44] is by far the one mostly used, along with its variations. Users can belong to more communities (called "home" communities), but can also have social relationships outside of their home communities (in "acquainted" communities). Communities are usually bound to a geographical space (static social communities), but they may also be represented by a group of people who happen to be in the same place at the same time (e.g. at a conference—temporal communities). According to this model, users spend their time in the locations of their home communities, but also visit areas where acquainted communities are located. As previously stated, a utility function may be used to decide which content objects must be fetched when two nodes are in range of each other. In a *community-based* approach, each community would be assigned a weight, and the utility of a data object would be computed according to the community its owner comes from and the community of the (potentially) interested nodes.

One step that has to be executed before designing a *community-based* dissemination algorithm is the *community detection*. There are several methods used for organizing nodes from an opportunistic network into communities. One way is to simply classify nodes based on the number of contacts and contact duration of a node pair according to a threshold value, while another approach would be to define k-CLIQUE communities as unions of all k-CLIQUEs that can be reached from each other through a series of adjacent k-CLIQUEs [48]. The phase following the detection of existing communities is the design of a *community structure*. All nodes in a community can be identical (from the perspective of behavior), but there are also situations where certain nodes are more important in the dissemination scheme. As previously described, some data dissemination algorithms use network overlays constructed using hubs or brokers (e.g. nodes with the highest centrality in a community). The advantage of such an approach is that only nodes having a high centrality transfer messages to other communities. When a node wants to send a content object, it transfers it to the hub (or to a node with a higher centrality, which has a better chance of reaching the hub). The hub then transfers the object to the hub of the destination's community, where it eventually reaches the desired destination. The structure of a community has a high relevance in classifying data dissemination techniques, because a well-structured community can speed up the dissemination process significantly.

3.3 Critical Analysis of Dissemination Algorithms

We now analyze the properties of four popular techniques for disseminating data in an opportunistic network, using the categories of the proposed taxonomy. The presented study evaluates the most relevant recent work in data dissemination algorithms. We also apply the proposed taxonomy to analyze and differentiate between the presented data dissemination techniques. The **Socio-Aware Overlay** algorithm [48] is a data dissemination technique that creates an overlay for an opportunistic network with publish/subscribe communication, composed of nodes having high centrality values that have the best visibility in a community. The data dissemination technique assumes the existence of a network *with infrastructure*, built by creating an overlay comprising of representative nodes from each community. The dissemination of subscriptions is done, together with the community detection, during the *node interaction* phase, through gossiping. The gossiping dissemination sends each message to a random group of nodes, so from a *node state* point of view, the Socio-Aware algorithm takes a *hybrid* approach. In order to choose an appropriate hub (or broker) in a network, the algorithm uses a measurement unit called node centrality.

Node discovery is performed through Bluetooth and WiFi devices, while there are two modes of *node interaction*, namely unicast and direct. The former is similar to Epidemic routing, while the latter provides a more direct communication mechanism like WiFi access points. From the standpoint of *content organization*, the Socio-Aware algorithm is based on a publish/subscribe approach. At the *data exchange* phase, subscriptions and un-subscriptions with the destination of community broker nodes are exchanged, as well as a list of centrality values with a time stamp. When a broker node changes upon calculation of its closeness centrality, the subscription list is transferred from the old one to the new one. Then, an update is sent to all the brokers. During the gossiping stage, subscriptions are propagated towards the community's broker. When a publication reaches the broker, it is propagated to all other brokers, and then the broker checks its own subscription list. If there are members in its community that must receive the publication, the broker floods the community with the information. The Socio-Aware algorithm is a socially-aware *community-based* algorithm, that has its own *community detection* method. This method assumes a *community structure* that is based on a classification of the nodes in an opportunistic network, from the standpoint of another node. A first type of node is one from the same community, having a high number of contacts of long/stable durations. Another type of node is called a familiar stranger and has a high number of contacts with the current node, but the contact durations are short. There are also stranger nodes, where the contact duration is short and the number of contacts is low, and finally friend nodes, with few contacts, but high contact durations.

In order to construct an overlay for publish/subscribe systems, *community detection* is performed in a decentralized fashion, because opportunistic networks do not have a fixed structure. Thus, each node must detect its own local community. The authors propose two algorithms for distributed community detection, named Simple and *k*-CLIQUE. In order to detect its own local community, a node interacts with encountering devices and executes the detection algorithm. The detection algorithm is done in the *data exchange* phase of the interaction between nodes. Each node accomplishes the *content identification* by maintaining information about the encountered nodes and contact durations (represented as a map called the familiar set) and the local community detected so far. When two nodes meet, a data exchange is performed, with each node acquiring information about the other's familiar set and local community. Each node then updates its local community and familiar set values, according to the algorithm used. As more nodes are encountered over time, the shape of the local community may be modified. The **Wireless Ad Hoc Podcasting** system [32] extends podcasting to ad-hoc domains. The purpose is the wireless ad-hoc delivery of content among mobile nodes. Assuming a network without infrastructure, the wireless podcasting service enables the distribution of content using opportunistic contacts whenever podcasting devices are in wireless communication range. From the standpoint of content organization, the Ad Hoc Podcasting service employs a publish/subscribe approach. Thus, it organizes content into channels, which allows the users to subscribe and automatically receive updates for the content they are interested in. However, the channels themselves are divided into episodes and enclosures. Furthermore, enclosures are also divided into chunks, which are transport-level small data units of a size that can typically be downloaded in an individual node encounter. The reason for this division is the need for improving efficiency in the case of small duration contacts. The chunks can be downloaded opportunistically from multiple peers, and they are further divided into pieces, which are the atomic transport units of the network.

For *node interaction*, when two nodes are within communication range they associate and start soliciting episodes from the channels they are subscribed to. Since data is not being pushed, the nodes have complete control over the content they carry and forward. *Node discovery* is done by using broadcast beacons sent periodically by each node. *Content identification* is performed to identify channels and episodes at the remote peer that the current node is subscribed to. Two nodes in range exchange a Bloom filter hash index that contains all channel IDs that each node offers. Then each node checks the peer's hash index for channels it is subscribed to. The *data exchange* phase begins if one of the nodes has found a matching channel. In this case, it starts querying for episodes. In order to perform *content analysis*, the Wireless Ad Hoc Podcasting system proposes three different types of queries, employed according to the channel policy: a node requests any random episodes that a remote peer offers, a node requests episodes from the peer that are newer than a given date starting with the newest episode, or a node requests any episodes that are newer than a given date starting with the oldest episode.

When two nodes meet, and neither has content from a channel the other is subscribed to, several solicitation strategies are employed [32]. They are used to increase the probability of a node having content to share with other nodes in future encounters. The solicitation strategies proposed are Most Solicited, Least Solicited, Uniform, Inverse Proportional and No Caching. The Most Solicited strategy fetches entries from feeds that are the most popular. The Least Solicited strategy does the opposite, by favoring less popular feeds. The Uniform strategy treats all channels equally, by soliciting entries in a random fashion, and has the advantage of being easy to implement. The Inverse Proportional strategy maintains a history list and solicits a feed with a probability which is inverse proportional to its popularity. Finally, No Caching is more of a benchmark for other strategies than a strategy itself, and assumes that a device has no public cache at all and that it stores or distributes only content from the fields it is subscribed to. Experiments show that the Uniform strategy has the best overall performance, while Inverse Proportional is the best one in regards to fairness.

Authors of [21] propose a probabilistic publish/subscribe-based multicast distribution infrastructure for opportunistic networks based on DTN (Delay Tolerant Networking). The protocol uses a push-based asynchronous distribution delivery model. The idea is that nodes in the opportunistic network replicate bundles to their neighbors in order to get the bundle delivered by multiple hops of store-carry-and-forward.

As its name states, **DPSP** has a *content organization* based on a channel subscription system, where users subscribe to channels and senders publish content. Although from the *network organization* standpoint, DPSP assumes *no infrastructure*, the nodes in the network are divided into three categories: sources, sinks and other nodes. Sources are the nodes that send content (in the form of bundles of data) to channels, while sinks subscribe to channels and receive information from them. The rest of the nodes are not interested in specific bundles, but they store, carry and forward bundles and subscriptions.

The *node interaction* phase has several steps. When two nodes meet, *content identification* is performed through the exchange of subscription lists. An entry in a subscription list contains the channel's URI, the subscription's creation time, its lifetime, the number of hops from the original subscriber to the current node, and an identifier for the subscription. Then, each node builds a queue of bundles to forward to the peer, and uses a set of filters to select the best. The selected bundles are subsequently sorted according to their priorities, and the *data exchange* stage is performed by sending the bundles one by one until the contact finishes or the queue becomes empty.

In this approach, a set of filters is used in order to select the best bundles in a queue. Because the DPSP protocol is *socially-unaware*, the filters used do not consider the organization of users into communities. There are three filters that handle the *content analysis* and that can be used in any combination: Known Subscription Filter, Hop Count Filter and Duplicate Filter. The Known Subscription Filter removes bundles nobody is interested in, the Hop Count Filter removes bundles that are too old, while the Duplicate Filter removes bundles that the peer has already received. *Content analysis* is also performed when the remaining bundles from a queue are sorted according to their priorities. Four heuristics are used to assign priorities to bundles: Short Delay, Long Delay, Subscription Hop Count and Popularity. Short

Delay prefers newer bundles, Long Delay prefers older bundles, Subscription Hop Count sorts bundles according to hop count, and the Popularity heuristic sorts bundles by the number of nodes subscribed to the bundle's channel. The authors noticed that the Short Delay heuristic performs better with respect to delivery rates than the other heuristics.

ContentPlace [4] deals with data dissemination in resource-constrained opportunistic networks, by making content available in regions where interested users are present, without overusing available resources. To optimize content availability, Content Place exploits learned information about users' social relationships to decide where to place user data. The design of ContentPlace is based on two assumptions: users can be grouped together logically, according to the type of content they are interested in, and their movement is driven by social relationships.

For performance issues, ContentPlace assumes a network *without infrastructure*. When a node encounters another node it decides what information seen on the other node should be replicated locally. When two nodes are in range, they have to discover each other. The *node discovery* is not specified, but since the nodes are mobile devices it is probably done by WiFi or Bluetooth periodic broadcasts. For *content identification*, nodes advertise the set of channels the local user is subscribed to upon encountering another node. ContentPlace defines a utility function by means of which each node can associate a utility value to any data object. When a node encounters another peer, it selects the set of data objects that maximizes the local utility of its cache. Due to performance issues, when two nodes meet, they do not advertise all information about their data objects, but instead they exchange a summary of data objects in their caches. Finally, the *data exchange* is accomplished when a user receives a data object it is subscribed to when it is found in an encountered node's cache. *Content organization* in ContentPlace is done through channels to which users can subscribe. Consequently, unsubscription messages are not necessary, so a *stateless* approach is used for the nodes. ContentPlace is a socially-aware, *community-based* data dissemination algorithm. To have a suitable representation of users' social behavior, an approach that is similar to the caveman model is used, that has a *community structure* which assumes that users are grouped into home communities, while at the same time having relationships in acquainted communities. For *content analysis* nodes compute a utility value for each data object. The utility is a weighted sum of one component for each community its user has relationships with. The utility component of a data object for a community is the product of the object's access probability from the community members, by its cost (which is a function of the object's availability in the community), divided by the object's size. *Community detection*, like at the Socio-Aware Overlay, uses the algorithms described in [27]. By using weights based on the social aspect of opportunistic networking, ContentPlace offers the possibility of defining different policies. There are five policies defined: Most Frequently Visited (MFV), Most Likely Next (MLN), Future (F), Present (P) and Uniform Social (US). MFV favors communities a user is most likely to get in touch with, while MLN favors communities a user will visit next. F is a combination between MLN and MFV, as it considers all the communities the user is in touch with.

Data Dissemination Technique	Network Organization	Node Characteristics					Content Characteristics		Social Awareness	
		Node State	Node Interaction				Content Analysis	Content Organization	Community Detection	Community Structure
			Node Discovery	Content Identification	Data Exchange					
Socio-Aware Overlay	Overlay Infrastructure	Hybrid	Bluetooth and WiFi	Encountered nodes and cont. duration	Subscriptions and list of centralities	N/A	Publish Subscribe	Simple and K-clique algorithms	Contact duration and no. of contacts
Ad Hoc Podcasting	No Infrastructure	N/A	Broadcast beacons	Bloom filter hash index	Episodes or chunks	Solicitation strategies	Publish Subscribe	N/A	N/A
DPSP	No Infrastructure	N/A	N/A	Subscription lists	Selection of bundles	Filters and priority heuristics	Publish Subscribe	N/A	N/A
ContentPlace	No Infrastructure	Stateless	Bluetooth and WiFi	Set of channels the node is subscribed to	Data objects	Utility function	Publish Subscribe	N/A	Caveman model

Fig. 4 Critical analysis of four dissemination techniques

In the case of P, users do not favor other communities than the one they are in, while at US all the communities the users get in touch with have equal weights.

A critical analysis of the four described protocols, according to the proposed taxonomy, is presented in Fig. 4.

According to our analysis of the four solutions, only one assumes that the network over which data dissemination is performed has an infrastructure. The Socio-Aware Overlay algorithm builds an overlay infrastructure using the nodes with the highest centrality from each community. However, opportunistic networks generally contain nodes with a high degree of mobility, which make the task of creating and maintaining an infrastructure very hard to accomplish. The reason for this is that nodes may change communities very often (or they may not belong to a community at all), thus complicating the community detection phase. Furthermore, a device that is considered to be the central node (or hub) of a community may be turned off (due to different circumstances, like battery depletion), leaving the nodes in the hub's community without an opportunity to send messages to other communities, until a new hub is elected. Given these reasons, we believe that an approach that does not assume the existence of an infrastructure should be further considered.

The characteristics of a node from an opportunistic network play an important role in the structure of a data dissemination algorithm. Node characteristics refer to the way a node's state is represented and the way nodes interact when they are in contact. As stated in Sect. 4, the approach a data dissemination algorithm can take in regard to node state can be either stateless, stateful, or hybrid. Of the protocols we analyzed, ContentPlace chooses a stateless approach, while the Socio-Aware Overlay uses a hybrid representation of a node's state. The authors of the other two algorithms do not specify the node state, but we assume a stateful approach, because of the way the content is represented (for example, DPSP maintains subscription lists, for which node state is required). According to [48], a hybrid approach is the preferred solution because it takes advantage of both stateful and stateless approaches. Such an approach would not suffer under frequent topology changes, while at the same it would not require a large amount of control traffic.

The interaction between nodes has three steps that have been presented in detail in Sect. 4: node discovery, content identification and data exchange. Node discovery is usually done in the same way for all algorithms analyzed, but it may differ according to the type of devices that are present in the network. In case of the Socio-Aware Overlay and ContentPlace, the discovery is performed by using the Bluetooth or WiFi capabilities. The Ad Hoc Podcasting algorithm uses broadcast beacons, while the authors of DPSP do not mention a particular discovery method. It is a good approach to use the existing capabilities from the wireless protocols, but a data dissemination algorithm should try to extend the battery's life as much as possible. For example, when the battery is low, the broadcast beacons should be sent at larger time intervals.

Content identification, meaning the way in which nodes represent the data internally, also has a big impact in the efficiency of a data dissemination technique. The Socio-Aware Overlay maintains information about the encountered nodes and the duration of contacts, Ad Hoc Podcasting uses a Bloom filter hash index that contains all channel IDs, DPSP exchanges subscription lists and ContentPlace advertises the set of channels a node is subscribed to. The most efficient method is using Bloom filters, because they are space efficient data structures of fixed size that avoid unnecessary transmissions of data that the receiver has already received [3].

Data exchange should also be performed in a manner that optimizes the duration of a transfer. The nodes from the Socio-Aware Overlay exchange subscriptions and lists of centrality values, Ad Hoc Podcasting exchanges episodes or chunks, DPSP uses bundles and ContentPlace nodes exchange data objects. The smaller the data unit, the bigger is the chance of a transmission to successfully finish, even in opportunistic networks where contact durations are very small. Therefore, one of the best approaches is the one employed by Ad Hoc Podcasting, where data is split into episodes and chunks.

The type of content organization that best suits opportunistic networks is the publish/subscribe pattern. The reason for this is that participants are decoupled from time, space and flow. Interested users simply subscribe to certain channels and receive data whenever the publishers post it. Publishers and subscribers do not have to be online at the same time, and it is not necessary that a direct connection exists between them. Consequently, all the analyzed data dissemination techniques organize their content according to a publish/subscribe approach. Content can also be analyzed in order for a node to decide what to download from an encountered peer. The Ad Hoc Podcasting technique uses five solicitation strategies that aim to increase the probability of a node having content to share with other nodes. DPSP has three filters used to select the best bundles in a queue and four heuristics that sort the remaining bundles. Finally, ContentPlace computes a utility function based on every community a node is in relationship with. The ContentPlace approach performs the best, because it takes advantage of the social aspect of opportunistic networking.

According to [11], human social structures are at the core of opportunistic networking. This is because humans carry the mobile devices, and it is the human mobility that generates communication opportunities when two or more devices come into contact. Social-based forwarding and dissemination algorithms reduce by about an order of magnitude the overhead, compared to algorithms such as Epidemic routing. Therefore, the social aspect has a very important role in the efficiency of a data dissemination technique in an opportunistic network. Social awareness is based on the division of users into communities, which are defined as groups of interacting individuals organized around common values within a shared geographical location. Thus, an important step for socially-aware dissemination algorithms is community detection. Of the techniques we studied, only the Socio-Aware Overlay proposes its own community detection algorithms, called Simple and k-CLIQUE. ContentPlace uses similar algorithms, while Ad Hoc Podcasting and DPSP are socially-unaware. As far as community structure goes, the Socio-Aware Overlay splits the nodes in a community from the standpoint of another node, according to the contact duration and number of contacts, while ContentPlace adopts a model similar to the caveman model. We consider that the future of data dissemination algorithms should be based on a socially-aware approach to take advantage of the human aspect of opportunistic networking. After analyzing the four data dissemination techniques, we can conclude that there is no single best approach, but each algorithm provides certain aspects that offer advantages over the other implementations. In the next phase we plan to extend this work and propose a dissemination algorithm that uses the advantages of all analyzed solutions for maximum efficiency.

4 Future Trends and Research Directions in Big Data Platforms for the Internet of Things

The evolution of the Internet of Things towards connecting every Thing on the planet in a very complex and large environment gives raise to high demanding requirements, which are subject of actual research. The continuously increasing volume of data collected from and exchanged among Things will require highly scalable environments able to support the high resulting network traffic, and offer the necessary storage capacity and computing power for data preservation and transformation. Communication protocols are needed to enable not only the high capacity traffic but also maintain the connectivity between Things even in case of transient disconnection of wired or wireless links. Also, new solutions should be found for efficiently store, search and fetch the data manipulated in these environments.

IoT is at the base of developing many applications that include things, people, economy, mobility, and governance. They enrich the urban environment with situational information, which can be rapidly exploited by citizens in their professional, social, and individual activities to increase city competitiveness. In addition, outdoor computing and user-centric approaches to services can improve the core urban systems such as public safety, transportation, government, agency administration and social services, education, healthcare. Context-aware applications use large amounts of input data, in various formats, collected from sensors or mobile users, from public open data sources or from other applications.

Internet of Things is not yet a reality, "but rather a prospective vision of a number of technologies that, combined together, could in the coming 5–15 years drastically modify the way our societies function" [37]. The evolution of the IoT on medium and long term unleashed a huge interest and gave raise to many re-search projects, workshops, conferences, reports and survey papers. In this section we discuss the aspects related to the IoT infrastructure and services with emphasis on the main challenges.

It is estimated [38] that IoT will have to accommodate over 50,000 billion objects of very diverse types. Standardization and interoperability will be absolute necessities for interfacing them with the Internet. New media access techniques, communication protocols and sustainable standards shall be developed to make thing communicate with each other and people. One approach would be the en-capsulation of smart wireless identifiable devices and embedded devices in web services. We can also consider the importance of enhancing the quality of service aspects like response time, resource consumption, throughput, availability, and reliability. The discovery and use of knowledge about services availability and of publish/subscribe/notify mechanisms would also contribute to enhancing the management of complex thing structures.

Enhanced monitoring facilities are needed to support informed decisions cooperatively adopted in collections of things. Also, increasing the quality of information collected from things will be the use of distributed bio-inspired approaches.

The huge number of things will make their management a very difficult task. New solutions are needed to enable things' adaptation, autonomous behavior, intelligence, robustness, and reliability. They could be based on new general centralized or distributed organizational architectures. Another solution will be endowing things with self-* capabilities in various forms: self-organization, self-configuration, self-Healing, self-optimization, and self-protection.

New services shall be available for persistent distributed knowledge storing and sharing, and new computational resources shall be used for complicated tasks execution. Actual forecasts indicate that in 2015 more than 220 EB of data will be stored [38]. At the same time, optimal distribution of tasks between smart objects with high capabilities and the IoT infrastructure shall be found.

New mechanisms and protocols will be needed for privacy and security issues at all IoT levels including the infrastructure. Solutions for stronger security could be based on models employing the context-aware capability of things.

New methods are required for energy saving and energy efficient and self-sustainable systems. Researchers will look for new power efficient platforms and technologies and will explore the ability of smart objects to harvest energy from their surroundings.

Obviously, new platform architectures and developing techniques are needed for the efficient storing, real-time processing, data placement and replication of Big Data in order to achieve higher real-time guarantee for data provisioning and increased data storage efficiency. One promising solution is Cloud computing for Big Data. Cloud technologies simplify building Big Data infrastructure, support massive growth of storage capacity, and guarantee the performance agreed with customers (SLA). New Cloud storage and compute components will be needed to allow global data availability and access over different types of networks, for various cooperative user communities, which exploit different facets of Big Data, in application with different service level requirements (batch, online, real-time, event-driven, collaborative, etc.), and with different security needs going up to highly trusted and secure environments.

The large variety of technologies and designs used in the production of Things is a main concern when considering the interoperability. One solution is the adoption of standards for Things inter-communication. Adding self-configuration and self-management properties could be necessary to allow Things inter-operate and, in addition, integrate within the surrounding operational environment. This approach is superior to the centralized management, which cannot respond to difficulties induced by the dimensions, dynamicity and complexity of the Internet of Things. The autonomic behavior is important at the operational level as well. Letting autonomic Things react to events generated by context changes facilitates the construction and structuring of large environments that support the Internet of Things.

Special requirements come from the scarcity of Things' resources, and are concerned with power consumption. New methods of efficient management of power consumption are needed and could apply at different levels, from the architecture level of things to the level of the network routing. They could substantially contribute to lowering the cost of Things, which is essential for the rapid expansion of the internet of Things.

Some issues come from the distributed nature of the environment in which different operations and decisions are based on the collaboration of Things. One issue is how Things convergence on a solution and how the quality of the solution can be evaluated. Another issue is how to protect against faulty Things including those exhibiting malicious behavior. Finally, the way Things can cope with security issues to preserve confidentiality, privacy, integrity and availability are of high interest.

5 Conclusions and Remarks

Actual evolution of the Internet of Things towards connecting every thing on the planet in a very complex and large environment gives raise to high demanding requirements, which challenge the actual and future research. The continuously increasing volume of data collected from and exchanged among things will require highly scalable environments able to support the high resulting network traffic, and offer the necessary storage capacity and computing power for data preservation and transformation. Communication protocols are needed to enable not only the high capacity traffic but also maintain the connectivity between things even in case of transient disconnection of wired or wireless links. Also, new solutions should be found for efficiently store, search and fetch the data manipulated in these environments.

The chapter addresses new research and scientific challenges in context-aware environments for IoT. They refer first to the identification, internal organization, provision of context information, intelligence, self-adaptation, and autonomic behavior of individual things. Then, actual research and main challenges related to IoT infrastructure are discussed, with emphasis on services for context awareness, inter-communication, interoperability, inter-cooperation, self-organization, fault tolerance, energy saving, compute and storage services, and management of things collections and structures.

We also analyzed existing relevant work in the area of data dissemination in opportunistic networks for IoT. We began by highlighting the use of ONs in real life and describing some potential scenarios where they can be applied. Then, we presented the categories of a proposed taxonomy that captures the capabilities of data dissemination techniques used in opportunistic networks. Moreover, we critically analyzed four relevant data dissemination techniques using the proposed taxonomy. The purpose of the taxonomy, aside from classifying dissemination methods, has been to analyze and compare the strengths and weaknesses of the analyzed data dissemination algorithms. Using this knowledge, we believe that an efficient data dissemination technique for opportunistic networks for future IoT platforms can be devised. We believe that the future of opportunistic networking lies in the social property of mobile networks, so a great importance should be given to this aspect.

Finally, future trends and research directions for the IoT infrastructure are discussed including performance, monitoring, reliability, safety, survivability, self-healing, transparency, availability, privacy, and others.

Acknowledgments The work was partially supported by the project "SideSTEP—Scheduling Methods for Dynamic Distributed Systems: a self-* approach", PN-II-CT-RO-FR-2012-1-0084.

References

1. Athanasopoulos, D., Zarras, A.V., Issarny, V., Pitoura, E., Vassiliadis, P.: Cowsami: interface-aware context gathering in ambient intelligence environments. Pervasive Mob. Comput. **4**(3), 360–389 (2008)
2. Bardram, J.E.: The java context awareness framework (jcaf)-a service infrastructure and programming framework for context-aware applications. In: Pervasive Computing, pp. 98–115. Springer (2005)
3. Bjurefors, F., Gunningberg, P., Nordstrom, E., Rohner, C.: Interest dissemination in a searchable data-centric opportunistic network. In: Wireless Conference (EW), 2010 European, pp. 889–895. IEEE (2010)
4. Boldrini, C., Conti, M., Passarella, A.: Contentplace: social-aware data dissemination in opportunistic networks. In: Proceedings of the 11th International Symposium on Modeling, Analysis and Simulation of Wireless and Mobile Systems, pp. 203–210. ACM (2008)
5. Chen, C., Helal, S.: A device-centric approach to a safer internet of things. In: Proceedings of the 2011 International Workshop on Networking and Object Memories for the Internet of things, pp. 1–6. ACM (2011)
6. Chen, L.J., Sun, T., Liang, N.C.: An evaluation study of mobility support in zigbee networks. J. Signal Process. Syst. **59**(1), 111–122 (2010)
7. Chourabi, H., Nam, T., Walker, S., Gil-Garcia, J.R., Mellouli, S., Nahon, K., Pardo, T.A., Scholl, H.J.: Understanding smart cities: an integrative framework. In: 2012 45th Hawaii International Conference on System Science (HICSS), pp. 2289–2297. IEEE (2012)
8. Christensen, H.B.: Using logic programming to detect activities in pervasive healthcare. In: Logic Programming, pp. 421–436. Springer (2002)

9. Ciobanu, R.I., Dobre, C.: Predicting encounters in opportunistic networks. In: Proceedings of the 1st ACM Workshop on High Performance Mobile Opportunistic Systems, pp. 9–14. ACM (2012)
10. Conti, M., Crowcroft, J., Giordano, S., Hui, P., Nguyen, H.A., Passarella, A.: Routing issues in opportunistic networks. In: Middleware for Network Eccentric and Mobile Applications, pp. 121–147. Springer (2009)
11. Conti, M., Giordano, S., May, M., Passarella, A.: From opportunistic networks to opportunistic computing. IEEE Commun. Mag. **48**(9), 126–139 (2010)
12. Costa, P.D., Guizzardi, G., Almeida, J.P.A., Pires, L.F., van Sinderen, M.: Situations in conceptual modeling of context. In: EDOC Workshops, p. 6 (2006)
13. Coutaz, J., Crowley, J.L., Dobson, S., Garlan, D.: Context is key. Commun. ACM **48**(3), 49–53 (2005)
14. Cristea, V., Dobre, C., Costan, A., Pop, F.: Middleware and architectures for space-based and situated computing. Int. J. Space-Based Situated Comput. **1**(1), 43–58 (2011)
15. Cugola, G., Migliavacca, M.: Multicar: Remote invocation for large scale, context-aware applications. In: Proceedings of the IEEE Symposium on Computers and Communications, ISCC '10, pp. 570–576. IEEE Computer Society, Washington, D.C., USA (2010). doi:10.1109/ISCC. 2010.5546718. http://dx.doi.org/10.1109/ISCC.2010.5546718
16. Denœux, T.: A k-nearest neighbor classification rule based on dempster-shafer theory. In: Classic Works of the Dempster-Shafer Theory of Belief Functions, pp. 737–760. Springer (2008)
17. Desta, M.S., Hyytiä, E., Ott, J., Kangasharju, J.: Characterizing content sharing properties for mobile users in open city squares. In: 10th Annual IEEE/IFIP Conference on Wireless On-Demand Network Systems and Services (WONS), Banff, Canada (2013)
18. Dey, A.K., Abowd, G.D., Salber, D.: A conceptual framework and a toolkit for supporting the rapid prototyping of context-aware applications. Hum. Comput. Interact. **16**(2), 97–166 (2001)
19. Dobre, C., Manea, F., Cristea, V.: Capim: A context-aware platform using integrated mobile services. In: 2011 IEEE International Conference on Intelligent Computer Communication and Processing (ICCP), pp. 533–540. IEEE (2011)
20. Doukas, C., Maglogiannis, I., Tragas, P., Liapis, D., Yovanof, G.: Patient fall detection using support vector machines. In: Artificial Intelligence and Innovations 2007: From Theory to Applications, pp. 147–156. Springer (2007)
21. Greifenberg, J., Kutscher, D.: Efficient publish/subscribe-based multicast for opportunistic networking with self-organized resource utilization. In: 22nd International Conference on Advanced Information Networking and Applications—Workshops, 2008. AINAW 2008, pp. 1708–1714. IEEE (2008)
22. Gu, T., Pung, H.K., Zhang, D.Q.: A service-oriented middleware for building context-aware services. J. Network Comput. Appl. **28**(1), 1–18 (2005)
23. Haddadi, H., Hui, P., Henderson, T., Brown, I.: Targeted advertising on the handset: privacy and security challenges. In: Pervasive Advertising, pp. 119–137. Springer (2011)
24. He, J., Zhang, Y., Huang, G., Cao, J.: A smart web service based on the context of things. ACM Trans. Internet Technol. (TOIT) **11**(3), 13 (2012)
25. Henricksen, K., Robinson, R.: A survey of middleware for sensor networks: state-of-the-art and future directions. In: Proceedings of the International Workshop on Middleware for Sensor Networks, pp. 60–65. ACM (2006)
26. Hui, P., Crowcroft, J., Yoneki, E.: Bubble rap: Social-based forwarding in delay-tolerant networks. IEEE Trans. Mobile Comput. **10**(11), 1576–1589 (2011)
27. Hui, P., Yoneki, E., Chan, S.Y., Crowcroft, J.: Distributed community detection in delay tolerant networks. In: Proceedings of 2nd ACM/IEEE International Workshop on Mobility in the Evolving Internet Architecture, p. 7. ACM (2007)
28. Kalyan, A., Gopalan, S., Sridhar, V.: Hybrid context model based on multilevel situation theory and ontology for contact centers. In: Third IEEE International Conference on Pervasive Computing and Communications Workshops, 2005. PerCom 2005 Workshops, pp. 3–7. IEEE (2005)

29. Kanda, T., Glas, D.F., Shiomi, M., Ishiguro, H., Hagita, N.: Who will be the customer? A social robot that anticipates people's behavior from their trajectories. In: Proceedings of the 10th International Conference on Ubiquitous Computing, pp. 380–389. ACM (2008)
30. Kim, H., Cho, Y.J., Oh, S.R.: Camus: A middleware supporting context-aware services for network-based robots. In: IEEE Workshop on Advanced Robotics and its Social Impacts, 2005, pp. 237–242. IEEE (2005)
31. Kortuem, G., Kawsar, F., Fitton, D., Sundramoorthy, V.: Smart objects as building blocks for the internet of things. IEEE Internet Comput. 14(1), 44–51 (2010)
32. Lenders, V., Karlsson, G., May, M.: Wireless ad hoc podcasting. In: 4th Annual IEEE Communications Society Conference on Sensor, Mesh and Ad Hoc Communications and Networks, 2007. SECON'07, pp. 273–283. IEEE (2007)
33. Lilien, L., Gupta, A., Yang, Z.: Opportunistic networks for emergency applications and their standard implementation framework. In: IEEE International Performance, Computing, and Communications Conference, 2007. IPCCC 2007, pp. 588–593. IEEE (2007)
34. McKeever, S., Ye, J., Coyle, L., Dobson, S.: Using dempster-shafer theory of evidence for situation inference. In: Smart Sensing and Context, pp. 149–162. Springer (2009)
35. Ngo, H.Q., Shehzad, A., Liaquat, S., Riaz, M., Lee, S.: Developing context-aware ubiquitous computing systems with a unified middleware framework. In: Embedded and Ubiquitous Computing, pp. 672–681. Springer (2004)
36. Pelusi, L., Passarella, A., Conti, M.: Opportunistic networking: data forwarding in disconnected mobile ad hoc networks. IEEE Commun. Mag. 44(11), 134–141 (2006)
37. Project, C.E.F.: RFID and the inclusive model for the internet of things. http://www.grifs-project.eu/data/File/CASAGRASFinalReport.pdf (2012). Accessed 15 July 2013
38. Society, E.C.I., DG, M.: Infso d.4 networked enterprise & rfid, working group rfid of the etp eposs. internet of things in 2020. roadmap for the future. http://www.iot-visitthefuture.eu/fileadmin/documents/researchforeurope/270808_IoT_in_2020_Workshop_Report_V1-1.pdf (2009). Accessed 15 August 2013
39. Tan, L., Wang, N.: Future internet: The internet of things. In: 2010 3rd International Conference on Advanced Computer Theory and Engineering (ICACTE), vol. 5, pp. V5–376. IEEE (2010)
40. Truong, H.L., Dustdar, S.: A survey on context-aware web service systems. Int. J. Web Inf. Syst. 5(1), 5–31 (2009)
41. Van Kasteren, T., Noulas, A., Englebienne, G., Kröse, B.: Accurate activity recognition in a home setting. In: Proceedings of the 10th International Conference on Ubiquitous Computing, pp. 1–9. ACM (2008)
42. Vergori, P., Ntanos, C., Gavelli, M., Askounis, D.: The webinos architecture: a developern++s point of view. In: Mobile Computing, Applications, and Services, pp. 391–399. Springer (2013)
43. Vermesan, O., Friess, P., Guillemin, P., Gusmeroli, S., Sundmaeker, H., Bassi, A., Jubert, I.S., Mazura, M., Harrison, M., Eisenhauer, M., et al.: Internet of things strategic research roadmap. In: Vermesan, O., Friess, P., Guillemin, P., Gusmeroli, S., Sundmaeker, H., Bassi, A., et al. (eds.) Internet of Things: Global Technological and Societal Trends, pp. 9–52 (2011)
44. Wu, J.: Small worlds: the dynamics of networks between order and randomness—Book review. SIGMOD Rec. 31(4), 74–75 (2002)
45. Yau, S.S., Karim, F.: A context-sensitive middleware for dynamic integration of mobile devices with network infrastructures. J. Parallel Distrib. Comput. 64(2), 301–317 (2004)
46. Yau, S.S., Liu, J.: Hierarchical situation modeling and reasoning for pervasive computing. In: The Fourth IEEE Workshop on Software Technologies for Future Embedded and Ubiquitous Systems, 2006 and the 2006 Second International Workshop on Collaborative Computing, Integration, and Assurance. SEUS 2006/WCCIA 2006, pp. 6–15. IEEE (2006)
47. Ye, J., Dobson, S., McKeever, S.: Situation identification techniques in pervasive computing: a review. Pervasive Mob. Comput. 8(1), 36–66 (2012)
48. Yoneki, E., Hui, P., Chan, S., Crowcroft, J.: A socio-aware overlay for publish/subscribe communication in delay tolerant networks. In: Proceedings of the 10th ACM Symposium on Modeling, Analysis, and Simulation of Wireless and Mobile Systems, pp. 225–234. ACM (2007)

Improving Data and Service Interoperability with Structure, Compliance, Conformance and Context Awareness

José C. Delgado

Abstract The Web has been continuously growing, in number of connected nodes and quantity of information exchanged. The Internet of Things (IoT), with devices that largely outnumber Internet users and have little computing power, has contributed to show the limitations of current Web technologies (namely, HTTP) for the integration of small systems. New protocols, such as CoAP and WebSockets, alleviate some of the problems but do not solve them at the origin. This chapter revisits the interoperability problem in the IoT context and proposes an architectural style, Structural Services, which combines the service modelling capabilities of SOA with the flexibility of structured resources in REST. The simplicity of JSON is combined with operation descriptions in a service-oriented distributed programming language that provides design time self-description without the need for a separate schema language. Interoperability, at the data and service levels, is enhanced with partial interoperability (by using compliance and conformance) and a context model that contemplates forward and backward contextual information. Efficiency and lower computing power requirements, aspects highly relevant for IoT systems, are supported by using a binary message format resulting from compilation of the source program and that uses self-description information only when needed.

J. C. Delgado (✉)
Department of Computer Science and Engineering,
Instituto Superior Técnico, Universidade de Lisboa, Av. Prof. Cavaco Silva,
Taguspark 2744-016, Porto Salvo, Portugal
e-mail: jose.delgado@ist.utl.pt; jose.delgado@tecnico.ulisboa.pt

N. Bessis and C. Dobre (eds.), *Big Data and Internet of Things:*
A Roadmap for Smart Environments, Studies in Computational Intelligence 546,
DOI: 10.1007/978-3-319-05029-4_2, © Springer International Publishing Switzerland 2014

1 Introduction

Usually, Big Data is characterized by three main properties, referred to as the three Vs [1]:

- Volume (data size and number);
- Velocity (the rate at which data are generated or need to be processed);
- Variety (heterogeneity in content and/or form).

Essentially, *Big* means too complex, too much and too many to apply conventional techniques, technologies or systems, since their capabilities are not enough to handle such extraordinary requirements.

In the context of the Internet of Things (IoT), Big Data also means Big Services, in the sense of the three Vs: too many services interacting and exchanging too many data, at too high rates in a too heterogeneous environment. A service can be a complex application in a server or a very simple functionality provided by a small sensor. Two additional Vs (Value and Veracity) have been proposed by Demchenko et al. [2], but are less relevant to the scope of this chapter.

Service interaction implies interoperability. The main integration technologies available today are based on the SOA and REST architectural styles, with Web Services and REST on HTTP as the most used implementations. Web Services are a closer modelling match for more complex applications, since they allow a service to expose an arbitrary interface. In turn, they lack structure (all services are at the same level) and are complex to use. REST is certainly simpler, since essentially it entails a set of best practices to use resources in a generic way, as heralded by HTTP. Although REST provides structure, it supports a single syntax interface and only a set of previously agreed data types. Also, its apparent simplicity hides the functionality that the interlocutors need to be endowed with. In addition, both SOA and REST rely, in practice and in most cases, on HTTP and text based data (XML or JSON).

These technologies stem from the epoch in which people were the main Internet drivers, text was king, document granularity was high, the distinction between client and server was clear, scalability was measured by the number of clients and the acceptable application reaction times could be as high as one second (human timescale).

Today, the scenario is much different. Cisco estimates [3] indicate that, in the end of 2012, the number of devices connected to the Internet was around 9 billion devices, versus around 2.4 billion Internet users in June 2012, according to the Internet World Stats site (http://www.internetworldstats.com/stats.htm). The European Commission estimates that the number of Internet-capable devices will be on the order of 50–100 billion devices by 2020 [4], with projections that at that time the ratio of mobile machine sessions over mobile person sessions will be on the order of 30. The number of Internet-enabled devices grows faster than the number of Internet users, since the worldwide population is estimated by the United Nations to be on the order of 8 billion by 2020 [5]. This means that the Internet is no longer dominated by human users, but rather by smart devices that are small computers and require technologies suitable to them, rather than those mostly adequate to fully-fledged servers.

These technologies include native support for binary data, efficient and full duplex protocols, machine-level data and service interoperability and context awareness for dynamic and mobile environments, such as those found in smart cities [6]. Today, these features are simulated on top of Web Services, RESTful libraries, HTTP, XML, JSON and other technologies, but the problem needs to be revisited to minimize the limitations at the source, instead of just hiding them with abstraction layers that add complexity and subtract performance.

The main goals of this chapter are:

- To rethink the interoperability problem in the context of data intensive applications and taking into account the IoT and smart environment requirements, regarding the characteristics of devices (low computing capabilities, power constraints, mobility, dynamicity, heterogeneity, and so on);
- To show that the good characteristics of the SOA and REST models can be combined by using a better designed interoperability model, based on active resources instead of passive document descriptions;
- To illustrate how the three Vs of Big Data (and Big Services) can be better supported by this interoperability model, namely in terms of:
  - Performance (Volume and Velocity), by using binary data and cached data interoperability mappings;
  - Heterogeneity (Variety), by using compliance and conformance (less restrictive than schema sharing) and context awareness (which increases the range of applicability).

This chapter is organized as follows. Section 2 discusses interoperability in the IoT context and presents some of the current interoperability technologies. Section 3 provides a generic model of resource interaction and a layered interoperability framework. Section 4 establishes the main requirements to improve interoperability in IoT environments. Section 5 proposes and describes a new architectural style, *Structural Services*. Sections 6 and 7 show how interoperability can be improved with structural typing and context awareness, respectively. Section 8 discusses the rationale for the work presented in this chapter, Sect. 9 present guidelines for future work and Sect. 10 draws some conclusions and summarizes the scientific contributions of this chapter.

2 Background

IoT scenarios should be implemented by a complex ecosystem of distributed applications, with nodes interacting seamlessly with each other, independently of platform, programming language, network protocol and node capabilities (from a fully-fledged computer, connected to a wide, fast and reliable network, down to a tiny sensor, connected to a local, slow and *ad hoc* wireless network).

This is not what we have today. Distributed programming always faced practical difficulties, caused by the lack of a generic and widely available way to exchange data.

The RPC (Remote Procedure Call) approach, such as Java RMI (Remote Method Invocation) and CORBA [7] provided a solution, but usually language or system dependent due to differences in protocols and message formats. In any case, one fundamental problem was the attempt to mimic a local environment (namely, with pointers) on top of a distributed one, instead of designing a system with a distributed mind-set from the start.

The advent of the Web changed this, by providing standardized mechanisms:

- To retrieve data from a remote node (a server), first in HTML and then by a format evolution towards XML and JSON;
- To remotely invoke behaviour in that node, first with the CGI (Common Gateway Interface) [8] and then with several technologies, such as Web Services [9] and REST [10], the most common today.

However, these mechanisms are supported on HTTP, a protocol conceived for the original Web problem [11], browsing remote hypermedia documents with text as the main media type. The main goal was information retrieval, from humans to humans, with client-server dichotomy and scalability as the main tenets. This justified many decisions taken at that time. Given the explosive success of the Web, these are still conditioning current technologies.

The IoT has different goals and needs from those of the Web, since the emphasis is now on machine to machine interactions, rather than human to human relationships.

Even the current Web, at human level, is very different from the original Web. Instead of a static Web of documents, we now have a Web of services, or at least a dynamic Web of documents, generated on the fly from databases or under control of server side applications. The classic client-server paradigm is also taking a turn with one of the latest developments, real-time communications in the Web [12] directly between browsers, by using WebRTC [13] in a peer-to-peer architecture.

This is being done in an evolutionary way, with essentially JavaScript and Web-Sockets [14], to circumvent some of the limitations of HTTP. The most important issue is that finally humans are also becoming direct Web producers and not merely clients of servers.

This is also the philosophy underlying IoT, in which all nodes, down right to very simple sensors, should be able to initiate transactions and to produce information. Since the IoT encompasses not only machines but also humans in a complete environment, a common architecture (or at least a common way to integrate architectures) should be the way to implement the vision of the IoT [4].

Typical implementations of the IoT use Web technologies for global interoperability. There are IoT applications based on SOA [15, 16], but the RESTful approach seems to be more popular [17–19], since it is simpler and more adequate to applications with smaller devices. Using basic HTTP is much simpler than using Web Services and even link usage is kept to a minimum. In fact, in typical systems such as those described in [17, 20], the only resource representations returned with links are the list of devices. This means that there is only application data structure exposed, not behaviour, which is essentially just a CRUD (Create, Read, Update and Delete) application [10].

In both SOA and REST, interoperability is achieved by using common data types (usually structured), either by sharing a schema (i.e., WSDL files) or by using a previously agreed data type (typical in RESTful applications). There is usually no support for partial interoperability and polymorphism in distributed systems. Compliance [21] and conformance [22] are concepts that support this, but they are not used in the SOA and REST contexts.

Context awareness [23] is very important to many IoT applications, but there is typically no support from current integration technologies for context-oriented services [24].

IoT devices can be integrated with the Web in the following ways:

- The device itself has Web capability (an HTTP server using SOA or REST) and can interact with the rest of the world directly. This implies connection to the Internet and support for the TCP/IP stack;
- Through a gateway, which connects to the Internet on one side and to the device's non-IP network (such as a ZigBee [25] wireless sensor network) on the other. The gateway needs to mediate not only network protocols but also the application level messages.

The protocols underlying the Web (TCP/IP and HTTP) are demanding in terms of the IoT. Hardware capabilities are increasingly better and efforts exist to reduce the requirements, such as IPv6 support on Low power Wireless Personal Area Networks (6LoWPAN) [26, 27] and the CoAP (Constrained Application Protocol) [28] to deal with constrained RESTful environments. However, the fact is that new applications, such as those based on nanotechnology [29, 30] and vehicular networks [31], prevent the universal adoption of an Internet backbone approach.

3 The Interoperability Problem

IoT scenarios and applications entail a complex ecosystem of resources interacting with each other in a big data environment, with many and heterogeneous data exchanged between many and heterogeneous resources.

To be valuable and meaningful, interactions require *interoperability*. This is one of the most fundamental characteristics of distributed systems. Although there is no universally accepted definition of interoperability, since its meaning can vary accordingly to the perspective, context and domain under consideration, probably the most cited definition is provided by the 24765 standard [32]. According to it, interoperability is "the ability of two or more systems or components to exchange information and to use the information that has been exchanged". In other words, merely exchanging information is not enough. Resources must also be able to understand it and to react to it according to each other's expectations. To detail what this means, we need to analyse the interoperability problem, starting by establishing a model of resources and of their interactions.

3.1 A Generic Model of Resource Interaction

We define *resource* as an entity of any nature (material, virtual, conceptual, noun, action, and so on) that embodies a meaningful, complete and discrete concept, which makes sense by itself and can be distinguished from, although able to interact with, other entities.

Resources are the main modelling artefacts and should mimic, as closely as possible, the entities in the problem domain. Physical resources range from simple sensors up to people, although the latter are represented in the computer-based realm by user interface devices. Non-physical resources typically correspond to software-based modules, endowed with interaction capabilities.

Resources are discrete and distinct entities, atomic (an indivisible whole) or structured (recursively composed of other resources, its components, with respect to which it performs the role of container). Each component can only have one direct container (yielding a strong composition model, in a tree shaped resource structure). Resources can also exhibit a weak composition model, by referring to each other by *references*. The reference itself is an atomic resource, a component of the resource that holds it (its container), but not the referenced resource. References support the existence of a directed graph of resources, superimposed on the resource tree.

Atomic resources can have state, either immutable or changeable. The state of a structured resource is, recursively, the set of the states of its components. Resources can migrate (move) from one container to another, changing the system's structure.

Resources are self-contained and interact exclusively by sending *messages*, i.e., resources moved from the sender to the receiver. Each resource implements a *service*, defined as the set of *operations* supported by that resource and that together define its behaviour (the set of reactions to messages that the resource exhibits). We also assume that resources are typically distributed and messages are exchanged through a *channel*, a resource specialized in relaying messages between interacting resources.

One resource, playing the role of *consumer*, sends a request message (over the channel) to another resource, playing the role of *provider*, which executes the request and eventually sends a response message back to the consumer. This basic interaction patterns constitutes a *transaction*. More complex interactions can be achieved by composing transactions, either temporally (the consumer invokes the provider several times) or spatially (the consumer X invokes the provider Y, which in turn invokes another provider, Z, and so on).

A *process* is a graph of all transactions that are allowed to occur, starting with a transaction initiated at some resource and ending at a transaction that neither provides a response nor initiates new transactions. A process corresponds to a use case of a resource and usually involves other resources, as transactions flow (including loops and recursion, eventually).

Figure 1 depicts this conceptual model in UML. Association classes describe relationships (such as `Transaction` and the message `Protocol`) or roles, such as `Message`, which describes the roles (specialized by `Request` and `Response`)

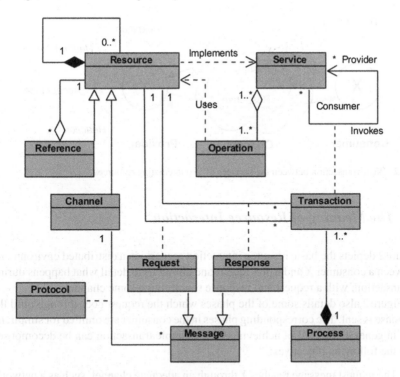

Fig. 1 A generic model of resource interaction

performed by resources when they are sent as messages. The interaction between resources has been represented as interaction between services, since the service represents the active part (behaviour) of a resource and exposes its operations. When we mention resource interaction, we are actually referring to the interaction between the services that they implement.

Therefore:

- Resources entail structure, state and behaviour;
- Services refer only to behaviour, without implying a specific implementation;
- Processes are a view of the behaviour sequencing and flow along services, which are implemented by resources.

This is a model inspired in real life, conceived to encompass not only computer-based entities but also humans (themselves resources) and any other type of physical resources, such as sensors.

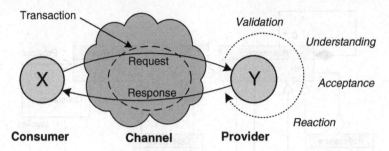

Fig. 2 Basic transaction between the services of two interacting resources

3.2 The Meaning of Resource Interaction

Figure 2 depicts the basic resource interaction scenario in a distributed environment, between a consumer X and a provider Y, and allows us to detail what happens during a transaction, with a request and response sent through some channel.

Figure 2 also details some of the phases which the request goes through until the response is sent. The corresponding phases in the consumer are omitted for simplicity but, in general, the goal of achieving such a simple transaction can be decomposed into the following objectives:

1. The request message reaches Y through an adequate channel, such as a network;
2. Y validates the request, according to its requirements for requests;
3. Y understands what X is requesting;
4. Y accepts to honour the request;
5. The reaction of Y to the request, including the corresponding effects, is consistent with the expectations of X regarding that reaction;
6. The response message reaches X (usually through the same channel as the request message);
7. X validates the response, according to its requirements for the response;
8. X understands what Y is responding;
9. X accepts the response;
10. X reacts appropriately, fulfilling the role of the response in a consistent manner with the overall transaction.

Therefore, just sending a request to a provider, or a response to the consumer, hoping that everything works, is not enough. We need to ensure that each message is validated and understood by the resource that receives it. In general, meaningfully sending a message (request or response) entails the following aspects:

- *Willingness* (objectives 4 and 9). Both sender and receiver are interacting resources and, by default, their services are willing to accept requests and responses. However, non-functional aspects such as security can impose constraints;

- *Intent* (objectives 3 and 8). Sending the message must have a given intent, inherent to the transaction it belongs to and related to the motivation to interact and the goals to achieve with that interaction;
- *Content* (objectives 2 and 7). This concerns the generation of the content of a message by the sender, expressed by some representation, in such a way that the receiver is also able to validate and interpret it;
- *Transfer* (objectives 1 and 6). The message content needs to be successfully transferred from the context of the sender to the context of the receiver;
- *Reaction* (objectives 5 and 10). This concerns the reaction of the receiver upon reception of a message, which should produce effects according to the expectations of the sender.

3.3 Abstraction Levels and Other Aspects of Interoperability

Each of these aspects corresponds to a different category of interoperability abstraction levels, which can be further refined. For example, the transfer aspects are usually the easiest to implement and lie at the lowest level. Transferring a message from the context of the sender to the context of the receiver requires *connectivity* over a channel, which can be seen at several levels, such as:

- Message protocol, which includes control information (for example, message type: request, response, etc.) and message payload (structured or just serialized as a byte stream);
- Routing, if required, with intermediate gateways that forward messages;
- Communication, the basic transfer of data with a network protocol;
- Physics, which includes communication equipment and physical level protocols.

Table 1 expresses a range of interoperability abstraction levels, ranging from very high level, governance management and coordination, to very low level, physical network protocols.

In practice, the higher levels tend to be dealt with *tacitly*, by assuming that what is missing has been somehow previously agreed or is described in the documentation available to the developer. Inferring intent and semantics from documentation, or undocumented behaviour stemming from programs, constitute examples of this.

In the same way, the lower levels tend to be tackled *empirically*, by resorting to some tool or technology that deals with what is necessary to make interoperability work. An example is using some data interoperability middleware, without caring for the details of how it works.

The syntactic category is the most common, with the pragmatic category used to express service interfaces or business processes. Semantics is lagging, normally dealt with tacitly by documentation, although in recent years the so called Semantic Web [33] has increased its relevance and visibility.

All these interoperability levels constitute an expression of resource coupling. On one hand, two uncoupled resources (no interaction between them) can evolve

Table 1 Interoperability abstraction levels

Category	Level	Description
Symbiotic (intent)	Coordination Alignment Collaboration Cooperation	Purpose and intent that motivate the interaction, with varying levels of mutual knowledge of governance, strategy and goals. Can range from a coordinated governance to mere outsourcing arrangement (cooperation)
Pragmatic (reaction and effects)	Contract Workflow Interface	Management of the effects of transactions at the levels of choreography (contract), process (workflow) and service (interface)
Semantic (meaning of content)	Inference Knowledge Ontology	Interpretation of a message in context, at the levels of rule (inference), relations (knowledge) and definition of concepts (ontology)
Syntactic (notation of content)	Structure Predefined type Serialization	Representation of resources, in terms of composition (structure), primitive resource (predefined types) and their serialization format in messages
Connective (transfer protocol)	Messaging Routing Communication Physics	Lower level formats and network protocols involved in transferring a message from the context of the sender to that of the receiver

freely and independently, which favours adaptability, changeability and even reliability. On the other hand, resources need to interact to cooperate towards common or complementary goals and some degree of previously agreed knowledge is indispensable. The more they share with the other, the more integrated they are and the easier interoperability becomes, but the greater coupling gets.

Therefore, one of the main problems of interoperability is to ensure that each resource knows enough to be able to interoperate but no more than that, to avoid unnecessary dependencies and constraints. This is an instance of the principle of least knowledge, also called the Law of Demeter [34].

Another important aspect is non-functional interoperability. It is not just a question of invoking the right operation with the right parameters. Adequate service levels, context awareness, security and other non-functional issues must be considered when resources interact, otherwise interoperability will not be possible or at least ineffective.

4 Improving on Current Solutions: From the Web to the Mesh

The Web is a rather homogeneous (due to HTTP), reliable (due to TCP/IP and to the Internet backbone) and static (due to DNS) environment. The nature of IoT is much more heterogeneous (due to lower-level networks, such as sensor and vehicular networks), unreliable (in particular, with mobile devices) and dynamic (networks are frequently *ad hoc* and reconfigurable in nature). This means that the Web, although offering a good and global infrastructure, is not a good match for the IoT.

The existence of a single, universal network and protocol is unrealistic. What we can do is to simplify the interoperability requirements so that the microcomputer level (such as in very small sensors) can be supported in a lighter and easier way than with current solutions (such as TCP/IP, HTTP, XML, JSON, SOA and REST). We simply cannot get rid of network gateways, but these should be transparent to applications.

Our approach is to revisit the interoperability problem in the IoT context and in the light of what we know today, both from the perspective of what the technology can offer and of what the applications require. Backward compatibility is important, but simply building more specifications on top of existing ones entails constraints that prevent using the most adequate solutions.

What we need is a global network of interconnected resources, but with interoperability technologies more adequate to humans, machines and dynamic networks. We designate this as the *Mesh*, to emphasize heterogeneity and dynamicity. We believe that it constitutes a natural match for the IoT, reflecting many of the characteristics of *mesh networks* [35].

Therefore, we propose to improve the IoT-on-Web scenario by adopting some basic principles that lead to an IoT-on-Mesh scenario, such as:

- *Structured references*. An URI is a mere string. References to resources should be structured data and include several addresses, so that a resource can reference another across various networks. In addition, a reference should not necessarily be limited to addresses, but should also be able to designate channels, directions (in mesh topologies) or even ID tokens, all in the context of the respective target networks or gateways, so that routing and indirect addressing are supported. The system should deal with the network in a virtual (or logical) way, with information on the actual network contained inside the structured references;
- *Efficiency*. Messages should be as short as possible and optimized for machine to machine interaction. This entails the possibility of sending only binary information, without names or other source information, whenever agreed, but maintaining the possibility of self-description information, whenever needed. Ideally, the level of source information sent should be automatically adjusted;
- *Human adequacy*. Description of resources need also to be adequate for people, the developers. JSON and REST are simpler than XML and SOA but less equipped for design-time support. The ideal would be to combine the simplicity of the former with the design-time support provided by the latter;
- *Layered protocol*. Protocols such as HTTP and SOAP include everything in one specification, placing the protocol at a relatively high level. The IoT is highly variable in this respect, which means that there should be a very simple protocol at the bottom, unreliable and asynchronous, with higher-level mechanisms separate and offered optionally or implemented by the application, in conjunction with exception handling;

- *Context-oriented approach.* This means that every message can have two components, functional (arguments to operations) and non-functional (context), corresponding to a separation between data and control planes in what could be designated a *software-defined IoT.* For example, context information can be used to specify the type of protocol to use when sending a message (reliable or unreliable) and also to get non-functional information in a response message;
- *Decoupling.* Web interaction at the service level is based on passive data resources (e.g., XML and JSON), with interoperability based on schema sharing (SOA, with WSDL) or pre-agreed media types (REST). This implies interoperability for all possible variants of service interaction and entails a higher level of coupling than required. We propose to use *partial interoperability*, based on the concepts of *compliance* and *conformance* (defined below).

These principles apply not only to machine-level but also to human-level interaction, although browsers would have to be adapted. The following sections describe the most relevant issues in further detail, establishing a relationship with currently available technologies.

5 The Structural Services Style

5.1 Getting the Best of SOA and REST

The main architectural styles [36] used today in distributed systems are SOA [9] and REST [10]. An architectural style can be defined as a set of constraints on the concepts and on their relationships [37].

Both SOA and REST have advantages and limitations, and our goal is to derive an architectural style that emphasizes the advantages and reduces the limitations:

- The SOA style allows any number of operations for each resource, whereas REST specifies that all resources must have the same set of operations. Real world entities (problem domain) have different functionality and capabilities. REST deals with this in the solution domain by decomposing each entity in basic components, all with the same set of operations, and placing the emphasis on structure. This simplifies access to components, making it uniform, but it also makes the abstraction models of entities harder to understand (hidden by the decomposition into basic components) and increases the semantic gap between the application's entities and REST resources. This is not adequate for all problems, namely those involving complex entities. We want SOA's flexibility in service variability, or the ability to model resources as close as possible to real world entities;
- The REST style allows structured resources, whereas SOA does not. Real world entities are usually structured and in many applications their components must be externally accessible. Just using a black-box approach (yielding a flat, one-level

Table 2 Main characteristics of architectural styles

Style	Brief description	Examples of characteristics or constraints
SOA	Style similar to object-oriented, but with distributed constraints	Resources have no structure Services are application specific (variable set of operations) No polymorphism Integration based on common schemas and ontologies
REST	Resources have structure but a fixed service	Client and server roles distinction Services have a fixed set of operations Stateless interactions
Structural services	Structured resources, variable services, minimal coupling, context awareness	Resources are structured (like REST) Services have variable interface (like SOA) Interoperability based on structural compliance and conformance (new feature) Functional and context-oriented interactions (new feature)

resource structure) limits the options for the architectural solution. We want REST's flexibility in resource structure.

Therefore, we propose the *Structural Servicesarchitectural style*, in which a resource has both structured state (resources as components) and service behaviour (non-fixed set of operations), all defined by the resource's designer. This means that a resource becomes similar to an object in object-oriented programming, but now in a distributed environment, with the same level of data interoperability that XML and JSON provide.

Table 2 summarizes the main characteristics of these architectural styles. The Structural Services style is a solution to several of the principles enunciated above, in particular the principles of human adequacy and structured references. The following sections illustrate how this style can be implemented.

5.2 The Need for New Implementation Mechanisms

A distributed system is, by definition, a system in which the lifecycles of the modules are not synchronized and, therefore, can evolve independently, at any time. Distribution does not imply geographical dispersion, although this is the usual case. Pointers become meaningless and distributed references, such as URIs, must be used. Assignment (such as in argument passing to operations) must have a copy semantics, since the reference semantics, common in modern programming languages, is not adequate for distributed systems.

Data description languages, such as XML and JSON, merely describe data and their structure. If we want to describe services, we can use WSDL (a set of conventions in XML usage), but the resulting verbosity and complexity has progressively turned away developers in favour of something much simpler, namely REST. If we want a programming language suitable for distributed environments we can use BPEL [38], but again with an unwieldy XML-based syntax that forces programmers to use visual programming tools that generate BPEL and increase the complexity stack. For example, a simple variable assignment, which in most programming languages would be represented as $x = y$, requires the following in BPEL:

```
<assign>
      <copy>
         <from variable="y" />
         <to variable="x" />
      </copy>
</assign>
```

JSON is much simpler than XML and, thanks to that, its popularity has been constantly increasing [39]. The evolution of the dynamic nature of the Web, as shown by JavaScript and HTML5 [14], hints that data description is not enough anymore and distributed programming is a basic requirement. In the IoT, with machine to machine interactions now much more frequent that human interaction, this is of even greater importance.

Current Web-level interoperability technologies are greatly constrained by the initial decision of basing Web interoperability on data (not services) and text markup as the main description and representation mechanism. This has had profound consequences in the way technologies have evolved, with disadvantages such as:

- Textual representation leads to parsing overheads (binary compression benefits communication but not processing overheads) and poor support for binary data;
- The flexibility provided by self-description (schemas) did not live up to its promise. XML-based specifications (WSDL, SOAP, and so on), although having a schema, are fixed by standards and do not evolve frequently. This means that schema overheads (in performance and verbosity of representation) are there but without the corresponding benefits;
- The self-description provided by schemas does not encompass all the interoperability levels described in Table 1, being limited mostly to the syntactic and in some cases semantic levels. Adequate handlers still need to be designed specifically for each specification, which means that data description languages take care of just some aspects and are a lot less general than they might seem at first sight;
- Interoperability is based on data schema sharing (at runtime or with previously standardized or agreed upon internet media types), entailing coupling for the entire range of data documents described by the shared schema, which is usually much more than what is actually used;
- Data description languages have no support for the pragmatic (service description and context awareness) and semantic (rules and ontologies) interoperability categories in Table 1, which must be built on top of them with yet other languages;

- The underlying protocol, usually HTTP, is optimized for scalability under the client-server model, whereas the IoT requires a peer to peer model with light and frequent messages.

Therefore, it is our opinion that different implementation mechanisms are needed to support the Structural Services style natively and efficiently in its various slants (instead of emulating them on top of a data-level description language such as it happens with WSDL, for example) and in various environments (such as the Web, IoT and non-IP networks).

5.3 Textual Representation of Resources

With respect to XML, JSON points the way in data representation simplicity, but it too has a compatibility load, in this case with JavaScript. Like XML, it is limited to data (no support for operations) and its schema is a regular JSON document with conventions that enable to describe the original JSON document. With both XML and JSON, a schema is a specification separate from the data (documents) it describes. A schema expresses the associated variability, i.e., the range of documents that are validated by that schema.

The basis of data interoperability with XML and JSON is schema sharing. Both the producer and consumer (reader) of a document should use the same schema, to ensure that any document produced (with that schema) can be read by the consumer. This means that both producer and consumer will be coupled by this schema.

This may be fine for document sharing, but not for services (in which the service description and the data sent as arguments are the documents), since typically a service provider should be able to receive requests from several consumers and each consumer must be able to invoke more than one provider.

Therefore, we use a different approach to represent resources, which can be described in the following way:

- The representation includes data, behaviour, context and self-description, in a format suitable and familiar for humans. In fact, it is a textual distributed programming language (SIL—Service Implementation Language), with an object-oriented look and feel;
- Self-description is done within the resource description itself (no separate schemas), through a design-time variability range. In fact, each resource has its own exclusive schema. At any time, each resource has a structured value, which can vary at runtime but must lie within the variability range, otherwise an error is generated;
- Interoperability is based on *structural typing* [40], in which a consumer can read a document, as long as it has a structure recursively interoperable (until primitive resources are reached) with what it expects. Section 6 details this aspect, including not only data but also service interoperability;

- Unlike XML and JSON, SIL has a dual representation scheme, source text and compiled binary, using TLV (Tag, Length and Value) binary markup. This is not mere data compression, but actual compilation by a compiler, which automatically synchronizes binary with source. The binary representation provides native support for binary data, has a smaller length and is faster to parse (the Length field enables breadth-first parsing), all very important for the small devices that pervade IoT applications.

Listing 1 shows a simple example of resources described in SIL. It includes a temperature controller (`tempController`), which is composed of a list of references to temperature sensors with history (`tempSensorStats`), each of which has a reference to a remote temperature sensor (`tempSensor`). The lines at the bottom illustrate how to use these resources and should be included in some resource that uses them.

```
tempSensor: spid {   // descriptor of a temperature sensor
    getTemp: operation (-> [-50.0 .. +60.0]);
};

tempSensorStats: { // temperature sensor with statistics
    sensor: @tempSensor;    // reference to sensor (can be remote)
    temp: list float;        // temperature history
    startStats <||;          // spawn temperature measurements
    getTemp: operation (-> float) {
        reply sensor@getTemp <--; // forward request to sensor
    };
    getAverageTemp: operation ([1 .. 24] -> float) {
        for (j: [temp.size .. temp.size-(in-1)])
            out += temp[j];
        reply out/in; // in = number of hours
    };
    private startStats: operation () {   // private operation
        while (true) {
            temp.add <-- (getTemp <--); // register temperature
            wait 3600;    // wait 1 hour and measure again
        }
    }
};

tempController: { // controller of several temperature sensors
    sensors: list @tempSensorStats;    //list of references to sensors
    addSensor: operation (@tempSensor) {
        t: tempSensorStats;    // creates a tempSensorStats resource
        t.sensor = in;         // register link to tempSensor
        sensors.add <-- @t;    // add sensor to list
    };
    getStats: operation (-> {min: float; max: float; average: float})
    {
        out.min = sensors[0]@getTemp <--;
        out.max = out.min;
        total: float := out.min;    // initial value
        for (i: [1 .. sensors.length-1] {    // sensor 0 is done
```

```
                    t: sensors[i]@getTemp <--; // dereference sensor i
                    if (t < out.min) out.min = t;
                    if (t > out.max) out.max = t;
                    total += t;
                };
                out.average = total/sensors.length;
                reply;    // nothing specified, returns out
            }
        };
    };
    // Using the resources
    // tc contains a reference to tempController
    tc@addSensor <-- ts1; // reference to a tempSensor resource
    tc@addSensor <-- ts2; // reference to a tempSensor resource
    x: tc@sensors[0]@getAverageTemp <-- 10;    // average last 10 hours
```

Listing 1. Describing and using resources, using SIL (Service Implementation Language)

This listing can be briefly described in the following way:

- The temperature sensor tempSensor) is remote and all we have is its SPID (SIL Public Interface Descriptor). This corresponds to a Web Service's WSDL, but much more compact and able to describe both structure and operations. It is automatically derived from the resource description by including only the public parts (public component resources and operation headings). The SPID of tempSensorStats, for example, is expressed by the following lines:

```
tempSensorStats: spid { // temperature sensor with statistics
    sensor: @tempSensor; // reference to sensor (can be remote)
    temp: list float; // temperature history
    getTemp: operation (-> float);
    getAverageTemp: operation ([1 .. 24] -> float);
}
```

- Resources and their components are declared by a name, a colon and a resource type, which can be primitive (such as integer, a range, e.g., [10 ⋯ 50], or float), or user defined (enclosed in curly brackets, i.e., "{...}"). Overall, there are some resemblances to JSON, but component names are not strings and operations are first class resources (in the line of REST, actually);
- The definition of a resource acts like a constructor, being executed only once, when a resource of this type is created, and can include statements. This is illustrated by the statement "startStats <||" in tempSensorStats. This is actually an asynchronous invocation ("<||") of private operation startStats, which is an infinite loop registering temperature measurements every hour. Synchronous invocation of operations is done with "<−−", followed by the argument, if any;
- Operations have at most one argument, but it can be structured (with "{...}"). The same happens with the operation's reply value, such as illustrated by operation getStats. Inside operations, the default names of the argument and the value to

return are in and out, respectively. The heading of operations specifies the type of the argument and of the reply value (inside parentheses, separated by "->");
- References to resources (indicated by the symbol "@") are not necessarily URIs, not even strings. They can be structured and include several addresses, so that a resource in a network (e.g., the Internet) can reference another in a different network (e.g., a ZigBee network). It is up to the middleware in the network nodes that support these protocols to interpret these addresses, so that transparent routing can be achieved, if needed;
- Resource paths, to access resource components, use dot notation, except if the path traverses a reference, in which case a "@" is used. Note, for example, the path used in the last line of Listing 1, which computes the average temperature, along the last 10 h, in sensor 0 of the controller.

5.4 Binary Representation of Resources

Listing 1 is what the programmer sees, a text program in a distributed programming language, with a syntax similar to usual and familiar object-oriented programming languages, instead of a verbose and cumbersome syntax of a data description language such as XML.

However, this is not necessarily what is used by the computing platform. A compiler transforms Listing 1 into binary code, like a Java compiler transforms Java source into bytecodes. This binary code is then interpreted, which guarantees its universality, as long as the interpreter is implemented in interacting computing nodes.

The binary representation uses a modified version of the TLV format (Tag, Length and Value) used by ASN.1 [41]. This not only supports the direct integration of binary information but also facilitates parsing, since each resource, primitive or structured, can be stepped over in a breadth-first traversal, thanks to the *Length* field (the resource size in bytes).

Resources (including messages) can be represented in three levels:

1. Binary, no variable names (components are referred to by their position index in the resource);
2. Variable names, in a dictionary that maps names to position indices;
3. Source text, when available. Typically, runtime messages are generated directly in binary, without source text. This is used mostly in design-time resources.

Levels 1 and 2 include the same information as level 3, apart from comments and formatting. Level 1 can be used without level 2 and is the most efficient and the most adequate for smaller devices in IoT. However, it requires a previous agreement on the types of the resources involved in the interaction. Hardwired solutions, typical of old binary systems, should be avoided and therefore it must be possible to establish a mapping between corresponding components (with the same name) of interacting resources, even if they are in different order or not all are present in both resources. This is structural type checking [40] and needs to be done in distributed systems,

Table 3 Sizes (in bytes) of several resource representations

	XML	JSON	SIL Source	SIL binary with names	SIL binary without names
Longer names	2472	1491	1317	927	358
Shorter names	1857	1153	979	589	358

in which the lifecycle of resources are not synchronized and a simple name is not enough to define a type.

To use just level 1 without losing type checking, both levels 1 and 2 should be sent in the first message, verify interoperability by checking resource types structurally (which takes time), establish a mapping between names and position indices and cache it, assigning an ID token to it. In subsequent messages with the same structure, this token can be used instead of performing a structural resource analysis again. This exploits the fact that many IoT devices, once installed, repeatedly use the same message structure for quite a long time. The interaction protocol must support this optimization and be able to recover from situations in which the token has changed, by signalling an error that forces retransmission of a message, now with components names, and generation of a new token. This mechanism bears resemblance to the Etag header of HTTP.

Table 3 gives an idea of the size of resource representations in various situations. The example refers to a data-only resource, so that XML and JSON can be used. The two lines refer to the same resource, but with longer and shorter component names. That is why the sizes in the last column are the same and the size reduction is greater when names are longer. Naturally, the sizes presented vary a lot with the concrete example, but this gives an idea of the values involved.

Compressed versions of XML and JSON [42] also accomplish relevant reductions in size, but time is needed to compress and to decompress messages, on top of which text parsing still needs to be done at the receiver. The compilation in SIL takes longer than a simple compression, but it does not need to be done at runtime. Compilation is a design time feature and messages sent are generated in binary directly. Messages received are parsed directly in binary, much faster than text parsing due to the navigable structure provided by the TLV scheme.

The interface between SIL and other languages is done in the same way as XML-based solutions, but with structure as the guiding mechanism. There is a mapping between each primitive data type in SIL and the corresponding one in another language. Structured resources in SIL are mapped onto classes. A dynamic load mechanism must be available to dynamically register new resources into a resource directory, so that their services become available in the network they connect to.

5.5 Message Protocol

The typical reliability of Web messaging provided by TCP is a luxury in many IoT applications, in terms of network characteristics and node processing capabilities.

Table 4 Main message types of the SIL message protocol

Message category/type	Description
Request	*Initial request in each transaction*
React	React to message, no answer expected
React and respond	React to message and answer/notify
React with context	React to message and context, no answer expected
React with context and respond	React to message and context and answer/notify
Assimilate	Merge the message into the receiver resource, subject to structural interoperability (only compatible components are replaced). It bears similarities to PUT in REST
Response	*Response to the request*
Answer	A (structured) value returned by a reply instruction
Answer with weak context	A (structured) value returned with context, handler at receiver optional
Answer with strong context	A (structured) value returned with context requiring a handler (includes exceptions)
Protocol fault	A status code resulting from a predefined protocol error (includes interoperability token mismatch)
Done	Request completed but has no value to reply

The CoAP [28] addresses this issue by assuming unreliability and resorting to UDP and lighter messages. CoAP separates the basic message transport protocol, which can be unreliable, from the message level protocol, which deals with requests and responses.

SIL also follows the layered protocol approach, with a basic binary message transport on top of which several types of messages can be defined. The control part of the protocol also uses the TLV scheme. There are no assumptions about the transport level, apart from addressability and capability of transferring the message from the sender to the receiver. It can be unreliable and unsecure and it can use any protocol, IP or non-IP based. Resource references are not necessarily URIs. Anything can be used, as long as the messaging middleware is able to understand it. If reliability or security are needed but not provided by the transport protocol, these must be implemented by the middleware or the application.

The main message types of the SIL message protocol are described in Table 4. The context aspect is discussed in Sect. 7.

6 Improving Interoperability with Compliance and Conformance

Although not an exclusive issue of the Structural Services style, the goal of reducing as much as possible the coupling between two interacting resources, while still providing the level of interaction required by the application, greatly benefits from a resource representation language such as SIL, which has native support for structural interoperability, based on *compliance* and *conformance*.

In a distributed system, any resource can change the name of the types it uses at any time, which means that name-based type checking to ensure interoperability requires lifecycle synchronization and hampers changeability. Yet, this is precisely what happens in current Web technologies, namely Web Services and RESTful applications:

- Schemas must be shared between interacting Web Services, establishing coupling for all the possible values satisfying each schema, even if they are not actually used. In this case, a reference to a schema acts like its name;
- REST also requires that data types (usually called media types) must have been previously agreed, either standardized or application specific;
- Searching for an interoperable Web Service is usually done by schema matching with similarity algorithms [43] and ontology matching and mapping [44]. This does not ensure interoperability and manual adaptations are usually inevitable.

The interoperability notion, as defined in this chapter, introduces a different perspective, stronger than similarity but weaker than commonality (sharing). The trick is to allow partial (instead of full) interoperability, by considering only the intersection between what the consumer needs and what the provider offers. If the latter subsumes the former, the degree of interoperability required by the consumer is ensured, regardless of whether the provider supports additional capabilities or not.

Interoperability of a consumer with a provider entails the following properties:

- *Compliance* [21]. The consumer must satisfy (*comply with*) the requirements established by the provider to accept requests sent to it, without which these cannot be honoured;
- *Conformance* [22, 45]. The provider must fulfil the expectations of the consumer regarding the effects of a request (including eventual responses), therefore being able to take the form of (*conform to*) whatever the consumer expects it to be.

In full interoperability, the consumer can use all the provider's capabilities. In partial interoperability, the consumer uses only a subset of those capabilities, which means that compliance and conformance need only be ensured for that subset.

These properties are not commutative. If *X* complies with *Y*, *Y* does not necessarily comply with *X*. However, they are transitive. For example, if *X* complies with *Y* and *Y* complies with *Z*, then *X* complies with *Z*.

Figure 3 illustrates this model. A resource *A*, in the role of consumer, has been designed for full interoperability with resource *B*, in the role of provider. *A* uses only the capabilities that *B* offers and *B* offers only the capabilities that *A* uses. Now, let us consider that we want to change the provider of *A* to resource *D*, which has been designed for full interoperability with resource *C*, in the role of consumer. The problem is that *A* was designed to interact with a provider *B* and *D* was designed to expect a consumer *C*. This means that, if we use *D* as a provider of *A*, *B* is how *A* views provider *D* and *C* is how *D* views consumer *A*.

There are two necessary conditions to ensure that *A* is interoperable with *D*:

Fig. 3 Partial interoperability
based on compliance and
conformance

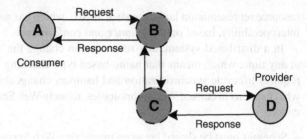

- *Compliance*. B must comply with C. Since A complies with B and C complies with
  D, this means that A complies with D and, therefore, A can use D as if it were B,
  as it was designed for;
- *Conformance*. C must conform to B. Since D conforms to C and B conforms to A,
  this means that D conforms to A and, therefore, D can replace (take the form of)
  B without A noticing it.

Partial interoperability has been achieved by *subsumption*, with the set of capabilities that A uses as a subset of the set of capabilities offered by D. This inclusion
relationship, without changing characteristics, is similar in nature to the *polymorphism* used in many programming languages, but here in a distributed context. It
constitutes the basis for transitivity in compliance and conformance.

Listing 2 illustrates these concepts by providing two additional temperature sensors (weatherSensor and airSensor) in relation to Listing 1. Only the additional and relevant parts are included here, with the rest as in Listing 1.

```
tempSensor: spid {
    getTemp: operation (->[-50.0 .. +60.0]);
};
weatherSensor: spid {
    getTemp: operation (->[-40.0 .. +50.0]);
    getPrecipitation: operation (-> integer);
};
airSensor: spid {
    getTemp: operation (->[-40.0 .. +45.0]);
    getHumidity: operation (-> [10 .. 90]);
};
// tc contains a reference to tempController
tc@addSensor <-- ts1; // reference to a tempSensor resource
tc@addSensor <-- ts2; // reference to a tempSensor resource
tc@addSensor <-- as1; // reference to an airSensor resource
tc@addSensor <-- ws1; // reference to an weatherSensor resource
tc@addSensor <-- ws2; // reference to an weatherSensor resource
tc@addSensor <-- as2; // reference to an airSensor resource
x: tc@sensors[0]@getAverageTemp <-- 10; // average last 10 hours
s: {max: float; average: float};
s = tc@getStats <--; // only max and average are assigned to s
temp: [-50.0 .. +50.0];
temp = ws1@getTemp <--; // complies. Variability ok
temp = ts1@getTemp <--; // does not comply. Variability mismatch
```

Listing 2. Example of partial interoperability, with structural compliance and conformance

Note that:

- `weatherSensor` and `airSensor` conform to `tempSensor`, because they offer the same operation and the result is within the expected variability. This means that they can be used wherever a `tempSensor` is expected, which is illustrated by adding all these types of sensors to `tempController` (through `tc`, a reference to it) as if they were all of type `tempSensor`. Non relevant operations do not matter;
- The result of invoking the operation `getStats` on `tempController` is a resource with three components (see Listing 1), whereas s has only two. But the assignment is still possible, thanks to structural interoperability (the extra component is ignored);
- The last statement in Listing 2 also triggers a compliance check by the compiler, which issues an error. The variability of the result value of operation `getTemp` in `tempSensor` (referenced by `ts1`) is outside the variability range declared for component `temp`.

In addition, it is important to note that compliance and conformance could also be implemented with XML and JSON (in what data are concerned, since they do not support operations). It would be just a matter of comparing descriptions. However, technology has not evolved that way. One of the reasons is that performing a structural compliance and conformance check in every message would be too slow. Therefore, they chose to stick to full interoperability, with schemas shared and media types previously agreed, but with more coupling that actually needed and lower changeability.

Practical implementation of partial interoperability requires an optimization mechanism, such as the token mentioned in Sect. 5.4, and a protocol to support it, as described in Sect. 5.5.

7 Improving Interoperability with Context Awareness

Non-functional information, or *context*, is very important in many applications, in particular in IoT environments with mobile applications that extract value from context awareness (such as location and local settings). If there is no support for it, functional parameters can become cluttered with contextual information.

Most context-oriented programming languages support context adaptation by activating and deactivating layers [46]. A context layer is a set of operations that deal with a specific context (or scenario) and that, when that layer is activated, automatically extend or replace operations that are already defined in various classes. The layer's operations can be scattered throughout the classes which they pertain to (*layer-in-class*) or declared together, in a specific construction separate from classes (*class-in-layer*), but achieving the same effect. When a layer is active, the behaviour of some operations is modified. Typically, this is a control-only feature, with no support for storing contextual state.

SIL includes support for context-oriented programming in a smaller granularity and with provision for context state, by allowing operations to have two parameters and two results (functional and non-functional). Exceptions have been integrated in the context-passing mechanism, which can be described in the following way:

- *Forward context.* A resource can be passed to an operation as non-functional parameter, together with a functional parameter, by using a *with* clause. This can be used to adapt the behaviour of the message recipient. If several types or values of context are expected, yielding different behaviour, there can be multiple implementations of the corresponding operation, each specifying which context type it accepts (with structural compliance). When a message with a specified context is received, the SIL environment invokes the first of the implementations of the target operation that declares a context with which the message's context complies;

- *Backward context.* SIL returns values by executing reply, reply...with and raise...with statements. The first returns a functional value and the others add a context value, weak and strong, respectively. The resource with receives the returned value can use a do...catch statement to handle context values. The difference between reply and raise is that the former invokes the catch handler (or ignores if the context does not comply with any catch handler) and resumes execution, as if there was no context, and the latter breaks execution and backtracks as much as needed until a handler is found (this includes exceptions).

Listing 3 provides a simple illustration of some of these features, in which sensors have been modified to exploit context information. In particular, it is possible in some cases to specify whether we want the temperature in degrees Celsius or Fahrenheit and in others to detect which temperature scale applies to returned values, simply be reading the backward context. Only the relevant parts are shown. The rest is as in Listings 1 and 2.

```
tempSensor: spid {
    getTemp: operation (-> [-50.0 .. +60.0] with ''celsius'');
};
weatherSensor: spid {
    getTemp: operation (-> [-40.0 .. +50.0] with ''celsius'');
    getTemp: operation (with ''celsius'' -> [-40.0 .. +50.0]);
    getTemp: operation (with ''fahrenheit'' -> [-58.0 .. +140.0]);
    getPrecipitation: operation (-> integer);
};
airSensor: spid {
    getTemp: operation (-> [-58.0 .. +140.0] with ''fahrenheit'');
    getHumidity: operation (-> [10 .. 90]);
};
// ts has a reference to a tempSensor
// ws has a reference to a weatherSensor
// as has a reference to a airSensor
//
```

```
do {
    tempC: float; // to hold a celsius temperature
    tempC = ts@getTemp <--;
    tempC = ts@getTemp <-- with ''celsius'';
    tempC = ts@getTemp <-- with ''fahrenheit''; // invokes catch
    tempC = ws@getTemp <--; // weatherSensor: returns celsius
    tempC = as@getTemp <--; // airSensor: invokes catch
}
catch (tf: float with ''fahrenheit'') {
    tc: [-50.0 .. +60.0];
    tc = tf * 1.8 + 32;
    proceed tc; // replaces value and type of message response
};
```

Listing 3. Example of forward and backward context awareness

tempSensor can only provide temperature in Celsius degrees (it does not accept indication of another scale), but at least informs the consumer (which invoked getTemp) of that fact, by returning the string "celsius" as a backward context.

weatherSensor is more elaborate and has three implementations of getTemp, allowing to specify the scale or to omit that indication (in which case it assumes Celsius degrees, by default). To obtain a temperature there is no need to specify a functional parameter, but the context can be specified. The adequate operation will be chosen at runtime, according to the forward context.

airSensor can only provide temperature values in degrees Fahrenheit.

The do...catch statement is similar to the try...catch statement in Java, but the try has been replaced by do because it is more encompassing than an exception handling mechanism. Inside the do clause, forward context can be specified by using a with clause, passing a value that must comply with the with clause of the invoked operation.

When the backward context complies with the catch clause (in this case, when the context string is "fahrenheit"), the value part of the reply is assigned to the specified component and the corresponding code executed. In this case, it computes a temperature value in degrees Celsius from the original value in degrees Fahrenheit and executes the proceed statement with the new value, which now acts as the reply and is finally assigned to tempC.

This shows how a backward context can be used to modify a reply value on the fly and increase context awareness and range of interoperability, since now more sensors can be used with tempController due to automatic context-oriented adaption. In particular, note the last statement in the do clause, in which no temperature scale was specified but the returned value could still be used and correctly converted to degrees Celsius because it was accompanied by contextual information.

8 Discussion and Rationale

There are several signs that show that global interoperability is shifting away from classical client-server HTTP-based interactions, initially conceived for human consumption in what has become known as the Web. Namely:

- The IoT is changing the emphasis to machine-to-machine interactions, using the peer to peer model. Although the most immediate solution is to use existing Web Technologies (in most cases REST, due to its simplicity), these are not the most adequate model for IoT. The mere development of CoAP [28] is an acknowledgment of this. The IoT is also bringing heterogeneity of resources and networks, and this is why we contend that the Web (rather homogeneous at the platform level) will evolve into a Mesh, as discussed in Sect. 4;
- The need for more dynamic applications than simple browsing have spurred the development of technologies such as AJAX [47] and Comet [48]. More recently, the WebSockets protocol [14] paved the way for full-duplex communication at the Web level. Even more recently, there is a movement towards implementing real-time communication directly between browsers, without the need for a server, with WebRTC [13];
- Cloud computing applications are changing the overall computing scenario, with an increasing emphasis on distributed programming and support for asynchrony, concurrency and handling of unreliability. Languages such as Orc [49] for distributed task orchestration, Jolie [50] for distributed service orchestration and, more recently, Orleans [51] for distributed programming in cloud environments, show that there is the need and the capability of implementing distributed applications in a more consistent way than merely resorting to current Web technologies, such as those used by SOA and REST.

Backward compatibility with existing technologies is an asset that usually is half way to success, but it can also be a straightjacket, with constraints and compromises that increase complexity and hamper performance. Therefore, we need to revisit the global interoperability problem, in the light of the needs of new applications (typically, based on services rather than on documents) and of the capabilities that new technologies can offer. Then, we can try to find a way to migrate from existing technologies to those that are found to be more suitable to tackle the new classes of problems.

This is our justification to conceive the Structural Services architectural style, trying to reap the benefits from both SOA and REST, and to design SIL not as a passive data description language but as an active service implementation language, with the corresponding execution platform [52].

This can be illustrated by considering how service discovery, or the equivalent of UDDI, is done in SIL. There is a directory service, which resources available to provide services can be registered with. This is a regular service, not a special mechanism, and it can register resources within a local computing node or available remotely. The directory maintains a list of references to registered resources. When we want to search for a service, we supply the directory with the resource SPID that describes the service that we want and the directory returns a list of references to the all registered resources that conform to that service. Conformance includes context information, so that we can specify both functionality and non-functional aspects and constraints, such as security mechanisms, SLA, cost, and so on. This is how the references used in Listings 1 through 3 can be obtained.

In terms of programming, the most used solutions are external programming languages (e.g., a Web Service or a REST resource implemented in Java) and WS-BPEL, an XML-based language. Services are described separately (e.g., in a WSDL file). Orc and Jolie are more concerned with scheduling and control of processing activities than with interoperability (for which they resort to conventional solutions, such as Web Services and XML). Several languages and specifications are needed to implement the full solution.

SIL, and its execution platform, is complete in the sense that one single language is enough to describe, implement and execute services, with their state structure and behaviour. A complete distributed application can be implemented in SIL. On the other hand, interoperability is supported, so that native operations can be implemented not in SIL but in a native language, such as Java or C#, and different native resources, implemented in different languages, are able to interoperate through a SIL interface (SPID).

Therefore, the integration benefits of Web Services and RESTful applications are not lost. Nevertheless, SIL is a new solution, not following an evolutionary approach. The migration path from current technologies can be established in two main ways:

- *Co-existence.* The SIL server receives binary messages and does not depend on any particular transport protocol, relying solely on message delivery. Therefore, any existing server can be used, based on HTTP, WebSockets or any other protocol. In fact, several servers can be used simultaneously, receiving messages that are handed over to message handlers. The SIL handler first checks if this is an identifiable SIL message (in TLV format). If not, the full message can then be inspected by other handlers, which implement current technologies;
- *Conversion.* Although space limitations have prevented us from discussing SIL in full detail, it has been conceived to support the features provided by XML, JSON and the corresponding schema languages, which means that any XML, WSDL or JSON file can be converted to SIL, allowing current clients to use a service in a SIL server.

9 Future Research Directions

Currently, there is a partial implementation of the SIL language and environment, with basic functionality [52]. Future work will be carried out along the main following lines:

- Semantics. SIL supports semantics through rule resources (a set of condition-then-action statements), which become active upon creation in the scope of their container resource and get evaluated each time a component affecting one of the conditions is changed. It is up to the compiler to establish dependencies between conditions and components. This mechanism needs to be optimized;
- The compliance and conformance algorithms are implemented in the compiler but these need to be optimized in the binary version, to increase the performance of dynamic type checking;
- The interface to native languages needs to be diversified and improved. At the moment, only Java is supported;
- A quantitative comparison assessment of the SIL platform versus Web Services and RESTful solutions needs to be carried out in IoT applications (particularly in non-IP networks).

10 Conclusions

The IoT is changing the scenario of global interoperability. From a rather homogeneous, reliable and static network of servers that we identify as the Web, built on top of the Internet, we are moving to what we call a *Mesh*, a heterogeneous, unreliable and dynamically reconfigurable network of computing nodes, ranging from stationary, generic and fully-fledged servers to mobile, application-specific and low-level devices, such as sensors. Usually, the simpler a node is, the larger the quantities in which it is used, which leads to an enormous increase of the number of messages exchanged and raises a big data problem.

This is shifting the emphasis of the rather human level of the Web to the rather machine level of the Mesh, built on top of the IoT. The current approach to implement the IoT vision [4] is to extend current Web technologies to small devices. We contend that, although evolutionary and faster to implement, this is not the most adequate to the nature of IoT.

In fact, even the "I" in IoT needs clarification. Our understanding of the IoT concept is not a network of devices connected by IP-based protocols, such as CoAP (on UDP) and HTTP (on TCP), but a network of networks, usually heterogeneous in terms of protocols. In this respect, a sensor should be reachable globally (from any other device connected to the IoT) and directly (which implies transparent gateways, not application specific, between networks).

The main tenets of our approach are:

- To reduce the lowest level of interoperability to binary messages, with a very basic protocol and no source information, to reduce the computing power, memory and energy requirements of simpler devices, which repeatedly use the same message types and patterns;
- To maintain the capability of source-level interoperability, for flexibility in more dynamic systems;

- To increase interoperability by basing it on the compliance and conformance concepts, which support partial interoperability, and context awareness, by automatic adaptation of behaviour;
- To reduce complexity (the main drive behind the success of JSON and REST), by designing a language that can describe, implement and execute services, without the need of a plethora of languages and specifications to implement distributed applications.

We have presented several examples that illustrate how this can be done, by supporting distributed resources and services natively, instead of emulating them on top of data description languages.

Our main scientific contributions are the following:

- A systematization of the resource interoperability problem, by providing a resource interaction model (Fig. 1) and an interoperability abstraction scale (Table 1);
- A new architectural style, Structural Services (Table 2), which combines the structural capability of REST with the functional variability of SOA, with a platform-independent execution model and a language well matched to this style;
- New ways of supporting Big Data in the IoT, namely in terms of efficiency (for volume and velocity) and decoupling (for variety, or heterogeneity), with:

  – Native support for binary data;
  – Variable source information (which can be omitted in frequent cases);
  – Coupling reduced to the bare minimum required by the application, due to the use of structural interoperability (compliance and conformance) instead of schema sharing;
  – Resource-based contextual information, which provides a better decoupling than the traditional approach of activating context layers since the context sent as part of a message is data also subjected to structural interoperability.

We hope that our approach is a small contribution to the goal of implementing the IoT vision in a simpler but more encompassing and efficient way.

References

1. Zikopoulos, P., et al.: Understanding Big Data. McGraw-Hill, New York (2012)
2. Demchenko, Y., Zhao, Z., Grosso, P., Wibisono, A., de Laat, C.: Addressing big data challenges for scientific data infrastructure. In: Proceedings of IEEE 4th International Conference on Cloud Computing Technology and Science, pp. 614–617. Taipei, Taiwan (2012)
3. Karen, T.: How many internet connections are in the world? Right now. http://blogs.cisco.com/news/cisco-connections-counter/ (2013). Accessed 28 Aug 2013
4. Sundmaeker, H., Guillemin, P., Friess, P., Woelffle, S.: Vision and challenges for realising the internet of things. European Commission Information Society and Media. http://bookshop.europa.eu/en/vision-and-challenges-for-realising-the-internet-of-things-pbKK3110323/ (2010). Accessed 28 Aug 2013
5. United Nations: World Population Prospects: The 2012 Revision, Key Findings and Advance Tables. Working Paper No. ESA/P/WP.227. Department of Economic and Social Affairs,

Population Division, United Nations (2013). http://esa.un.org/unpd/wpp/Documentation/pdf/ WPP2012_%20KEY%20FINDINGS.pdf. Accessed 28 August 2013

6. Schaffers, H., et al.: Smart cities and the future internet: towards cooperation frameworks for open innovation. In: Domingue, J., et al. (eds.) The Future Internet, pp. 431–446. Springer, Berlin Heidelberg (2011)
7. Bolton, F.: Pure Corba. SAMS Publishing, Indianapolis (2001)
8. Venkitachalam, G., Chiueh, T.: High performance common gateway interface invocation. In: Proceedings of IEEE Workshop on Internet Applications, pp. 4–11. San Jose, California (1999)
9. Earl, T.: Service-Oriented Architecture: Concepts, Technology, and Design. Prentice Hall PTR, Upper Saddle River, NJ (2005)
10. Webber, J., Parastatidis, S., Robinson, I.: REST in Practice: Hypermedia and Systems Architecture. O'Reilly Media, Sebastopol (2010)
11. Berners-Lee, T.: Weaving the Web: The Original Design and Ultimate Destiny of the World Wide Web by Its Inventor. HarperCollins Publishers, New York (1999)
12. Loreto, S., Romano, S.: Real-time communications in the web: issues, achievements, and ongoing standardization efforts. IEEE Internet Comput. **16**(5), 68–73 (2012)
13. Johnston, A., Yoakum, J., Singh, K.: Taking on WebRTC in an Enterprise. IEEE Commun. Mag. **51**(4), 48–54 (2013)
14. Lubbers, P., Albers, B., Salim, F.: Pro HTML5 Programming: Powerful APIs for Richer Internet Application Development. Apress, New York (2010)
15. Priyantha, N., Kansal, A., Goraczko, M., Zhao, F.: Tiny web services: design and implementation of interoperable and evolvable sensor networks. In: Proceedings of 6th ACM Conference on Embedded Network Sensor Systems, pp. 253–266 (2008). doi:10.1145/1460412. 1460438
16. Akribopoulos, O., Chatzigiannakis, I., Koninis, C., Theodoridis, E.: A web services-oriented architecture for integrating small programmable objects in the web of things. In: Proceedings of Developments in E-systems Engineering Conference, pp. 70–75 (2010). doi:10.1109/DeSE. 2010.19
17. Gupta, V., Udupi, P., Poursohi, A.: Early lessons from building Sensor. Network: an open data exchange for the web of things. In: Proceedings of Conference on Pervasive Computing and Communications Workshops, pp. 738–744 (2010). doi:10.1109/PERCOMW.2010.5470530
18. Taherkordi, A., Eliassen, F., Romero, D., Rouvoy, R.: RESTful service development for resource-constrained environments. In: Wilde, E., Pautasso, C. (eds.) REST: From Research to Practice. Springer Science+Business Media, New York (2011)
19. Guinard, D., Trifa, V., Wilde, E.: A resource oriented architecture for the Web of Things. In: Proceedings of 2nd International Internet of Things Conference, pp. 1–8 (2010). doi:10.1109/ IOT.2010.5678452
20. Guinard, D., Trifa, V., Mattern, F., Wilde, E.: From the internet of things to the web of things: resource oriented architecture and best practices. In: Uckelmann, D., Harrison, M., Michahelles, F. (eds.) Architecting the Internet of Things, pp. 97–129. Springer, Berlin (2011)
21. Kokash, N., Arbab, F.: Formal behavioral modeling and compliance analysis for service-oriented systems. In: Boer, F., Bonsangue, M., Madelaine, E. (eds.) Formal Methods for Components and Objects. Lecture Notes in Computer Science, vol. 5751, pp. 21–41. Springer, Berlin, Heidelberg (2009)
22. Adriansyah, A., van Dongen, B., van der Aalst, W.: Towards robust conformance checking. In: Proceedings of Business Process Management Workshops, pp. 122–133. Springer, Berlin Heidelberg (2010)
23. Perera, C., Zaslavsky, A., Christen, P., Georgakopoulos, D.: Ca4iot: context awareness for internet of things. In: Proceedings of IEEE International Conference on Green Computing and Communications, pp. 775–782. Besançon, France (2012)
24. Kapitsaki, G., Prezerakos, G., Tselikas, N., Venieris, I.: Context-aware service engineering: a survey. J. Syst. Softw. **82**(8), 1285–1297 (2009)
25. Gislason, D.: Zigbee Wireless Networking. Elsevier, UK (2008)
26. Hui, J., Culler, D.: IPv6 in low-power wireless networks. Proc. IEEE **98**(11), 1865–1878 (2010)

27. Jacobsen, R., Toftegaard, T., Kjærgaard, J.: IP connected low power wireless personal area networks in the future internet. In: Vidyarthi, D. (ed.) Technologies and Protocols for the Future of Internet Design: Reinventing the Web, pp. 191–213. IGI Global, Hershey (2012)

28. Castellani, A., Gheda, M., Bui, N., Rossi, M., Zorzi, M.: Web services for the internet of things through CoAP and EXI. In: Proceedings of International Conference on Communications Workshops, pp. 1–6 (2011). doi:10.1109/iccw.2011.5963563

29. Balasubramaniam, S., Kangasharju, J.: Realizing the internet of nano things: challenges, solutions, and applications. IEEE Comp. 46(2), 62–68 (2013)

30. Akyildiz, I., Brunetti, F., Blázquez, C.: Nanonetworks: a new communication paradigm. Comp. Netw. 52(12), 2260–2279 (2008)

31. Moustafa, H., Zhang, Y.: Vehicular Networks: Techniques, Standards, and Applications. Auerbach publications, Boston (2009)

32. ISO/IEC/IEEE: Systems and software engineering—Vocabulary. International Standard ISO/IEC/IEEE 24765:2010(E), 1st edn, p. 186. Geneva, Switzerland (2010)

33. Sheth, A., Henson, C., Sahoo, S.: Semantic sensor web. IEEE Internet Comput. 12(4), 78–83 (2008)

34. Palm, J., Anderson, K., Lieberherr, K.: Investigating the relationship between violations of the law of demeter and software maintainability. In: Proceedings of Workshop on Software-Engineering Properties of Languages for Aspect Technologies. http://www.daimi.au.dk/eernst/splat03/papers/Jeffrey_Palm.pdf (2003). Accessed 28 Aug 2013

35. Bruno, R., Conti, M., Gregori, E.: Mesh networks: commodity multihop ad hoc networks. IEEE Commun. Mag. 43(3), 123–131 (2005)

36. Dillon, T., Wu, C., Chang, E.: Reference architectural styles for service-oriented computing. In: Li, K., et al. (eds.) IFIP International Conference on Network and Parallel Computing. Lecture Notes in Computer Science 4672, pp. 543–555. Springer, Berlin Heidelberg (2007)

37. Fielding, R., Taylor, R.: Principled design of the modern web architecture. ACM Trans. Internet Technol. 2(2), 115–150 (2002)

38. Juric, M., Pant, K.: Business Process Driven SOA using BPMN and BPEL: From Business Process Modeling to Orchestration and Service Oriented Architecture. Packt Publishing, Birmingham, UK (2008)

39. Severance, C.: Discovering JavaScript Object Notation. IEEE Comp. 45(4), 6–8 (2012)

40. Gil, J., Maman, I.: Whiteoak: introducing structural typing into Java. ACM Sigplan Not. 43(10), 73–90 (2008)

41. Dubuisson, O.: ASN.1 Communication Between Heterogeneous Systems. Academic Press, San Diego, CA (2000)

42. Sumaray, A., Makki, S.: A comparison of data serialization formats for optimal efficiency on a mobile platform. In: Proceedings of 6th International Conference on Ubiquitous Information Management and Communication (2012). doi:10.1145/2184751.2184810

43. Jeong, B., Lee, D., Cho, H., Lee, J.: A novel method for measuring semantic similarity for XML schema matching. Expert Syst with Appl 34, 1651–1658 (2008)

44. Euzenat, J., Shvaiko, P.: Ontology matching. Springer, Berlin (2007)

45. Kim, D., Shen, W.: An approach to evaluating structural pattern conformance of UML models. In: Proceedings of ACM Symposium on Applied Computing, pp. 1404–1408. ACM Press (2007)

46. Salvaneschi, G., Ghezzi, C., Pradella, M.: Context-oriented programming: a software engineering perspective. J. Syst. Softw. 85(8), 1801–1817 (2012)

47. Paulson, L.: Building rich web applications with Ajax. IEEE Comp. 38(10), 14–17 (2005)

48. McCarthy, P., Crane, D.: Comet and Reverse Ajax: The Next-Generation Ajax 2.0. Apress, New York (2008)

49. Kitchin, D., Cook, W., Misra, J.: A language for task orchestration and its semantic properties. In: Proceedings of 17th International Conference on Concurrency Theory, pp. 477–491. Springer, Berlin Heidelberg (2006)

50. Montesi, F., Guidi, C., Lucchi, R., Zavattaro, G.: Jolie: a java orchestration language interpreter engine. Electron. Notes Theor. Comp. Sci. 181, 19–33 (2007)

51. Bykov, S., et al.: Orleans: cloud computing for everyone. In: Proceedings of 2nd ACM Symposium on Cloud Computing, p. 16. ACM Press (2011)
52. Delgado, J.: Service Interoperability in the Internet of Things. Internet of Things and Intercooperative Computational Technologies for Collective Intelligence, pp. 51–87. Springer, Berlin Heidelberg (2013)

Big Data Management Systems for the Exploitation of Pervasive Environments

Alba Amato and Salvatore Venticinque

Abstract The amount of available data has exploded significantly in the past years, due to the fast growing number of services and users producing vast amounts of data. The Internet of Things (IoT) has given rise to new types of data, emerging for instance from the collection of sensor data and the control of actuators. The explosion of devices that have automated and perhaps improved the lives of all of us has generated a huge mass of information that will continue to grow exponentially. For this reason the need to store, manage, and treat the ever increasing amounts of data that comes via the Internet of Things has become urgent. In this context, Big Data becomes immensely important, making possible to turn into this amount of data in information, knowledge, and, ultimately, wisdom. The aim of this chapter is to provide an original solution that uses Big Data technologies for redesigning an IoT context aware application for the exploitation of pervasive environment addressing problems and discussing the important aspects of the selected solution. The chapter also provides a survey of Big Data technical and technological solutions to manage the amounts of data that comes via the Internet of Things.

1 Introduction

The amount of available data has exploded significantly in the past years, due to the fast growing number of services and users producing vast amounts of data. The Internet of Things (IoT) has given rise to new types of data, emerging for instance from the collection of sensor data and the control of actuators. The explosion of devices that have automated and perhaps improved the lives of all of us has generated

A. Amato (✉) · S. Venticinque
Department of Industrial and Information Engineering, Second University of Naples,
Naples, Italy
e-mail: alba.amato@unina2.it

S. Venticinque
e-mail: salvatore.venticinque@unina2.it

N. Bessis and C. Dobre (eds.), *Big Data and Internet of Things:*
A Roadmap for Smart Environments, Studies in Computational Intelligence 546,
DOI: 10.1007/978-3-319-05029-4_3, © Springer International Publishing Switzerland 2014

a huge mass of information that will continue to grow exponentially. For this reason the need to store, manage, and treat the ever increasing amounts of data that comes via the Internet of Things has become urgent. In this context, Big Data becomes immensely important, making possible to turn into this amount of data in information, knowledge, and, ultimately, wisdom. An interesting view of what are the Big Data has been exposed to Gartner that defines Big Data as "high volume, velocity and/or variety information assets that demand cost-effective, innovative forms of information processing that enable enhanced insight, decision making, and process automation" [1]. In fact the huge size is not the only property of Big Data. Only if the information has the characteristics of Volume, Velocity and/or Variety we can talk about Big Data [2] as shown in Fig. 1. Volume refers to the fact that we are dealing with ever-growing data expanding beyond terabytes into petabytes, and even exabytes (1 million TB). Variety refers to the fact that Big Data is characterized by data that often come from heterogeneous sources such as machines, sensors and unrefined ones, making the management much more complex. Finally, the third characteristic, that is velocity that, according to Gartner [3], "means both how fast data is being produced and how fast the data must be processed to meet demand". In fact in a very short time the data can become obsolete. Dealing effectively with Big Data "requires to perform analytics against the volume and variety of data while it is still in motion, not just after" [2]. IBM [4] proposes the inclusion of veracity as the fourth Big Data attribute to emphasize the importance of addressing and managing the uncertainty of some types of data. Striving for high data quality is an important Big Data requirement and challenge, but even the best data cleansing methods cannot remove the inherent unpredictability of some data, like the weather, the economy, or a customer's actual future buying decisions. The need to acknowledge and plan for uncertainty is a dimension of Big Data that has been introduced as executives seek to better understand the uncertain world around them. Big Data are so complex and large that it is really difficult and sometime impossible, to process and analyze them using traditional approaches. In fact traditional relational database management systems (RDBMS) can not handle Big Data sets in a cost effective and timely manner. These technologies may not be enabled to extract, from large data set, rich information that can be exploited across of a broad range of topics such as market segmentation, user behavior profiling, trend prediction, events detection, etc and in many fields like public health, economic development and economic forecasting. Besides Big Data have a low information per byte, and, therefore, given the vast amount of data, the potential for great insight is quite high only if it is possible analyze the whole dataset [2]. In fact data is the raw material that is processed into information. Individual data by itself is not very useful, but volumes of it can identify trends and patterns. This and other sources of information come together to form knowledge. In the simplest sense, knowledge is information of which someone is aware. Wisdom is then born from knowledge plus experience [5]. So, the challenge is to find a way to transform raw data into valuable information. To capture value from Big Data, it is necessary an innovation in technologies and techniques that will help individuals and/organizations to integrate, analyze, visualize different types of data at different spatial and temporal scales. The aim of this chapter is to provide

Fig. 1 Big data characteristics

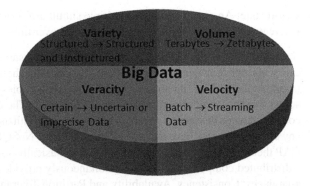

an original solution that uses Big Data technologies for redesigning an IoT context aware application for the exploitation of pervasive environment addressing problems and discussing the important aspects of the selected solution. The chapter also provides a survey of Big Data technical and technological solutions to manage the amounts of data that comes via the Internet of Things. The paper is organized as follows: in Sect. 2 a survey of technical and technological solutions related to Big Data is presented, Sects. 4 and 5 describe an example of application context that is suitable for the utilization Big Data solution. Sect. 6 briefly describes the proposed solution; conclusions are drawn in Sect. 7.

2 NoSQL

The term NoSQL (meaning 'not only SQL') is used to describe a large class of databases which do not have properties of traditional relational databases and which are generally not queried with SQL (structured query language). NoSQL data stores are designed to scale well horizontally and run on commodity hardware. Also, the 'one size fit's it all' [6] notion does not work for all scenarios and it is a better to build systems based on the nature of the application and its work/data load [7].

NoSQL data stores come up with following key features [8]:

- the ability to horizontally scale simple operation throughput over many servers,
- the ability to replicate and to distribute (partition) data over many servers,
- a simple call level interface or protocol (in contrast to a SQL binding),
- a weaker concurrency model than the ACID transactions of most relational (SQL) database systems,
- efficient use of distributed indexes and RAM for data storage, and
- the ability to dynamically add new attributes to data records.

Most of the NoSQL databases, has as the main objective, the achievement of scalability and higher performance. To provide this they do not guarantee all ACID properties, but use the a relaxed set of properties named BASE:

- Basically available: Allowance for parts of a system to fail

- Soft state: An object may have multiple simultaneous values
- Eventually consistent: Consistency achieved over time.

Because the data does not have to be 100 percent consistent all the time, applications can scale out to a much greater extent. By relaxing the consistency requirement, for example, NoSQL databases can have multiple copies of the same data spread across many servers or partitions in many locations. The data is instead eventually consistent when the servers are able to communicate with one another and catch up on any updates one may have missed [9]. Proponents of NoSQL often cite Eric Brewer's CAP theorem [10], formalized in [11], which basically states that is impossible for a distributed computing system to simultaneously provide all three of the following guarantees: Consistency, Availability and Partition Tolerance (from these properties the CAP acronym has been derived). Where:

- Consistency: all nodes see the same data at the same time
- Availability: a guarantee that every request receives a response about whether it was successful or failed
- Partition Tolerance: the system continues to operate despite arbitrary message loss or failure of part of the system that create a netwrok partition

Only two of the CAP properties can be ensured at the same time. Therefore, only CA systems (consistent and highly available, but not partition-tolerant), CP systems (consistent and partition tolerant, but not highly available), and AP systems (highly available and partition tolerant, but not consistent) are possible and for many people CA and CP are equivalent because loosing in Partitioning Tolerance means a lost of Availability when a partition takes place. So, the trade-offs are complex and the NoSQL databases, acting as distributed systems, must choose between either support: AP or CP.

Common concepts in NoSQL databases are [12]:

- Sharding, also referred to as horizontal scaling or horizontal partitioning. It is a partitioning mechanism in which records are stored on different servers according to some keys. Data is partitioned in such a way that records, that are typically accessd/updated together, reside on the same node. Data shards may also be replicated for reasons of reliability and load-balancing and it may be either allowed to write to a dedicated replica only or to all replicas maintaining a partition of the data. To allow such a sharding scenario there has to be a mapping between data partitions (shards) and storage nodes that are responsible for these shards. This mapping can be static or dynamic, determined by a client application, by some dedicated 'mapping-service-component' or by some network infrastructure between the client application and the storage nodes. The downside of sharding scenarios is that joins between data shards are not possible, so that the client application or proxy layer inside or outside the database has to issue several requests and postprocess (e. g. filter, aggregate) results instead.
- Consistent hashing [13]. The idea behind consistent hashing is to use the same hash function, used to generate fixed-length output data that acts as a shortened reference to the original data, for both the object hashing and the node hashing.

This is advantageous to both objects and machines. The machines will get an interval of the hash function range and the neighboring machines can take over portions of the interval of their adjacent nodes if they leave and can assign parts of their interval if some new member node joins and gets mapped to a nearby interval. Another advantage of consistent hashing is that clients can easily determine the nodes to be contacted to perform read or write operations.

- MapReduce [14] is a programming model and an associated implementation by Google for processing and generating large data sets. Users specify a map function that processes a keyvalue pair to generate a set of intermediate keyvalue pairs, and a reduce function that merges all intermediate values associated with the same intermediate key. MapReduce is typically used to do distributed computing on clusters of computers. When applied to databases, MapReduce processes a set of keys by submitting the process logic (map- and reduce-function code) to the storage nodes which locally apply the map function to keys that should be processed and that are in their set. The intermediate results can be consistently hashed just as regular data and processed by the following nodes in clockwise direction, which apply the reduce function to the intermediate results and produce the final results. It should be noted that due to the consistent hashing of the intermediate results there is no coordinator needed to direct the processing nodes to find the intermediate results. A popular open source implementation is Apache Hadoop [15], a framework that allows for the distributed processing of large data sets across clusters of computers using simple programming models. It is designed to scale up from single servers to thousands of machines, each offering local computation and storage and to execute queries and other batch read operations against massive datasets that can be tens or hundreds of terabytes and even petabytes in size. A commercial closed source model is Dryad [16]. Originated by Microsoft Research, Dryad, DryadLINQ and the Distributed Storage Catalog (DSC) are currently available as community technology previews. DryadLINQ allows programmers to use Microsoft technologies such as Microsoft .NET and LINQ to express their algorithms. Dryad is a general purpose runtime for execution of data parallel applications allowing the execution of the algorithms across large quantities of unstructured data distributed across clusters of commodity computers. DSC is a storage system that provides the bottom layer to the stack. It ties together a large number of commodity machines to store very large (i.e., Bing-level) quantities of data. These are commercial versions of the same technology used by the Bing search engine for large, unstructured data analysis.
- Versioning of Datasets in Distributed Scenarios. If Datasets are distributed among nodes, they can be read and altered on each node and no strict consistency is ensured by distributed transaction protocols. Questions arise on how concurrent modifications and versions are processed and to which values a dataset will eventually converge to. There are several options to handle these issues and, the most used are Multiversion Concurrency Control (MVCC), whose aim is to avoid the problem of writers blocking readers and viceversa, by making use of multiple versions of data, and vector clocks, an algorithm for generating a partial ordering of events in a distributed system and detecting causality violations.

However, NoSQL databases are not the solution to every problem of data management [17]. In fact, they completely miss a common query language like SQL in RDBMS. SQL is based on Relational Algebra that ensures completeness of the query language and that offers many optimization techniques to support query execution. It represents one of the main reasons why the RDBMS systems have acquired increasing importance. A developer can move from one database system to another with reduced effort at least for basic operations. Another lack of NoSQLs is the extreme heterogeneity of the existing solutions for what concerns the organization of the data model, the query model and the data access recommended patterns. This forces the developer to handle manually low-level data management issues like indexing, query optimizing data structures, relations between objects, and so on. This results in a higher complexity of NoSQL compared to RDBMS solutions for what concerns programmability and management of the data store.

3 NoSQL Classification

According to [8] NoSQL Data Models can be classified in:

- Key-value data stores (KVS). They store values associated with an index (key). KVS systems typically provide replication, versioning, locking, transactions, sorting, and/or other features. The client API offers simple operations including puts, gets, deletes, and key lookups. Notable examples include: Amazon DynamoDB, Project Voldemort, Memcached, Redis and RIAK
- Document data stores (DDS). DDS typically store more complex data than KVS, allowing for nested values and dynamic attribute definitions at runtime. Unlike KVS, DDS generally support secondary indexes and multiple types of documents (objects) per database, as well as nested documents or lists. Notable examples include Amazon SimpleDB, CouchDB, MembaseCouchbase, MongoDB and RavenDB
- Extensible record data stores (ERDS). ERDS store extensible records, where default attributes (and their families) can be defined in a schema, but new attributes can be added per record. ERDS can partition extensible records both horizontally (per-row) or vertically (per-column) across a datastore, as well as simultaneously using both partitioning approaches. Notable examples include Google BigTable, HBase, Hypertable and Cassandra

Another important category is constituted by Graph data stores. They [18] are based on graph theory and use graph structures with nodes, edges, and properties to represent and store data. Key-Value, Document based and Extensible record categories aim at the entities decoupling to facilitate the data partitioning and have less overhead on read and write operations, whereas Graph-based take the modeling the relations like principal objective. Therefore techniques to enhancing schema with a Graph-based database may not be the same as used with Key-Value and others. The graph data model fits better to model domain problems that can be represented by graph as ontologies, relationship, maps etc. Particular query languages allow querying the data bases by using classical graph operators as neighbor, path, distance etc. Unlike the NoSQL systems we presented, these systems generally provide ACID transactions. Different Graph Store products exist in the market today. Some provide custom API's and Query Languages and many support the W3C's RDF standard. Notable examples include neo4j, AllegroGraph and InfiniteGraph.

Amazon Dynamo [19] is a key-value distributed storage system that is developed and used by Amazon. Dynamo is used to manage the state of services that have very high reliability requirements and need tight control over the tradeoffs between availability, consistency, cost-effectiveness and performance. Dynamo is a structured overlay based on consistent hashing with maximum one-hop request routing. Dynamo uses a synthesis of well known techniques to achieve scalability and availability: Data is partitioned and replicated using consistent hashing, and consistency is facilitated by object versioning [20]. The consistency among replicas during updates is maintained by a quorum-like technique and a decentralized replica synchronization protocol. Dynamo employs a gossip based distributed failure detection and membership protocol. Dynamo is a completely decentralized system with minimal need for manual administration. Storage nodes can be added and removed from Dynamo without requiring any manual partitioning or redistribution.

Project Voldemort [21] is an open source distributed key-value data store used by LinkedIn for high-scalability storage. It represents an open source implementation of the basic parts of Dynamo's distributed key-value storage system. Like Dynamo, Voldemort uses consistent hashing for data partitioning and supports virtual nodes. Data is automatically replicated over multiple servers and is automatically partitioned so each server contains only a subset of the total data. Server failure is handled transparently. Pluggable serialization is supported to allow for rich keys and values including lists and tuples with named fields, as well as to integrate with common serialization frameworks like Protocol Buffers, Thrift, Avro and Java Serialization. Data items are versioned to maximize data integrity in failure scenarios without compromising availability of the system and each node is independent of other nodes with no central point of failure or coordination. It also supports pluggable data placement strategies to support geographically separated datacenters. Voldemort single node performance is at the range of 10–20k operations per second depending on the machines, the network, the disk system, and the data replication factor. Voldemort provides eventually consistency, just like Amazon Dynamo [22].

Memcached [23] is a free and open source, high-performance, distributed memory object caching system, generic in nature, but intended for use in speeding up dynamic web applications by alleviating database load. Memcached is an in-memory key-value store for small chunks of arbitrary data (strings, objects) from results of database calls, API calls, or page rendering. Keys are hashed and ash table span across an arbitrary number of servers. Its API is available for most popular languages.

Redis [24] Redis is an open source, BSD licensed, advanced key-value store. It is often referred to as a data structure server since keys can contain strings, hashes, lists, sets and sorted sets. It is possible to run atomic operations on these types, like appending to a string; incrementing the value in a hash; pushing to a list; computing set intersection, union and difference; or getting the member with highest ranking in a sorted set. In order to achieve better performance, Redis works with an in-memory dataset and it is possible to persist it either by dumping the dataset to disk every once in a while, or by appending each command to a log. Redis in particular does not offer fault-tolerance and as data is held in memory it will be lost if a server crashes.

RIAK [25] is another open source distributed key-value data store, developed by Basho Technologies, that provides tunable consistency. Consistency is tuned by specifying how many replicas must respond for a successful read/write operation and can be specified per-operation. It provides a decentralized key-value store that supports standard Get, Put and Delete operations. Riak is a distributed, highly scalable and fault-tolerant store with map/reduce, HTTP, JSON and REST queries. RIAK relies on consistent hashing for data partitioning and vector clocks for versioning, like Dynamo and Voldemort. Riak also includes a MapReduce mechanism for non-key-based querying. MapReduce jobs can be submitted through the RIAK's HTTP API or the protobufs API. To this end, the client makes a request to RIAK node which becomes the coordinating node for the MapReduce job.

Amazon SimpleDB [26] is a proprietary document data store offered as a service in Amazon's AWS cloud portfolio. Amazon SimpleDB automatically manages infrastructure provisioning, hardware and software maintenance, replication and indexing of data items, and performance tuning. Amazon SimpleDB automatically creates multiple geographically distributed copies of each data item you store. This provides high availability and durability in the unlikely event that one replica fails, Amazon SimpleDB can failover to another replica in the system. Amazon SimpleDB supports two read consistency options: eventually consistent reads and consistent reads. The eventual consistency option (default) maximizes the read performance (in terms of low latency and high throughput). However, an eventually consistent read might not reflect the results of a recently completed write. Consistency across all copies of data is usually reached within a second; repeating a read after a short time should return the updated data. So in addition to eventual consistency, Amazon SimpleDB also gives to the user the flexibility and control to request a consistent read if the application requires it. A consistent read returns a result that reflects all writes that received a successful response prior to the read. Unlike with key-value datastores, SimpleDB supports more than one grouping in one database: documents are put into domains, which support multiple indexes. SimpleDB data model is comprised of domains, items, attributes and values. Domains are collections of items that are described by attribute-value pairs. SimpleDB constrains individual domains to grow up to 10 GB each, and currently has a limit of 100 active domains.

Apache CouchDB [27] is an open source database that focuses on ease of use and on being "a database that completely embraces the web". It stores JSON objects that consist of named fields without predefined schema. Field values can be strings, numbers, or dates; but also ordered lists and associative arrays. CouchDB uses JavaScript for MapReduce queries, and regular HTTP for an API and provides ACID semantics at the document level, but eventual consistency otherwise. To support ACID on document level, CouchDB implements a form of Multi-Version Concurrency Control (MVCC) in order to avoid the need to lock the database file during writes. Conflicts are left to the application to resolve. CouchDB structure stores data into views and each view is constructed by a JavaScript function that acts as the map phase in MapReduce. CouchDB was designed with bi-directional replication (or synchronization) and off-line operation in mind. Namely, CouchDB can replicate to devices (like smart-phones) that can go offline and later sync back the device.

Couchbase [28], originally known as Membase, is an open source, distributed document oriented data store that is optimized for interactive applications. These applications must serve many concurrent users; creating, storing, retrieving, aggregating, manipulating and presenting data. In support of these kinds of application needs, Couchbase is designed to provide easy-to-scale key-value or document access with low latency and high sustained throughput. It is designed to be clustered from a single machine to very large scale deployments. Couchbase has initially grown around memcached, by adding to it features like persistence, replication, high availability, live cluster reconfiguration, re-balancing, multi-tenancy and data partitioning. Couchbase supports fast fail-over with multi-model replication support for both peer-to-peer replication and master-slave replication. Every Couchbase node is architecturally identical consisting of a data manager and cluster manager component. The cluster manager supervises the configuration and behavior of all the servers in a Couchbase cluster. It configures and supervises internode behavior like managing replication streams and rebalancing operations. It also provides metric aggregation and consensus functions for the cluster, and a RESTful cluster management API. The cluster manager is built atop Erlang/OTP, a proven environment for building and operating fault-tolerant distributed systems. The data manager is responsible for storing and retrieving documents in response to data operations from applications. Couchbase has only recently migrated from a key-value store to a document data store, with version 2.0 bringing features like JSON document store, incremental MapReduce and cross datacenter replication. Couchbase stores JSON objects with no predefined schema.

MongoDB [29], is an open source document-oriented data storebase system. MongoDB stores structured data as JSON-like documents with dynamic schemas. MongoDB supports queries by field, range queries, regular expression searches. Queries can return specific fields of documents and also include user-defined JavaScript functions. Any field in a MongoDB document can be indexed and secondary indices are also available. MongoDB supports master-slave replication. A master can perform reads and writes, whereas a slave copies data from the master and cannot be used for writes. MongoDB scales horizontally using sharding and can run over multiple servers, balancing the load and/or duplicating data to keep the system up and running in case of hardware failure. MongoDB supplies a file system function, called GridFS, taking advantage of load balancing and data replication features over multiple machines for storing files. MapReduce can be used in MongoDB for batch processing of data and aggregation operations.

RavenDB [30], is a transactional, open source Document Database written in .NET, and offering a flexible data model designed to address requirements coming from real-world systems. RavenDB allows you to build high-performance, low-latency applications quickly and efficiently. Data in RavenDB is stored schema-less as JSON documents, and can be queried efficiently using Linq queries from your .NET code or using RESTful API using other tools. Internally, RavenDB makes use of indexes which are automatically created based on your usage, or created explicitly by the consumer. RavenDB is built for web-scale, and offers replication and sharding support out-of-the-box.

Big Table [31], Bigtable is a distributed storage system designed by Google for managing structured data that is designed to scale to a very large size: petabytes of data across thousands of commodity servers. Google initially designed BigTable as distributed data storage solution for several applications (like Google Earth and Google Finance), aiming at providing flexible, high-performance solution for different application requirements. It is designed for storing items such as billions of URLs, with many versions per page; over 100 TB of satellite image data; hundreds of millions of users; and performing thousands of queries a second. Bigtable is designed with semi-structured data storage in mind. It is a large map that is indexed by a row key, column key, and a timestamp. Each value within the map is an array of bytes that is interpreted by the application. Every read or write of data to a row is atomic, regardless of how many different columns are read or written within that row.

Hbase [32] is a distributed, column-oriented, data storage system offering strict consistency designed for data distributed over numerous nodes. It provides strongly consistent reads and writes versus an eventually consistent data store. HBase is largely inspired by Google's BigTable and is designed to work well with Hadoop which is an open source implementation of Google's MapReduce framework. The default distributed files system for Hadoop (HDFS) is designed for sequential reads and writes of large files in a batch manner. This strategy disallows the system to offer close to real-time access which requires efficient random accesses of the data. HBase is an additional layer on top of HDFS that efficiently supports random reads— and in general access—on the data, using a sparse multi-dimensional sorted map. HBase does not support a structured query language like SQL. HBase applications are written in Java much like a typical MapReduce application. HBase does support writing applications in Avro, REST, and Thrift.

Hypertable [33] is a high performance, open source, massively scalable database modeled after Bigtable, Google's proprietary, massively scalable database. Its goal is to set the open source standard for highly available, scalable, database systems. Hypertable runs on top of a distributed file system such as the Apache Hadoop DFS, GlusterFS, or the Kosmos File System (KFS). It is written almost entirely in C++. Hypertable is similar to a relational database in that it represents data as tables of information, with rows and columns, but the main differences are that row keys are UTF-8 strings, there is no support for data types, values are treated as opaque byte sequences. Regarding its data model it supports all abstractions available in Bigtable; in contrast to Hbase column-families with an arbitrary numbers of distinct columns are available in Hypertable. Tables are partitioned by ranges of row keys (like in Bigtable) and the resulting partitions get replicated between servers. The data representation and processing at runtime is also borrowed from Bigtable; updates are done in memory and later flushed to disk. Hypertable has its own query language called HQL (Hypertable Query Language) and exposes a native C++ as well as a Thrift API.

Cassandra [34] is a distributed data storage system developed by Facebook which, similarly to BigTable, is designed for managing very large amounts of structured data spread out across many commodity servers, providing a key-value store with tunable consistency. Cassandra lets the application developer dial in the appropriate level of consistency versus scalability or availability for each transaction. This tunable consistency is a level of sophistication that other databases such as HBase do not always offer. However, the extra sophistication comes with more of a learning curve for the developer [9]. The main goal of Cassandra is to provide a highly available service with no single point of failure. The Cassandra API [35] consists of three very simple methods (insert, get, delete) and it allows the user to manipulate data using a distributed multi-dimensional map indexed by the key. The different attributes (columns) of the data stored by Cassandra are grouped together into sets (called "column families"). Cassandra exposes two kinds of such families: "simple column families" and "super column families", where the latter are a column family within a column family. This allows a key to map to multiple values. Cassandra also has an in-memory cache to speed access to the most important data on designated servers in the cluster. In terms of scalability, Cassandra achieves the highest throughput for the maximum number of nodes in all experiments [36].

Neo4j [37], is an open source, robust (fully ACID) transactional property graph database. Due to its graph data model, Neo4j is highly agile and blazing fast. For connected data operations, Neo4j runs a thousand times faster than relational databases. Nodes store data and edges represent relationships. The data model is called property graph to indicate that edges could have properties. Neo4j provides a REST interface or a Java API. The core engine of Neo4j supports the property graph model. This model can easily be adapted to support the LinkedData RDF model, consisting of Triples. Besides it is possible to add spatial indexes to already located data, and perform spatial operations on the data like searching for data within specified regions or within a specified distance of a point of interest. In addition classes are provided to expose the data to geotools and thereby to geotools enabled applications like geoserver and uDig.

AllegroGraph [38] is a modern, high-performance, persistent graph database. AllegroGraph uses efficient memory utilization in combination with disk-based storage, enabling it to scale to billions of quads while maintaining superior performance. AllegroGraph supports SPARQL, RDFS++, and Prolog reasoning from numerous client applications. AllegroGraph is a proprietary product of Franz Inc., which markets a number of Semantic Web products and claims Pfizer, Ford, Kodak, NASA and the Department of Defense among its AllegroGraph customers.

InfiniteGraph [39] InfiniteGraph is a proprietary graph database currently available in both free and paid license versions produced by Objectivity, a company that develops data technologies supporting large-scale, distributed data management, object persistence and relationship analytics. Its goal is to create a graph database with "virtually unlimited scalability". InfiniteGraph is used in applications including real-time and location-aware web and mobile advertising platforms, military

Table 1 Data store comparison

Name	Classification	License	Data storage
Dynamo	KVS	Proprietary	Plug-in
Voldemort	KVS	Open source	RAM
Memcached	KVS	Open source	RAM
Redis	KVS	Open source	RAM
RIAK	KVS	Open source	Plug-in
SimpleDB	DDS	Proprietary	S3 (simple storage service)
CouchDB	DDS	Open source	Disk
Couchbase	DDS	Open source	Disk
MongoDB	DDS	Open source	Disk
RavenDB	DDS	Open source	Disk
Google BigTable	ERDS	Proprietary	GFS
HBase	ERDS	Open source	Hadoop
Hypertable	ERDS	Open source	Disk
Cassandra	ERDS	Open source	Disk
Neo4J	Graph	Open source	Disk
AllegroGraph	Graph	Proprietary	Disk
InfiniteGraph	Graph	Proprietary	Disk

operations planning and mission assurance, and advanced healthcare and patient records management.

Table 1 provides a comparison of all the examples given in terms of Classification, Licence and Storage System. Comparison based on several issues are available at [40].

4 Context Awareness in Pervasive Environments

Pervasiveness of devices provides to application and services the possibility for using peripherals and sensors as their own extensions to collect about the user and the environment, but also to improve service delivery. In the last years, context awareness has widely demonstrated its crucial role to achieve optimized management of resources, systems, and services in many application domains, from mobile and pervasive computing to dynamically adaptive service provisioning. Furthermore the explosion of devices that have automated and perhaps improved the lives of all of us has generated a huge mass of information that will continue to grow exponentially. In fact if we take in consideration, that by 2020 there will be more than 30 billion devices connected to the Internet, it is possible to understand how much data this devices could produce if they are connected 24/7. Data can be collected from various sources such as social networks, data warehouse, web applications, networked machines, virtual machines, sensors over the network, mobile devices, etc. It is necessary to think about how and where to store them. It is necessary a scalable, distributed storage systems, a set of flexible data models that allow for an easy interaction with programming languages. The need to store, manage, and treat the ever increasing amounts of data is becoming increasingly felt and contextualisation can be an attractive paradigm to combine heterogeneous data streams to improve quality of a mining process or classifier. Another issue concerns the integration of multiple data sources in an automated, scalable way to aggregate and store these heterogeneous and unbelievable amounts of data to conduct deep analytics on the combined data set. The traditional storage can represent a low cost solution to handle this kind of information. Also the effort spent in redesigning and optimizing data warehouse for analysis requests could result in poor performance. In fact current databases and management tools are inadequate to handle complexity, scale, dynamism, heterogeneity, and growth of such systems.

For these reasons context awareness in pervasive environments represent an interesting application field of big data technologies. In fact, according to IBM [2]:

- Big Data solutions are ideal for analyzing not only raw structured data, but semi-structured and unstructured data from a wide variety of sources.
- Big Data solutions are ideal when all, or most, of the data needs to be analyzed versus a sample of the data; or a sampling of data is not nearly as effective as a larger set of data from which to derive analysis.
- Big Data solutions are ideal for iterative and exploratory analysis when measures on data are not predetermined.

Big data technologies can address the problems related to the collection of data streams of higher velocity and higher variety. They allow for building an infrastructure that delivers low, predictable latency in both capturing data and in executing short, simple queries; that is able to handle very high transaction volumes, often in a distributed environment; and supports flexible, dynamic data structures [41]. With such a high volume of information, it is relevant the possibility to organize data at its original storage location, thus saving both time and money by not moving around large volumes of data. The infrastructures required for organizing big data are able to process and manipulate data in the original storage location. This capability provides very high throughput (often in batch), which are necessary to deal with large data processing steps and to handle a large variety of data formats, from unstructured to structured [41]. The analysis may also be done in a distributed environment, where some data will stay where it was originally stored and be transparently accessed for required analytics such as statistical analysis and data mining, on a wider variety of data types stored in diverse systems; scale to extreme data volumes; deliver faster response times driven by changes in behavior; and automate decisions based on analytical models. Most importantly, the infrastructure must be able to integrate analysis on the combination of big data and traditional enterprise data. New insight comes not just from analyzing new data, but from analyzing it within the context of the old to provide new perspectives on old problems [41]. Context-aware Big Data solutions could focus only on relevant information by keeping high probability of hit for all application-relevant events, with manifest advantages in terms of cost reduction and complexity decrease [42].

5 The M.A.R.A. Case Study

Exploitation of archaeological sites can be very difficult because of a lack of supporting infrastructures and because of the complex recognition and comprehension of
the relevant ruins, artworks and artifacts. The availability of personal devices can be
used to plan the visit and to support the tourist by suggesting him the itineraries, the
points of interest and by providing multimedia contents in the form of digital objects
which can semantically augment the perceived reality. In this context relevant issues
are the profiling of the user, the discovery and the delivery of the contents that can
improve the user's satisfaction, new models of interactions with reality. The Second
University of Naples is engaged on a multidisciplinary project with both cultural and
a technological aims [43]. Three case studies have been chosen to test the approach
and the framework. The S. Angelo in Formis Basilica, in Campania, near S. Maria
Capua Vetere, the ancient town of Norba and the amphitheater of Capua. In these sites
we cannot install complex infrastructures and they are difficult to understand without
a tourist guide. For the presented case studies we need to provide a technological
solution that does not need infrastructures for letting the software know the user
location and his feeling about the environment [44]. It means that Bluetooth, NFC,
GPS, electronic compass, camera, network connection and others technologies have
to be used, together or independently, to get information about the user perceptions
and to augment his exploitation of the archaeological site. In particular we will deal
with context aware recommender systems that are achieving widespread success in
nowadays in a lot of fields. The aim of those systems is making personalized recommendations during a live user interaction. A recommendation system learns from
user's behavior and makes recommendation that can be interesting for users. The
key component of a context aware recommendation system is data, often extremely
diverse in format, frequency, amount and provenance. This heterogeneous data will
serve as the basis for recommendations obtaining using algorithms to find similarities and affinities and to build suggestions for specific user profiles. One of the most
important application field of recommender systems is cultural heritage. Archaeological sites become pervasive environments if personal devices like tablets and
smart-phones are able to detect surrounding ruins, artifact and other kind of points of
interest by their on-board peripherals. In this context pervasiveness offers to software
applications the possibility to interact with the reality by the device, in order to perceive the information surrounding the users, and to adapt their own behavior and the
environment itself. By modeling an archaeological site as a pervasive environment
we are able to improve its exploitation by the visitors. In a pervasive computing environment, the intelligence of context-aware systems will be limited if the systems are
unable to represent and reason about context. In domains like tourism, the notion of
preferences varies among users and strongly depends on factors like users' personalities, parameters related to the context like location, time, season, weather and others
elements like user's feedback, so it is necessary to provide users with many other
kinds of personalized experiences, based on data of many kinds. The growth of visitors and interactive archaeological sites in recent years poses some key challenges

for recommender systems. In fact it is necessary to introduce recommender system technologies that can quickly produce high quality recommendations, even for very large-scale problems, so that users can benefit of context awareness in services exploitation and mobile services can became really useful and profitable. Besides, it is necessary to address the problem of the variety of data from sensors, RFID, devices, annotation tools, GIS (Geospatial, GPS), web. So the problem is both to capture data quickly and to store them quickly in structured form. The structure of the data, then, allows to identify a pattern based strategy for the extraction of consistent, comparable and updated information. In Fig. 2 an high level representation of our general case study is shown. We model the environment where the user is moving and to reconstruct the perceptions of the user himself in order to get his particular vision about what is surrounding him. A real representation of the environment is necessary to identify landmarks and possibilities of intervention using pervasive actuators and sensors whose input will be updated as the environment changes. The environment has been modeled as a geo-referred map with itineraries, landmarks and point of interest. The user can download the map of the area to be visited at home, before to leave, or on site, if the network will be available. The map includes all the points of interest that identify the relevant objects of that area and different cultural itineraries which could be exploited on site. Also contents can be discovered and downloaded in advance according to the device capabilities. Software on user's mobile device and remote services will assist the cultural visit by augmenting the reality by the user's personal device. A software agent executes on the user's device to support services exploitation. It percepts the surrounding environment using the on-board peripherals and executes plans which are chosen by an ad-hoc reasoning to optimize the user's satisfaction. The knowledge of the environment acquired by the agent represents part of its own beliefs. Another set of believes describes the user profile by recording and evaluating hiher actions or explicitly asking feedbacks. Some examples are user's position, interest, nearby objects, landscape, etc. Of course the way of localization of users and objects depends on the device technology, the available infrastructures and the kind of environment. Indoor or outdoor localization can be implemented using heterogeneous technologies, and often absolute localization can not be performed, but it is only possible to detect nearby landmarks or objects [45]. These techniques can result quite intensive in terms of computational requirements, and so the needed resources can exploit a distributed infrastructure. Part of the computation will be performed locally on the smartphone and expensive tasks will be off-loaded remotely. Besides the limited energy and data storages of the user's device can affect agent's capability. Remote services allows to the agent to move on remote more complex reasoning on a wider knowledge base. In order to augment user's knowledge and capability to interact with the environment, services have to choose, according to their context awareness; (1) what content and application it has to deliver; (2) when it needs to present the content; (3) how this should be done.

In Fig. 3 the architectural solution of the MARA framework is shown. The framework is composed of different tools and applications [44]. Tools for experts in the domain of the Cultural Heritage are used to augment the archaeological site with a set of multimedia contents. They include a map editor, a semantic annotator and a content

Fig. 2 Problem model

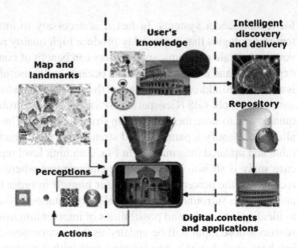

Fig. 3 Architecture and roles

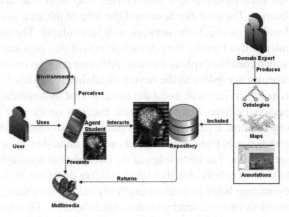

manager. A set of context aware services are available for intelligent multimedia discovery and delivery. A tourist guide supports the user in visiting an archaeological site, detects and suggests points of interest, provides context awareness of remote service, allows for the utilization of remote services and plays multimedia contents. On the left side of Fig. 3, the user is using his device that hosts a light agent that is able to perceive information from the field by pervasive sensors. The agent executes autonomously and proactively in order to support the user's activity within the environment where he is moving. It discovers surrounding objects, it uses them to update the representation of the user's knowledge, reacts using the local knowledge to organize and propose the available contents and facilities by an interactive interface. If the connection works the device can access remote services, which can access a wider knowledge and have greater reasoning capabilities to look for additional contents and applications. Experts of the application domain define the ontology for the specific case study described in [46]. They use or design a map to represent the environment.

They add POIs to the map to geo-refer multimedia contents and can link them to a concept of the ontology. Furthermore they select relevant contents and annotate them using concepts and individuals of the ontology. Remote applications implement context aware services. They use personal devices to collect perceptions and for content delivery. An ontology implements the representation of the global knowledge that is necessary to share a common dictionary and to describe the relationships among the entity eobjects, which are part of the model. In our model a common ontology include all the general concepts, which are useful for a description of a pervasive environment where mobile users are moving, using their devices and interacting with available facilities and other users. The general ontology is complemented with a domain ontology that is designed by an expert of the specific application field. Concepts of the ontology are used on client side to describe a representation of the reality as it is perceived by the user. On the back-end the ontology is used to annotate digital resources like point of interests, contents, applications. It is also used to support reasoning. User's behaviors, information from pervasive devices or from other users, device properties, external events are heterogeneous data that are perceived by the device and that are used to build a dynamic changing representation of the user knowledge about the reality, within which he is moving. The applications are knowledge driven. The user's knowledge can be used by the application that is running on the device to adapt its logic locally, an is updated remotely to improve the awareness of services at server side. Application are events based. Events are triggers for reactive activity by the agent. An event may update beliefs, trigger plans or modify goals. Events may be generated externally and received by sensors or integrated systems. Additionally, events may be generated internally to trigger decoupled updates or plans of activity. Events can be updates of the user's knowledge or can be explicit service requests raised by the user. At each invocations semantic queries, that depend on the user's knowledge, are built and processed to get the action to be performed and the contents to be delivered. Results of the query are individuals of the ontology that are described by semantic annotation. The user's knowledge is composed of many semantic concepts with *static* and *dynamic* properties. Semantic techniques are used for intelligent content and application discovery and delivery. An ontology has been designed to describe the sites of interest and to annotate the related media. A general part includes the concepts which are common to all the class of applications that can be modeled according to the proposed approach. Among the others the *Time* class and his properties (*CurrentTime, AvailableTime, ElapsedTime, Exploitation-Time*) allow to organize and assist the visit taking into account time information and handling time constraints. *Position* class and its properties allow to localize the user and objects around him. An application specific part of the ontology includes the concepts that belong to the domain of the cultural heritage and additional classes and individual which are proper of the case studies introduced in the previous section. The ontology is used also for annotating the multimedia contents. To annotate texts, images and any kind of contents we chose the AktiveMedia tool. In Fig. 4 a picture of the Amphitheater of S. Maria Capua Vetere is annotate with the *Column* and the *Arc* classes which are part of this kind of building. The perceptions are automatically communicated by a client application for Android Device and recommendations are

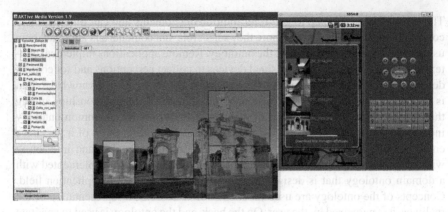

Fig. 4 The annotator

asincronously notified to the user by the tourist guide that periodically search for the most relevant items. The output produced by the annotator is an RDF file with concepts and properties of the AktiveMedia ontology and of the domain ontology.

5.1 Digital Repository and Semantic Discovery

The Fedora repository is used to store digital objects and supports their retrieval. Into the Fedora repository a digital object is composed of a set of files which are:

- *object metadata*: used by the client application to understand how to deliver the content;
- *binary streams*: which are images, video, text ... any kind of raw information to be delivered;
- *RDF annotation*: that describe the semantic of the object according to the ontology;
- *disseminations*: filters to be eventually used for adapting the object according to the target client.

We loaded the Aktive-Media ontology and the domain ontology into the Fedora repository in order to exploit its embedded SPARQL engine that is used to select the optimal set of individuals (i.e. contents). Multimedia contents are automatically stored into the repository after the annotation phase. The RDF output is automatically processed using an XSL transformation to make it compliant with the model used by the Fedora repository. Each content can be linked position.

5.2 Content Types

Different types of content models have been defined and simple examples have been produced.

- Multiple images whose transparency can be graduated by the user to compare changes in different periods. In the same way real picture can be compared with paintings. Old picture can be compared with what is seen by the camera.
- Part of the image acquired by the camera are recognized and linked to related multimedia contents.
- Virtual reconstructions which are synchronized with the camera output or the detected RFIDs.
- Text, audio, video and composite media.

A content descriptor is attached to every digital object. It is used by the device when the content is being delivered. The descriptor defines the right player for that media, configuration parameter and necessary input.

The semantic discovery service of MARA returns a set of digital objects related to POIs in the pervasive environments. Each content is annotated by concepts from the ontology and can be discovered by a SPARQL query to the content repository. The result of the query is a set of N instances of digital objects whose relevance to the user context is calculated as described in [47].

6 Moving to Big Data

Performance figures discussed in [48] demonstrate feasibility of the proposed solution implemented with a classical RDBMS. However a number of limitations have been assumed. First of all the amount of data are limited to the proprietary knowledge base with a limited number of archaeological sites. If we aim at handling all the national cultural heritage or the world one volume of data will be not supported. Besides the coverage of an increasing number of sites, eventually wider, will affect the amount of geographical information and the number of connected mobile users. Data continuously received from thousands of devices scattered in the environment handling queries and providing perception will augment volume, velocity and variety of information. Finally the exploitation of the user feedback could be used to improve the expertise of the system building a social network of visitors and enriching the knowledge base with semantic annotation inferred by a social network of visitors who become authors and editors themselves. The new vision of M.A.R.A. could not be implemented without considering the Big Data requirements and solutions that are ideal [2] to deal with raw data from a wide variety of sources that must be analyzed in toto. At this point we wonder understand what would be the best choice for re-designing and developing the framework to satisfy the new requirements. Understanding what NoSQL data models and technology could be more effective among the available alternatives needs further insides.

First of all maps, contents and users are represented by documents and profiles which are structured XML documents, but containing attributes like concepts, keyvalue pairs, RDF triples, which can dynamically change.

In current implementation of the MARA knowledge base all this documents are indexed and stored in a relational databases. Ontology, annotations, maps and profiles are processed when they are uploaded and updated by new inserting new record or updating the existing one. For this reason the adoption of a document data store supporting the map reduce paradigm seems the most rational alternative. I would not impact on the original design and supports the distribution of both the data and the computational effort for indexation and retrieval. Also users and data locality could be exploited to optimize performance. In fact points of interest and contents close to the users will be eventually more relevant to the others. About the drawback of such a choice, we have to consider that NOSQL DDS usually do not support ACID transactions, than there is the lack of a SPARQL interface for reasoning and semantic retrieval of relevant information, which is currently used by the MARA application layer. Furthermore some technological solution limit the maximum allowed size for a digital object. In fact MARA uses ontologies defined using RDFOWL that can be naturally represented using a labeled, directed, multi-graph. RDF [49] extends the linking structure of the Web to use URIs to name the relationship between things as well as the two ends of the link (this is usually referred to as a "triple"). Using this simple model, it allows structured and semi-structured data to be mixed, exposed, and shared across different applications. This linking structure forms a directed, labeled graph, where the edges represent the named link between two resources, represented by the graph nodes. This graph view is the easiest possible mental model for RDF and is often used in easy-to-understand visual explanations. To enables inferences and as query language, is used SPARQL that is a subgraph pattern matching query language. Besides, in MARA project we need to integrate multiple independent schemas that evolve separately depending on the different conditions correlated to the archaeological site. Semantic terms link instance data with other resources and apps and linked resources enable interoperability between apps as shown in Fig. 5. As an additional factor, it is necessary to bridge the structured and unstructured worlds and to make queries that integrate across data. It is a complex task because the different searches need to follow links between the data elements across schemas to find relationships. Graph databases [18] may overcome these limitation of the as they generally provide ACID transactions. Adoption of such solution will provide custom API's and Query Languages and many support the W3C's RDF standard, including a SPARQL engine. This model can easily integrated in MARA paltform as it supports the LinkedData RDF model. Besides it is possible to add spatial indexes to already located data, and perform spatial operations on the data like searching for data within specified regions or within a specified distance of a point of interest. In addition classes are provided to expose the data to geotools and thereby to geotools enabled applications like geoservers. As shown in Fig. 6, the second choice requires the re-desing of the knowledge base and of applications and tools for production, collection and indexing of data. In fact the new knowledge base can be modeled as a unique ontology dynamically augmented and updated with new information.

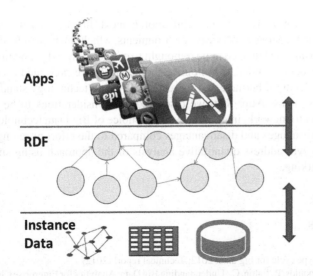

Fig. 5 RDF utilization

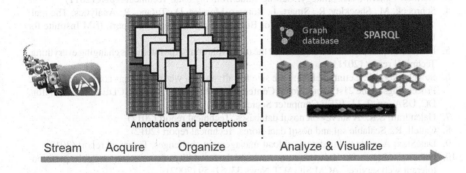

Fig. 6 Graph database utilizzation

7 Conclusion

In this chapter we have presented a survey of technical and technological solutions and analyzed critically the utilization of Big Data to redesign a context aware application for the exploitation of pervasive environment addressing problems and discussing the important aspects of the selected solution. The requirements of many applications are changing and require the adoption of these technologies. NoSQL databases ensure better performance than RDBMS systems in various use cases, most notably those involving big data. But the choice of the one that best fits the application requirements is a challenge for the programmers that decide to develop a scalable application. There are many differences among the available products and also among the level of maturation on them. From a solution point of view there needs to be a clear analysis of the application context. In particular we focused on technologies that operate in pervasive environments, which can benefit from the huge information available but need to be rethought to extract knowledge and improve the context awareness in order to customize the services. We presented a case study in the field of cultural heritage and its realization using technology standards such as RDBMS systems. After that, we discussed the considerations to be made for upgrading this framework, in particular on the choice of Big Data technologies considering its advantages and disadvantages compared to the effort of re-engineering. Future works will address quantitative analysis of the approach using simulations and benchmarking.

References

1. Gartner: Hype cycle for big data, 2012. Technical report (2012)
2. IBM, Zikopoulos, P., Eaton, C.: Understanding Big Data: Analytics for Enterprise Class Hadoop and Streaming Data. 1st edn. McGraw-Hill Osborne Media, New York (2011)
3. Gartner: Pattern-based strategy: Getting value from big data. Technical report (2011)
4. Schroeck, M., Shockley, R., Smart, J., Romero-Morales, D., Tufano, P.: Analytics: The real-world use of big data. IBM Institute for Business Value—executive report, IBM Institute for Business Value (2012)
5. Evans, D.: The internet of things—how the next evolution of the internet is changing everything. Technical report (2011)
6. Stonebraker, M., Cetintemel, U.: One size fits all: an idea whose time has come and gone. In: Proceedings of the 21st International Conference on Data Engineering. ICDE'05, Washington, DC, USA, pp. 2–11. IEEE Computer Society (2005)
7. Gajendran, S.K.: A survey on nosql databases. Technical report (2012)
8. Cattell, R.: Scalable sql and nosql data stores. Technical report (2012)
9. DataStax: A guide to big data workload-management challenges. Technical report (2012)
10. Gilbert, S., Lynch, N.: Brewer's conjecture and the feasibility of consistent, available, partition-tolerant web services. ACM SIGACT News 33, 51–59 (2002)
11. Brewer, E.A.: Towards robust distributed systems (abstract). In: Proceedings of the Nineteenth Annual ACM Symposium on Principles of Distributed Computing, PODC'00, p. 7. ACM, New York (2000)
12. Strauch, C.: Nosql databases (2011) (Online; 26 July 2013)
13. Karger, D., Lehman, E., Leighton, T., Panigrahy, R., Levine, M., Lewin, D.: Consistent hashing and random trees: distributed caching protocols for relieving hot spots on the world wide web. In: Proceedings of the Twenty-ninth Annual ACM Symposium on Theory of Computing STOC'97, pp. 654–663. ACM, New York (1997)
14. Dean, J., Ghemawat, S.: Mapreduce: simplified data processing on large clusters. Commun. ACM 51, 107–113 (2008)
15. Apache: Hadoop (2012) (Online 26 July 2013)
16. Jo Foley, M.: Microsoft drops dryad; puts its big-data bets on hadoop. Technical report (2011)
17. Locatelli, O.: Extending nosql to handle relations in a scalable way models and evaluation framework (2012012)
18. Robinson, I., Webber, J., Eifrem, E.: Graph Databases. O'Reilly Media, Incorporated (2013)
19. DeCandia, G., Hastorun, D., Jampani, M., Kakulapati, G., Lakshman, A., Pilchin, A., Siva-subramanian, S., Vosshall, P., Vogels, W.: Dynamo: amazon's highly available key-value store. SIGOPS Oper. Syst. Rev. 41, 205–220 (2007)

20. Lamport, L.: Time, clocks, and the ordering of events in a distributed system. Commun. ACM **21**, 558–565 (1978)
21. Sumbaly, R., Kreps, J., Gao, L., Feinberg, A., Soman, C., Shah, S.: Serving large-scale batch computed data with project voldemort. (2009)
22. Voldemort: Project voldemort a distributed database. (2012) (Online; 26 July 2013)
23. Memcached: Memcached (2012) (Online; 26 July 2013)
24. Redis: Redis (2012) (Online; 26 July 2013)
25. Riak: Riak (2012) (Online; 26 July 2013)
26. Amazon: Simpledb (2012) (Online; 26 July 2013)
27. Apache: Couchdb (2012) (Online; 26 July 2013)
28. Couchbase: Couchbase (2012) (Online; 26 July 2013)
29. MongoDB: Mongodb (2012) (Online; 26 July 2013)
30. RavenDB: Ravendb (2012) (Online; 26 July 2013)
31. Chang, F., Dean, J., Ghemawat, S., Hsieh, W.C., Wallach, D.A., Burrows, M., Chandra, T., Fikes, A., Gruber, R.E.: Bigtable: A distributed storage system for structured data. ACM Trans. Comput. Syst. **26**, 4:1–4:26 (2008)
32. HBase: Hbase (2012) (Online; 26 July 2013)
33. Hypertable: Hypertable (2012) (Online; 26 July 2013)
34. Cassandra: Cassandra (2012) (Online; 26 July 2013)
35. BigFoot: Current practices of big data analytics. Technical report (2013)
36. Rabl, T., Gómez-Villamor, S., Sadoghi, M., Muntés-Mulero, V., Jacobsen, H.A., Mankovskii, S.: Solving big data challenges for enterprise application performance management. Proc. VLDB Endow. **5**, 1724–1735 (2012)
37. Neo Technology, I.: Neo4j, the world's leading graph database. (2012) (Online; 26 July 2013)
38. AllegroGraph: Allegrograph (2012) (Online; 26 July 2013)
39. InfiniteGraph: Infinitegraph (2012) (Online; 26 July 2013)
40. findthebest.com: Compare nosql databases (2012) (Online; 26 July 2013)
41. Oracle: Big data for the enterprise. Technical report (2013)
42. Nessi: Nessi white paper on big data. Technical report (2012)
43. Amato, A., Di Martino, B., Venticinque, S.: Bdi intelligent agents for augmented exploitation of pervasive environments. In: WOA, pp. 81–88. (2011)
44. Amato, A., Di Martino, B., Venticinque, S.: Semantically augmented exploitation of pervasive environments by intelligent agents. In: ISPA, pp. 807–814. (2012)
45. Aversa, R., Di Martino, B., Venticinque, S.: Distributed agents network for ubiquitous monitoring and services exploitation. **2**, 197–204 (2009)
46. Renda, G., Gigli, S., Amato, A., Venticinque, S., Martino, B.D., Cappa, F.R.: Mobile devices for the visit of "anfiteatro campano" in santa maria capua vetere. In: EuroMed, pp. 281–290. (2012)
47. Amato, A., Di Martino, B., Venticinque, S.: A semantic framework for delivery of context-aware ubiquitous services in pervasive environments, pp. 412–419. (2012)
48. Amato, A., Di Martino, B., Scialdone, M., Venticinque, S.: Personalized recommendation of semantically annotated media contents. In: Intelligent Distributed Computing VII, vol. 511, pp. 261–270. Springer International Publishing, Switzerland (2013)
49. RDF: Rdf (2012) (Online; 26 July 2013)

On RFID False Authentications

Kevin Chiew, Yingjiu Li and Congfu Xu

Abstract Many reader/tag authentication protocols are proposed to effectively authenticate tags and readers. However, we demonstrate with YA-TRAP as an example how false authentications that a legitimate tag could be wrongly rejected by a reader may arise from these protocols when they are applied to C1G2 (class 1 generation 2) passive RFID tags. In this chapter, we identify a protocol pattern of which the implementation on C1G2 passive tags leads to false authentications, and further identify three types of the existing protocols that can bring with false authentications due to containing this pattern. Moreover, we give a necessary and sufficient condition for false authentications prevention, and propose a naive semaphore-based solution which revises the pattern by adding semaphore operations so as to avoid false authentications. Our experiments demonstrate the arising of false authentications and verify the effectiveness of our solution.

1 Introduction

RFID technology has been used in real applications such as supply chain management [7, 11, 16, 17], in which RFID tags are attached to products so that they can be conveniently identified by tag readers. Such RFID applications may encounter insecure situations like duplication of tag IDs, invalid or counterfeit tags and readers, or

K. Chiew (✉)
Provident Technology Pte. Ltd., Singapore, Singapore
e-mail: kev.chiew@gmail.com

Y. Li
School of Information Systems, Singapore Management University, Singapore, Singapore
e-mail: yjli@smu.edu.sg

C. Xu
College of Computer Science, Zhejiang University, Hangzhou 310027, P. R. China
e-mail: xucongfu@zju.edu.cn

N. Bessis and C. Dobre (eds.), *Big Data and Internet of Things:*
A Roadmap for Smart Environments, Studies in Computational Intelligence 546,
DOI: 10.1007/978-3-319-05029-4_4, © Springer International Publishing Switzerland 2014

---

1. Tag ⟵ Reader: T_r
2. Tag:
    2.1. **if** $(T_r - T_t <= 0$ or $T_r > T_{max})$ **then** $H_r = PRNG_i^j$
    2.2. **else** $T_t = T_r$, $H_r = HMAC_{K_i}(T_t)$
3. Tag ⟶ Reader: H_r
4. Reader ⟶ Server: T_r, H_r
5. Server:
    5.1. **let** $s = LOOKUP(HASH\_TABLE_{T_r}, H_r)$
    5.2. **if** $(s == -1)$ **then** MSG =TAG-ERROR
    5.3. **else** $MSG = G(Ks)$ (or MSG ="VALID")
6. Server ⟶ Reader: MSG

Fig. 1 Tsudik's YA-TRAP protocol

even malicious attacks, hence in practice it is critical to authenticate the legitimacy of tags and readers. Aiming at this purpose, many reader/tag authentication protocols and secure frameworks [1, 2, 5, 6, 12, 16, 18] have been proposed for authenticating the legitimacy of a reader or a tag by integrating hashing [18] or encryption [5] operations. Ideally, these protocols are expected to be effective for all kinds of tags. However, for C1G2 (class-1 generation-2) passive RFID tags [4, 8], this expectation cannot come to the truth. The C1G2 standard [4] has been the only de facto standard adopted by ISO as ISO 18000-6C [8] and has been massively used in real applications such as supply chain management. Our study shows that, false authentications that a legitimate tag is wrongly verified illegitimate and rejected by readers may arise from some of the protocols if they are working in an environment where one or more readers are interacting with the same C1G2 passive tag. False authentications may bring with disaster in real applications because in a supply chain for example, if a legitimate tag is attached to a container and is wrongly rejected due to false authentication, then the container could be delayed or even denied to pass through the supply chain, leading to economic losses of both the container customer and the supply chain service provider.[1]

1.1 Example of False Authentications

Taking YA-TRAP [18] (shown in Fig. 1) as an example, we demonstrate how such false authentications may arise. Step 3 of the protocol is denoted as "Tag ⟶ Reader: H_r", meaning that the tag returns to the reader a response H_r w.r.t. challenge T_r earlier sent by the reader. Although it is not clearly stated either in [18] or in other papers [1, 2, 12] how returning response H_r is accomplished, we can assume that theoretically it can be carried out by many approaches two of which are suggested

---

[1] This false authentication may also arise under other situations such as (1) two or more readers interacting with one tag, (2) one reader interacting with two or more tags, or (3) two or more readers interacting with two or more tags, in which all tags involved are not C1G2 passive tags. However, it is one of the directions for our future study.

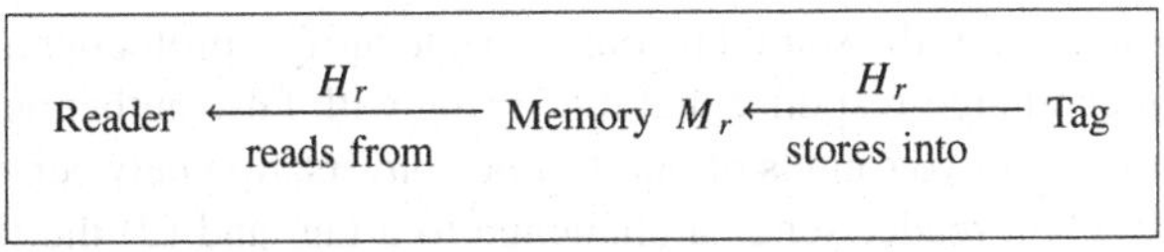

Fig. 2 Returning a response by two steps

as follows: (a) returning response H_r by the tag's initiatively sending it to the reader; and (b) returning response H_r by two steps as illustrated in Fig. 2: (1) the tag first stores response H_r into memory M_r inside itself and waits for the reader to read it, followed by (2) the reader reads response H_r from memory M_r where it is stored.

Approach (a) requires a tag to possess the capability of sending a response to a reader on its own initiative, whereas approach (b) works for a tag either with or without such capability. Based on the current ISO standard 18000-6C for RFID [8], C1G2 passive RFID tags do not possess such capability. For an application with C1G2 passive tags[2] involved, so far it seems that approach (b) is the only option for a tag to return a response. As a matter of fact, we are using approach (b) as a solution in our R&D project on RFID security in which IAIK tag emulators [3] together with other C1G2 tags are used.

Given the understandings, we observe what the reader reads from memory M_r is *not* the expected response H_r in either of the following two cases: (1) the reader reads memory M_r *before* it has been updated with response H_r; and (2) the reader reads memory M_r *after* it has been overwritten by another response w.r.t. a latter challenge from another reader. If response H_r is correct (meaning a legitimate tag), then in either situation, the tag is rejected as an illegitimate tag by the reader, and a false authentication arises.

1.2 Objective and Contribution of this Study

The above example of false authentications motivates us to conduct an intensive study on (1) why false authentications may arise, (2) what kinds of protocols may bring with false authentications, and (3) effective solutions for preventing such false authentications from arising. All of these, we would like to state once more however, are based on the assumption that all tags are C1G2 passive tags and returning a response is carried out by the suggested approach (b) as shown in Fig. 2.

In this chapter, we show that the major reason leading to false authentications is the contention of the shared resource, i.e., memory M_r, which is used to store non-shared data (i.e., responses) for readers. If a reader fails to read its data from this shared resource within a given time slot, then it may read unexpected data and wrongly reject a legitimate tag, leading to a false authentication.

---

[2] We henceforth use terms *tags*, *C1G2 tags* or *C1G2 passive tag* interchangeably to refer to *C1G2 passive RFID tags* unless otherwise specified.

From the contention of the shared resource, we identify a protocol pattern known as the pattern of challenge-response that may bring with false authentications. This pattern is comprised of two rounds of one-to-one conversation between a reader and a tag as follows: (1) a reader sends a challenge to a tag, and (2) the tag returns a response to the reader. We further come out a necessary and sufficient condition for false authentications prevention under this protocol pattern.

Based on whether a protocol contains the component of either reader or tag authentication, we classify the existing protocols into *four* types, namely (1) reader authentication only (RAO) protocols, (2) tag authentication only (TAO) protocols, (3) tag-then-reader authentication (TRA) protocols, and (4) reader-then-tag authentication (RTA) protocols. The first two types of protocols are one-way authentications, i.e., either a reader authenticates itself to a tag or a tag authenticates itself to a reader; whereas the next two types of protocols are mutual authentications, i.e., a reader and a tag have to authenticate to each other mutually. Based on the classification, we show that three out of four types of protocols, i.e., TAO, TRA and RTA protocols, contain the above identified pattern of challenge-response, and false authentications may arise from them.

To prevent false authentications from arising, we propose a naive semaphore-based solution which suggests to revise the protocol pattern by adding several steps of semaphore operations. A semaphore is a predefined memory inside a tag, just as the memory for storing responses. It can be set with different values indicating different states of the shared resource. A semaphore operation is to set the semaphore with a value so as to signal whether the shared resource is ready for access by a user (either the tag itself or a reader). With semaphore operations, the aforementioned necessary and sufficient condition for false authentications prevention can be well satisfied, thus false authentications can be prevented.

Our experimental results show that (1) false authentications resulting from the protocol pattern can easily happen in practice and can be captured; and (2) the proposed solution can effectively eliminate such false authentications.

In summary, our major contributions are three-fold: (1) we point out that false authentications may arise from the existing protocols in the applications with C1G2 passive tags involved, (2) we identify a protocol pattern under which false authentications may arise, and further identify three types of the existing protocols that contain this pattern and may bring with false authentications; and (3) we give a necessary and sufficient condition for false authentications prevention. In addition, other contributions are: (1) we propose a semaphore-based solution to prevent false authentications from arising; and (2) we demonstrate with experiments the arising of false authentications and verify the effectiveness of our solution.

1.3 Chapter Organisation

The chapter is organised as follows. We review the related work in Sect. 2, following which we give assumptions for this study together with the primitive operations of the conversations between a tag and a reader. After analysing the reason of false authentications in Sect. 4, we identify the protocol pattern that may incur false authentications and show how. We then give a classification for the existing protocols in Sect. 5 and identify those that contain the pattern and may lead to false authentications. Lastly in Sect. 6 we present a semaphore-based solution for false authentications prevention, and show the experimental results in Sect. 7 before concluding the paper in Sect. 8.

2 Related Work

Many reader/tag authentication protocols have been proposed to authenticate readers and tags. Based on the application environments, some protocols only fulfill one-way authentications [5, 18, 21], i.e., either tags authenticate to readers or the reverse, while some other protocols fulfill mutual authentications [1, 2, 14, 22], i.e., both readers and tags authenticate to each other mutually.

To prevent unauthorised readers from accessing legitimate tags, Weis et al. proposed the hash-locking protocol [21] that enables a tag to authenticate the legitimacy of a reader first before allowing the reader to access the tag's memory. In real applications, this type of protocols only takes a very small portion of all reader/tag authentication protocols.

In most applications, readers are guaranteed legitimate but tags could be counterfeit. Some protocols have been proposed to prevent unauthorised tags from being accepted by readers. The major difference amongst these protocols is how a response is generated. One of the representatives of such protocols is YA-TRAP [18] under which a response from a tag is a hashing result of the secret key held by the tag and the challenge from a reader. Feldhofer et al. [5] implemented such a protocol under which a response is generated with AES. Vajda and Buttyán [19] proposed five protocols of challenge-response with different ways of response generating for a tag to authenticate itself to a reader.

In some circumstances there could be counterfeit readers, making it necessary for a reader and a tag to mutually authenticate to each other. Mutual authentications are combinations of the above one-way authentications with two possible orders, i.e., either a reader first authenticates itself to a tag [1, 14] or the reverse [2, 12, 22]. Alomair et al. [1] proposed a mutual authentication protocol which they claimed unconditionally secure. The protocol achieves low-cost computations on tags. Peris-Lopez et al. proposed M^2AP protocol [14] which contains only two rounds of conversation between a reader and a tag and uses primitive operations. They claimed it as the minimalist mutual-authentication protocol for low-cost RFID tags, and studied the security property that Yang et al. analysed for their protocol proposed in [22]. Other properties such as time complexity and privacy-preserving of lightweight protocols have been studied in [2, 12].

3 Preliminary

In the next, we give some assumptions based on which we conduct this study, and classify the conversations between a reader and a tag into two classes of primitive operations.

3.1 Assumption

(1) Free of attacks. Tags and readers are assumed interacting with each other under no attacks.

(2) Legitimate tags and readers. All tags are assumed legitimate and their responses to any challenge are always assumed correct unless otherwise specified. We use Roman font T_1, T_2, ... to denote tag 1, tag 2, ... Similarly, all readers are assumed legitimate. We use Roman font R_1, R_2, ... to denote reader 1, reader 2, ...

(3) C1G2 tags. All tags are assumed C1G2 passive RFID tags by default. Returning a response is carried out by the suggested approach (b) as shown in Fig. 2. Likewise, sending a challenge is carried out by a reader's writing the challenge into a memory inside a tag.

(4) $R(\cdot)$ and $\hat{R}(\cdot)$. A response from a legitimate tag w.r.t. challenge c is always assumed correct and is denoted as $R(c)$, in which $R(\cdot)$ is a single-valued function $R : \mathbb{N} \rightarrow \mathbb{N}$ such that $R(x) = R(y)$ iff $x = y$; whereas a response from a legitimacy-unknown tag is denoted as $\hat{R}(c)$, and the tag is assumed legitimate iff $\hat{R}(c) = R(c)$. In other words, notation $R(\cdot)$ implies a legitimate tag, whereas notation $\hat{R}(\cdot)$ implies a legitimacy-unknown tag.

3.2 Primitive Operation

Based on the current ISO standard 18000-6C for C1G2 tags [8], we classify the conversations between a tag and a reader into the following two classes of primitive operations, namely reader to tag operations and tag to tag operations.

• Reader to tag operations. This class of operations are carried out by readers. They contains three commands that enable a reader to access a tag by either reading data from or writing data into a memory inside the tag. A set of commands, including these three commands, are normally supported by readers commercially obtainable from the market [15].

(1) Inventory( ). This operation is a one-to-many conversation between a reader and tags, enabling a reader to read IDs of tags within the range of its effective radio wave so as to identify these tags. As a broadcasting command though, it is achieved by many rounds of communications between a reader and nearby tags due to collisions [10, 13, 20], and is normally used in tag identification phase of a reader/tag authentication protocol.

(2) *rWrite(TagID, Memory, Data)*. This operation is a one-to-one conversation, enabling a reader to write data into a specified memory inside a tag identified by its ID TagID. Sending a challenge to a tag is an execution of this command.

(3) *rRead(TagID, Memory)*. This operation is a one-to-one conversation. It enables a reader to receive by initiatively reading data from a specified memory inside a tag identified by its ID TagID. Returning a response is an execution of such a command.

• *Tag to tag operations*. This class of operations are carried out by tags. They contain two commands that allow a tag to access a memory inside itself. These two commands are supported by the IAIK tag emulators known as DemoTags [3].

(1) *tRead(Memory, Data)*. This command is normally executed after a prior *rWrite()* command run by a reader. It enables a tag to read the data from a specified memory followed by running a function with the data as input. The function can be generating a response w.r.t. a challenge, or other actions triggered by the *rWrite()* command.

(2) *tWrite(Memory, Data)*. This operation enables a tag to store the data into a specified memory inside itself. It is normally executed right after a prior *tRead()* command of which there is a result to be sent to a reader.

4 Protocol Pattern

In this section, we first identify a protocol pattern that may bring with false authentications, and show how false authentications arise from this pattern, followed by a necessary and sufficient condition for false authentications prevention.

4.1 Pattern of Challenge-Response

For the false authentications as demonstrated in Sect. 1, we assume that the major reason is what a reader reads from memory M_r is not what is prepared for it. From Fig. 2, we can see that memory M_r works as a shared resource accessible to several users, i.e., the tag itself and readers. The tag uses it for storing responses into it and readers use it for reading responses from it. Similar to the concept of critical section in operating systems, there is a contention amongst all users for the shared memory M_r. If a reader fails to read the response for it from this shared resource within a critical time slot, then it may read a response for another reader and wrongly reject this tag if it is legitimate. One the other hand, the contention of shared resource is because there lacks a coordinator that can inform each user whether the resource is ready for it to exclusively access—a semaphore can play the role of such a coordinator (see details in Sect. 6).

Given the above reason, we further identify a protocol pattern causing false authentications. This pattern, referred to as the pattern of challenge-response as shown in

Fig. 3 Pattern of challenge-response and its implementation. **a** Pattern of challenge-response. **b** Implementation of the pattern

(a)

1. R → T: c
2. R ← T: $R(c)$

(b)

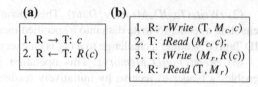

1. R: $rWrite$ (T, M_c, c)
2. T: $tRead$ (M_c, c);
3. T: $tWrite$ (M_r, $R(c)$)
4. R: $rRead$ (T, M_r)

Fig. 3a, contains two rounds of one-to-one conversation between a reader and a tag as follows: (1) a reader sends a challenge c to a tag, and (2) the tag returns a response $R(c)$ to the reader. When implemented with the primitive operations, the pattern can be presented in another form as shown in Fig. 3b in which R denotes the reader, T the tag, c the challenge, $R(c)$ the response, M_c the memory to which challenge c is sent, and M_r the memory into which response $R(c)$ is stored. The conversations between the reader and the tag in the pattern are carried out in a line-by-line top-down order as indicated by the numbers, while in Fig. 3b the notation R or T in front of each command highlights (somehow redundantly) which reader or tag executes the command.

4.2 False Authentications Arising from Protocol Pattern

Figure 3b illustrates an ideal execution sequence of the commands for the pattern however, in practice these commands may not follow the sequence in their executions. Based on when a reader executes a read command, there are two possible cases as identified in what follows.

Definition 1 (Early-read) *An early-read, a.k.a. an early-reader on reader R, is a read action such that a reader R reads a tag's memory* before *the expected response is stored into it.*

Definition 2 (Lagged-read) *A lagged-read, a.k.a. a lagged-reader on reader R, is a read action such that a reader R reads a tag's memory* after *another value is stored into it.*

Either an early-read or a lagged-read can bring with false authentications as identified below.

Definition 3 (False Negative Authentication) *Given a protocol, an authentication is known as a false negative authentication (FNA for short) iff a reader wrongly rejects a legitimate tag which it should accept.*

There is another possible false authentication as follows.

Definition 4 (False Positive Authentication) *Given a protocol, an authentication is known as a false positive authentication (FPA for short) iff a reader accepts an illegitimate tag which it should reject.*

An FPA only arises when there is an attack. In this study however, FPAs will never arise given the assumption of no attacks. A further discussion on how FPAs may arise is out of the scope of this paper. In what follows, false authentications are referred to as FNAs if there is no confusion.

Given these definitions, we have the following claim.

Claim 1 *Either an early-read or a lagged-read is a sufficient condition under which an FNA arises.*

Proof: By definitions, an early-read results in that a reader reads a value left in the memory by the tag in its last write operation, while a lagged-read results in that a reader reads a response w.r.t. a latter challenge coming from another reader. In either situation, what the reader reads is not the response that it should read, leading to the arising of an FNA.   ∎

Given the above analysis, we have the following conclusions about false authentications arising from the pattern when there is one or more readers interacting with the same tag.

Theorem 1 *FNAs may arise from the pattern if there is only one reader interacting with one tag.*

Proof: Given that there is only one reader interacting with one tag, from Fig. 3b we observe that there are totally three possible execution sequences of the commands as follows: $1 \to 2 \to 3 \to 4$, $1 \to 2 \to 4 \to 3$ and $1 \to 4 \to 2 \to 3$. An early-read is found in either of the latter two sequences, and consequently an FNA arises according to Claim 1.   ∎

Theorem 2 *FNAs may arise from the pattern if there are two or more readers interacting with the same tag.*

Proof: Given two readers R_1 and R_2 interacting with the same tag T, assume (1) reader R_2 sends challenge c_2 to tag T *after* reader R_1 has sent challenge c_1, and (2) reader R_1 reads tag's memory M_r *after* it has been updated with response $R(c_1)$ by the tag.[3] Thus the first *four* commands are executed in a line-by-line top-down order as follows:

$$1.R_1: rWrite(T, M_c, c_1)$$
$$1.R_2: rWrite(T, M_c, c_2)$$
$$2^1.T: tRead(M_c, R(c_1))$$
$$3^1.T: tWrite(M_r, R(c_1))$$

in which the superscript 1 (or 2 in the next) of a step such as "$2^1.T: tRead(M_c, R(c_1))$" means to which reader the tag is responding. Based on the pattern shown in Fig. 3b, there are *twelve* possible execution sequences (line-by-line top-down order) for the next *four* commands as shown in Fig. 4 from which we have the following observations:

---

[3] Theorem 1 has covered the case that reader R_1 reads memory M_r *before* it has been updated with response $R(c_1)$.

Fig. 4 Twelve possible execution sequences of four commands

Case 1	Case 2
4. R_1: $rRead$ (T, M_r)	2^2. T: $tRead$ (M_c, $R(c_2)$)
2^2. T: $tRead$ (M_c, $R(c_2)$)	4. R_1: $rRead$ (T, M_r)
3^2. T: $tWrite$ (M_r, $R(c_2)$)	3^2. T: $tWrite$ (M_r, $R(c_2)$)
4. R_2: $rRead$ (T, M_r)	4. R_2: $rRead$ (T, M_r)
Cases 3–8	
4. R_1: $rRead$ (T, M_r)	4. R_1: $rRead$ (T, M_r)
2^2. T: $tRead$ (M_c, $R(c_2)$)	4. R_2: $rRead$ (T, M_r)
4. R_2: $rRead$ (T, M_r)	2^2. T: $tRead$ (M_c, $R(c_2)$)
3^2. T: $tWrite$ (M_r, $R(c_2)$)	3^2. T: $tWrite$ (M_r, $R(c_2)$)
4. R_2: $rRead$ (T, M_r)	4. R_2: $rRead$ (T, M_r)
2^2. T: $tRead$ (M_c, $R(c_2)$)	4. R_1: $rRead$ (T, M_r)
4. R_1: $rRead$ (T, M_r)	2^2. T: $tRead$ (M_c, $R(c_2)$)
3^2. T: $tWrite$ (M_r, $R(c_2)$)	3^2. T: $tWrite$ (M_r, $R(c_2)$)
2^2. T: $tRead$ (M_c, $R(c_2)$)	2^2. T: $tRead$ (M_c, $R(c_2)$)
4. R_1: $rRead$ (T, M_r)	4. R_2: $rRead$ (T, M_r)
4. R_2: $rRead$ (T, M_r)	4. R_1: $rRead$ (T, M_r)
3^2. T: $tWrite$ (M_r, $R(c_2)$)	3^2. T: $tWrite$ (M_r, $R(c_2)$)
Cases 9–10	
2^2. T: $tRead$ (M_c, $R(c_2)$)	2^2. T: $tRead$ (M_c, $R(c_2)$)
3^2. T: $tWrite$ (M_r, $R(c_2)$)	3^2. T: $tWrite$ (M_r, $R(c_2)$)
4. R_1: $rRead$ (T, M_r)	4. R_2: $rRead$ (T, M_r)
4. R_2: $rRead$ (T, M_r)	4. R_1: $rRead$ (T, M_r)
Cases 11–12	
4. R_2: $rRead$ (T, M_r)	2^2. T: $tRead$ (M_c, $R(c_2)$)
2^2. T: $tRead$ (M_c, $R(c_2)$)	4. R_2: $rRead$ (T, M_r)
3^2. T: $tWrite$ (M_r, $R(c_2)$)	3^2. T: $tWrite$ (M_r, $R(c_2)$)
4. R_1: $rRead$ (T, M_r)	4. R_1: $rRead$ (T, M_r)

(1) The execution sequence in case 1 is ideal, while in case 2 the sequence is acceptable. In either case, there is neither an early-read nor a lagged read on any reader, and tag T is accepted by both readers R_1 and R_2.

(2) In any of cases 3–8, step 4 of reader R_2 (i.e., 4. R_2:) is carried out *prior to* step 3^2, i.e., reader R_2 reads memory M_r *before* it has been updated with response $R(c_2)$. Therefore there is an early-read on reader R_2, leading to an FNA such that tag T is rejected by reader R_2.

(3) In any of cases 9–10, step 4 of reader R_1 (i.e., 4. R_1:) is carried out *after* step 3^2, i.e., reader R_1 reads memory M_r *after* it has been updated with response $R(c_2)$. Therefore there is a lagged-read on reader R_1, leading to an FNA such that tag T is rejected by reader R_1.

(4) In any of cases 11–12, there are an early-read on reader R_2 and a lagged-read on reader R_1, leading to two FNAs such that tag T is rejected by both readers R_1 and R_2.

In conclusion, *ten* out of twelve cases lead to FNAs when *two* readers interacting with one tag. Given two cases in which no FNAs arise, the scenario of n (where $n > 2$)

readers can be reduced to the scenario of $n - 1$ readers by assuming that one of the readers is lucky to have no FNAs, and the conclusion holds true for more than two readers interacting with one tag.   ∎

From the above conclusions, we come out the necessary and sufficient condition for false authentications prevention.

Theorem 3 *Under the protocol pattern of challenge-response, FNAs can be prevented iff there is neither an early-read nor a lagged-read on any reader.*

Proof: (a) Necessity. If there is either an early-read or a lagged-read on any reader, then by Claim 1 an FNA arises. (b) Sufficiency. If there is neither an early-read nor a lagged-read on any reader, then it is guaranteed what a reader reads is exactly what it should read, and no FNAs will arise.   ∎

5 Classification of Authentication Protocols

With the above protocol pattern, we give a classification for the existing protocols and investigate what kinds of protocols contain this pattern and may bring with FNAs.

A typical authentication protocol works by two phases, namely, a tag identification phase followed by an authentication phase [1, 10, 21]. There are many reader/tag authentication protocols with different rounds of conversations in the authentication phase though, the tag identification phase is the same amongst all. If we ignore the tag identification phase, the existing protocols can be classified into *four* types based on the reader-tag conversations in their authentication phases. They can be simplified as illustrated in Fig. 5. The first two types of protocols are one-way authentications, i.e., either a tag authenticates itself to a reader or a reader authenticates itself to a tag; whereas the latter two types of protocols are mutual authentications by which tags and readers authenticate to each other mutually. They are briefly described as follows.

(1) The first type is reader authentication protocols under which only readers authenticate themselves to tags, referred to as RAO (reader authentication only) as shown in Fig. 5a. Challenge c_1 from the reader is a function (e.g., hashing) of key k held by the tag. The tag runs the same function with key k as the input and compares the result with c_1. If both are equal, i.e., c_1 is correct as illustrated in the figure, then it accepts the reader; otherwise it rejects the reader. With hash-locking protocol [21] as a representative, this type of protocols prevent unauthrised readers from accessing tags.

(2) The second type is tag authentication protocols under which only tags authenticate themselves to readers, referred to as TAO (tag authentication only) as shown in Fig. 5b. Upon receiving challenge c_1, the tag (of which the legitimacy is unknown) returns response $\hat{R}(c_1)$ which is a function of challenge c_1. The reader compares it with $R(c_1)$ on the back-end server. If $\hat{R}(c_1) = R(c_1)$, then the reader accepts the tag; otherwise it rejects the tag. This type of protocols allow readers to prevent

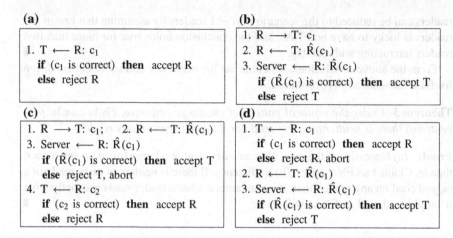

Fig. 5 Illustration of simplified four types of authentication protocols. **a** RAO protocol. **b** TAO protocol. **c** TRA protocol. **d** RTA protocol

unauthorised tags from being accepted. Representatives of this type of protocols include YA-TRAP and five protocols of challenge-response proposed in [19].

(3) The third type of protocols such as the protocol of Avoine et al. [2], the protocol of Yang et al. [22] and the protocol of Molner and Wagner [12], referred to as TRA (tag-then-reader authentication) in Fig. 5c, allow tags to authenticate themselves to readers first followed by readers authenticating themselves to tags. TRA protocols work as a combination of TAO and RAO. The reader first sends challenge c_1 and verifies response $\hat{R}(c_1)$ on the back-end server. If response $\hat{R}(c_1) = R(c_1)$, then it accepts the tag and proceeds to authenticating itself to the tag by sending challenge c_2; otherwise, it rejects the tag and aborts the authentication process. Once the tag passes the authentication to the reader, it verifies challenge c_2 and accepts the reader if correct or otherwise rejects the reader.

(4) The fourth type of protocols such as the protocol of Alomari et al. [1] and M^2AP [14], referred to as RTA (reader-then-tag authentication) in Fig. 5d, allow readers to authenticate themselves to tags first and then tags authenticate themselves to readers. Similar to TRA protocols, RTA protocols work as a combination of RAO and TAO. The difference is, under TRA protocols, the tag responds challenge c_1 without verifying its validity; whereas under RTA protocols, the tag first verifies challenge c_1 before responding with response $\hat{R}(c_1)$. If c_1 is correct, then the tag accepts the reader; otherwise, it rejects the reader and aborts the authentication. Once the reader passes the authentication to the tag, it starts to authenticate the tag by verifying response $\hat{R}(c_1)$. It accepts the tag if $\hat{R}(c_1) = R(c_1)$ or otherwise rejects the tag.

From Theorems 1 and 2 we have the following conclusion.

Corollary 1 *Under any of TAO, TRA and RTA protocols, FNAs may arise if there is one or more readers interacting with the same tag.*

Proof: From the above classification, we observe that the pattern of challenge-response exists in all TAO, TRA and RTA protocols. By Theorems 1 and 2, the conclusion follows. ∎

6 Semaphore-Based Solution

In the above we have analysed that the major reason of false authentications is due to the contention of the shared resource. If there is a coordinator that can inform each user when the shared resource is available for it to access, the contention can be resolved. Enlightened by the idea of semaphore as a solution to critical-section problem in operating systems, we assume that the role of such a coordinator can be taken up by semaphores. In the next, we propose to add semaphore operations to the protocol pattern for false authentications prevention. This is a plain solution without considering attacks though, it is simple and effective. Improved solutions against attacks deserve further study in separate papers.

6.1 Semaphore

We define three semaphores s_0, s_1 and s_2, each of which is a predefined memory inside a tag. They play the following roles.

(1) Semaphore s_0 indicates the availability of a tag. If $s_0 = 0$, then the tag is ready for processing a challenge; if $s_0 = S(c)$, where $S(c)$ known as a token is supposedly a single-valued function w.r.t. challenges such that $S(x) = S(y)$ iff $x = y$, then the tag is generating response $R(c)$. At the reader side, the reader keeps checking semaphore s_0 until $s_0 = 0$ and sends a challenge; whereas at the tag side, the tag updates the state of semaphore s_0 from 0 to token $S(c)$ when it starts to generate response $R(c)$ w.r.t. challenge c from the reader.

(2) Semaphore s_1 indicates the readiness of response $R(c)$. The response is ready iff $s_1 = S(c)$ where $S(c)$ is the token currently held by semaphore s_0. At the reader side, upon sending challenge c, the reader keeps checking semaphore s_1 until $s_1 = S(c)$ and reads response $R(c)$; whereas at the tag side, the tag updates semaphore s_1 with token $S(c)$ when response $R(c)$ is ready.

(3) Semaphore s_2 indicates that the reader has read response $R(c)$. It is set to $S(c)$ iff the reader finishes reading response $R(c)$, in which token $S(c)$ is currently held by both semaphores s_0 and s_1. At the reader side, upon reading response $R(c)$, the reader updates semaphore s_2 with token $S(c)$; whereas at the tag side, the tag keeps checking semaphore s_2 until $s_2 = S(c)$ and updates semaphore s_0 with 0.

All three semaphores are readable and updatable to the tag though, semaphores s_0 and s_1 are only readable to readers and semaphore s_2 is updatable to readers. We reserve extra memory units in memory M_r for semaphores s_0 and s_1 besides the units for response $R(c)$, and use memory M_c for semaphore s_2 and a challenge from a

```
1. Reader ⟶ Tag: c
2. Tag:
    2.1. if (s₀ = 0) then  s₀ ← S(c), s₁ ← S(c)
    2.2. Tag ⟶ Reader: s₀, s₁, R(c)
3. Reader:
    3.1. if (s₀ ≠ S(c)) then  goto 1
    3.2. Reader ⟶ Tag:  S(c)
4. Tag: if (s₂ = S(c)) then  s₀ ← 0
```

Fig. 6 Revised protocol pattern with semaphores

```
1. R: rRead (T, Mᵣ, s₀)
   if (s₀ ≠ 0) then  goto 1              // tag busy
2. R: rWrite (T, M_c, c)     // another challenge   c′ at the same time
3. T: tRead (Mᵣ, s₀)
   if (s₀ = 0) then   // available, otherwise ignoring other challenges
      3.1. T: tRead (M_c, c)
      3.2. T: tWrite (Mᵣ, S(c))     // s₀ ← S(c), indicating busy now
      3.3. T: tWrite (Mᵣ, R(c), S(c)) // s₁ ← S(c), writing R(c) done
4. R: rRead (T, Mᵣ, s₀)
   if (s₀ = 0) then   goto 4           // awaiting tag running step 3
   if (s₀ ≠ S(c)) then   goto 1        // challenge   c′ earlier than   c
5. R: rRead (T, Mᵣ, R(c), s₁)          // reading   s₁ together with   R(c)
   if (s₁ ≠ S(c)) then   goto 5        // R(c) not yet ready
6. R: rWrite (T, M_c, S(c))            // s₂ ← S(c), reading R(c) done
7. T: tRead (M_c, s₂)
   if (s₂ ≠ S(c)) then   goto 7        // awaiting reader to read   R(c)
8. T: tWrite (Mᵣ, 0)                   // s₀ ← 0, acknowledgement
```

Fig. 7 Implementation of protocol pattern with semaphores

reader. This arrangement of memory allows a reader to read response $R(c)$ together with semaphores s_0 and s_1 by a single read command.

6.2 Solution with Semaphore Operations

The revised protocol pattern with semaphore operations is shown in Fig. 6 in which two steps are added as compared with the original pattern shown in Fig. 3a. Accordingly the implementation of the revised pattern is shown in Fig. 7, in which we can also merge steps 4 and 5 into one command as "R: $rRead(T, M_r, s_0, s_1, R(c))$". It works by several phases that are carried out concurrently at both reader and tag sides.

- At the reader side, there are *two* phases as follows.

(1) Send challenge c (steps 1, 2 and 4). Reader R keeps querying the availability of tag T in step 1, and runs step 2 of sending challenge c iff tag T is available. However, it is very likely that reader R′ sends challenge $c′$ prior to reader R. Thus in

step 4, semaphore s_0 tells reader R about (a) whether tag T has started to generate a response (if yes, then $s_0 \neq 0$), and (b) whether tag T is generating a response w.r.t. its challenge c or w.r.t. challenge c' from reader R'. If $s_0 = S(c')$, meaning that the tag is processing challenge c' rather than challenge c, then reader R has to start its try from step 1; otherwise, it proceeds to the next phase.

(2) Read response $R(c)$ (step 5) and acknowledge (step 6). In step 5 reader R keeps reading semaphore s_1 together with response $R(c)$. If $s_1 \neq S(c)$, meaning that the response is not yet ready, then it knows what it reads is not $R(c)$ and ignores it; otherwise, it knows from $s_1 = S(c)$ what it reads is exactly $R(c)$ and updates semaphore s_2 with token $S(c)$ to acknowledge finish of reading response $R(c)$.

- At the tag side, there are *two* phases as follows.

(1) Read challenge c (step 3.1) and store response $R(c)$ (step 3.3). The tag responds a challenge iff it is available indicated by semaphore $s_0 = 0$. When starting to process challenge c (step 3.1), the tag updates semaphore s_0 with token $S(c)$ (step 3.2) and does not respond any latter challenge. It updates semaphore s_1 with token $S(c)$ to indicate its finish of storing response $R(c)$ into memory M_r (step 3.3).

(2) Receive acknowledgement (step 7). Once it finishes storing response $R(c)$ (step 3.3), the tag keeps querying sema-phore s_2 (step 7) until $s_2 = S(c)$ which means the reader's completion of reading the response, then the tag resets s_0 to 0, indicating ready to process a new challenge.

Given the above solution, we have the following conclusion.

Theorem 4 *No FNAs will arise from the solution when one or more readers are interacting with the same tag.*

Proof: By Claim 1, we only need to prove that there is neither early-read nor lagged-read on any reader. From the above analysis, we can see that there are several phases of operations carried out by the reader or the tag. At each side, either the reader or the tag can control by itself the execution sequence of the operations; whereas both sides may run respective operations concurrently. We discuss the following two cases.

(1) One reader. An early-read will arise iff the reader reads response $R(c)$ (step 5) *before* it is stored into memory M_r (step 3.3). From Fig. 7, the reader keeps checking sema-phore s_1 until $s_1 = S(c)$ before accepting response $R(c)$ that it reads together with semaphore s_1, while setting sema-phore s_1 to $S(c)$ is run by the tag together with storing response $R(c)$ in step 3.3. Even if the reader goes to step 5 earlier than the tag going to step 3.3, it would be blocked by semaphore s_1 until the tag finishes running step 3.3 which sets semaphore s_1 to $S(c)$ indicating the completion of storing response $R(c)$. Therefore, the potential early-read is prevented by semaphore s_1. On the other hand, the tag keeps response $R(c)$ unchanged in memory M_r until the reader sets s_2 to $S(c)$ indicating end of reading $R(c)$, a lagged-read on the reader is hence prevented.

(2) Two or more readers. Given tag T and reader R_1 together with several other readers, it is possible that several readers send respective challenges to tag T at the same time after verifying semaphore $s_0 = 0$. From Fig. 7, any of other readers will

be blocked at step 4 if $s_0 = S(c_1)$, which indicates that challenge c_1 from reader R_1 arrives to tag T earlier than any other challenges and the tag is responding to challenge c_1. Thus there is no way for other readers to access memory M_r hence either early-read and lagged-read will not arise on these readers. From step 5 onwards, only reader R_1 is allowed to interact with the tag, and either early-read or lagged-reader on reader R_1 is prevented as analysed in the first case. ∎

6.3 Memory Cost of Semaphores

The memory cost of a semaphore is determined by token $S(\cdot)$ which is supposedly a single-valued function w.r.t. challenges, denoted as $S : \mathbb{N}_c \rightarrow \mathbb{N}_s$ where $\mathbb{N}_c$ is the space of c-bit length unsigned numbers and $\mathbb{N}_s$ the space of s-bit length unsigned numbers. In practice, token $S(\cdot)$ is a hashing operation of challenges, whereas a challenge is normally a pseudo-random number of at least 128-bit length for security consideration [9, 12]. Given s-bit length of $S(\cdot)$ and t challenges, the probability of collision that any two challenges are hashed to the same token is $1 - 2^s!/(2^{st} \cdot (2^s - t)!)$.

Normally the minimum block size of a read/write operation by a tag/reader is 8-bit (one-byte) or 16-bit (two-byte) [3, 15]. Therefore, if we use 8-bit length of token, then the probability of collision is less than 0.04 for $t = 5$ (5 challenges); whereas if we use 16-bit length of token, the probability is less than 0.015 for $t = 40$ (40 challenges). Both probabilities of collision are negligibly low in real applications.

Not all three semaphores cost memory because semaphore s_2 shares memory M_c with challenges as aforementioned. Moreover, to further save memory, semaphore s_1 can be just 1-bit of length with *one* for response ready and *zero* for otherwise. Thus the memory cost of semaphores are acceptable as compared with the memory costs of challenges and responses.

If we let $s = 8$, then the probability of collision is less than 0.04 for $t = 5$; if we let $s = 16$, then the probability is less than 0.015 for $t = 40$, 1.5×10^{-4} for $t = 5$, or less than 7×10^{-4} for $t = 10$. Therefore, for no more than five readers interacting with the same tag, 8-bit length of token is enough to guarantee negligibly low probability of collision, while 16-bit length is enough for 40 readers.

7 Experiment

We conduct three experiments to verify our conclusions. In the first experiment, we verify that early-read could arise from one reader interacting with a tag; in the second experiment, we verify that lagged-read could arise from two readers interacting with the same tag; while in the third experiment, we verify that semaphore operations can effectively prevent either early-read or lagged-read from arising.

The tags and readers that we use in our experiments are (1) IAIK UHF tag emulators [3] and (2) CAEN A828 readers [15], and the back-end servers are laptop PCs running Microsoft Windows XP and Windows 7. The CPU speeds of laptops do not

Table 1 Nth round of read from which onwards the reader reads the correct response

Challenge length	Length of response (bytes)									
	2	4	6	8	10	12	14	16	18	20
1 block	3	3	4	4	5	5	6	7	7	8
2 blocks	3	4	4	5	5	6	6	7	7	8
3 blocks	4	4	5	5	6	6	7	7	8	8

affect the results because the speed of reader-tag communications is determined by tags and readers. Any challenge c is a random number, and response $R(c)$ is part of the hashing result of SHA-256 with challenge c as the input[4]. Embedded with an ATMega128 micro-controller, an IAIK tag emulator can run a hashing operation of SHA-256 or write a byte of data into a memory within ten milliseconds, while a CAEN A828 reader can finish a read/write operation within ten milliseconds.

In the first experiment, we run 20 rounds of *rRead*( ) command to read a response right after a *rWrite*( ) command which sends a challenge, and verify what the reader reads. We vary the challenge length from *one* to *three* blocks of input size[5], and vary the length of response from 2 bytes to 20 bytes. Our experimental result in Table 1 shows from which round onwards the reader reads the correct response; in other words, before this round of read, any read action that the reader takes leads to an early-read with unexpected data read by the reader. The table also shows, either longer length of challenges or longer length of responses leads to longer time of response generating or storing and results in a later round of read from which onwards the reader reads the correct response. This experiment demonstrates that we have captured early-read arising from one reader interacting with one tag.

In our second experiment, we run three consecutive steps as follows: (1) reader R_1 sends challenge c_1, (2) reader R_2 sends challenge c_2, and (3) reader R_1 runs 20 rounds of command *rRead*( ) to read a response. We keep the challenge length within one block of input size but vary the response length from 2 bytes to 20 bytes. Table 2 shows from which round onwards reader R_1 reads response $R(c_2)$ which is destined for reader R_2 rather than reader R_1. The result tells that to avoid early-read reader R_1 may postpone reading response $R(c_1)$, however, it has to finish its read action before the round of read from which onwards it reads response $R(c_2)$; otherwise, lagged-read on reader R_1 arises. This experiment demonstrates that we have captured lagged-read arising from two reader interacting with one tag.

In the third experiment, the tag and readers are equipped with semaphores. We take the first byte of challenge c as token $S(c)$ because we use only two readers in the experiment, and let challenge c_2 from reader R_2 be sent after challenge c_1 from

---

[4] The hashing result of SHA-256 is 256 bits (i.e., 32 bytes) though, we take part of the result as a response in our experiments. The time for the tag to write a response into a memory increases with the response length.

[5] The input size of SHA-256 is $64n - 9$ bytes for n blocks. The execution time of SHA-256 grows with the input size though, it is a constant for any bytes of length within the same input size.

Table 2 Nth round of read from which onwards reader R_1 reads response $R(c_2)$

Challenge length	Length of response (bytes)									
	2	4	6	8	10	12	14	16	18	20
1 block	5	6	7	7	9	9	11	13	14	15

Table 3 Rounds of reading semaphore s_1 before response $R(c_1)$ ready

Challenge length	Length of response (bytes)									
	2	4	6	8	10	12	14	16	18	20
1 block	2	2	3	3	4	4	5	5	6	7

reader R_1. Similar to the second experiment, we keep the challenge length within one block of input size but vary the length of responses from 2 bytes to 20 bytes. Table 3 shows how many rounds of read action that reader R_1 has taken to read semaphore s_1 in step 5 (Fig. 7) before $s_1 = S(c_1)$ indicating the readiness of response $R(c_1)$. The result complies with that in the first experiment. This experiment demonstrates that with semaphore operations, there is neither early-read nor lagged-read arising from the process of two readers interacting with one tag, because reader R_2 does not send any challenge to the tag before the tag receives acknowledgement from reader R_1 and sets semaphore s_0 to 0 indicating its readiness to receive challenges.

From the results of the above three experiments, in summary, we come out the conclusion that false authentications resulting from the protocol pattern can be easily captured in experiments and the semaphore-based solution can effectively prevent such false authentications from arising.

8 Conclusion

In conclusion, we have made the following major contributions in this chapter. (1) We have pointed out that false authentications may arise from the existing protocols when they are applied to C1G2 passive tags; (2) we have identified a protocol pattern under which false authentications may arise, and further identified three types of the existing protocols that contain this pattern and may bring with false authentications; and (3) we have given a necessary and sufficient condition for false authentications prevention. In addition, we have made some other contributions as follows. (1) We have proposed a semaphore-based solution to prevent false authentications from arising; and (2) we have demonstrated with experiments the arising of false authentications and verify the effectiveness of our solution. For the next stage of study, it would be interesting to study secure solutions against various attacks.

Acknowledgments This work was partly supported by National Natural Science Foundation of China (No. 61272303) and China National Program on Key Basic Research Projects (973 Program, No. 2010CB327903).

References

1. Alomair, B., L. Lazos, Poovendran, R.: Towards securing low-cost RFID systems: an unconditionally secure approach. In: Proceedings of the 2010 Workshop on RFID Security (RFIDsec'10 Asia), Singapore, pp. 1–17, 22–23 Feb 2010
2. Avoine, G., Dysli, E., Oechslin, P.: Reducing time complexity in RFID systems. In : Proceedings of the 12th Annual Workshop on Selected Areas in Cryptography (SAC'05), Kingston, pp. 291–306, 11–12 Aug 2005
3. DemoTag. http://www.iaik.tugraz.at/content/research/rfid/tag_emulators/
4. EPCGlobal, EPC Radio-Frequency Identity Protocols Class-1 Generation-2 UHF RFID Protocol for Communications at 860 MHz–960 MHz Version 1.2.0. Available at http://www.epcglobalinc.org/standards/uhfc1g2/uhfc1g2_1_2_0-standard-20080511.pdf
5. Feldhofer, M., Dominikus, S., Wolkerstorfer, J.: Strong authentication for RFID systems using the AES algorithm. In: Proceedings of the Sixth International Workshop on Cryptographic Hardware and Embedded Systems (CHES 2004), Cambridge, pp. 357–370, 11–13 Aug 2004
6. Fu, G., Li, Y.: A role-based authorization framework for RFID-enabled supply chain networks. In: Proceedings of the 16th International Conference on Transformative Science, Engineering, and Business Innovation, Jeju Island, 12–16 June 2011
7. He, W., Li, Y., Chiew, K., Li, T., Lee, E.W.: A solution with security concern for RFID-based track and trace services in epcglobal-enabled supply chains. In: Turcu, C. (ed.) Designing and Deploying RFID Applications, pp. 95–108. InTech, UK (2011). Chapter 7
8. International Organization for Standards (ISO): ISO/IEC 18000–6: radio frequency identification for item management—part 6: parameters for air interface communications at 860 MHz to 960 MHz. http://www.iso.org/iso/iso_catalogue/catalogue_tc/catalogue_detail.htm?csnumber=34117
9. Juels, A., Weis, S. A.: Authenticating pervasive devices with human protocols. In: Proceedings of the 25th Annual International Cryptology Conference (Crypto 2005), Santa Barbara, pp. 293–308, 14–18 Aug 2005
10. Lai, Y.-C., Lin, C.-C.: Two blocking algorithms on adaptive binary splitting: single and pair resolutions for RFID tag identification. IEEE/ACM Trans. Networking 17(3), 962–975 (2009)
11. Melski, A., Müller, J., Zeier, A., Schumann, M.: Improving supply chain visibility through RFID data. In: Proceedings of the the IEEE 24th International Conference on Data Engineering Workshop (ICDEW'08), Cancun, pp. 102–103, 7–12 April 2008
12. Molner, D., Wagner, D.: Privacy and security in library RFID: issues, practices, and architectures. In: Proceedings of the 11th ACM Conference on Computer and Communications Security (CCS'04), Washington, pp. 210–219, 25–29 Oct 2004
13. Myung, J., Lee, W., Srivastava, J., Shih, T.K.: Tag-splitting: adaptive collision arbitration protocols for RFID tag identification. IEEE Trans. Parallel Distrib. Syst. 18(6), 763–775 (2007)
14. Peris-Lopez, P., Hernández-Castro, J. C., Estévez-Tapiador, J. M., Ribagorda, A.: M^2AP: a minimalist mutual-authentication protocol for low-cost RFID tags. In: Proceedings of the 3rd International Conference on Ubiquitous Intelligence and Computing (UIC06), Wuhan, China, pp. 912–923, 3–6 September 2006
15. RFID, C.: CAENRFIDLib: ansi C functions library—technical information manual. http://www.caen.it/rfid/index.php
16. Shi, J., Li, Y., Deng, R.H.: A secure and efficient discovery service system in epcglobal network. Comput. Secur. 31(8), 870–885 (2012)
17. Shi, J., Li, Y., He, W., Sim, D.: Sectts: a secure track & trace system for RFID-enabled supply chains. Comput. Indus. 63(6), 574–585 (2012)
18. Tsudik, G.: YA-TRAP: yet another trivial RFID authentication protocol. In: Proceedings of the 4th IEEE Annual International Conference on Pervasive Computing and Communications Workshops (PerComW 2006), Pissa, Italy, pp. 643–646, 13–17 Mar 2006
19. Vajda, I., Buttyán, L.: Lightweight authentication protocols for low-cost RFID tags. In: Proceedings of the 5th International Conference on Ubiquitous Computing (UbiComp 2003), Seattle, WA, USA, 12–15 Oct 2003

20. Wang, C., Daneshmand, M., Sohraby, K., Li, B.: Performance analysis of RFID generation-2 protocol. IEEE Trans. Wireless Commun. **8**(5), 2592–2601 (2009)
21. Weis, S. A., Sarma, S. E., Rivest, R. L., Engels, D. W.: Security and privacy aspects of low-cost radio frequency identification systems. In: Proceedings of the 1st International Conference on Security in Pervasive Computing (SPC 2003), Boppard, Germany, pp. 201–212, 12–14 Mar 2003
22. Yang, J., Park, J., Lee, H., Ren, K., Kim, K.: Mutual authentication protocol for low-cost RFID. In: Proceedings of the Workshop on RFID and Lightweight Crypto, Graz, Austria, pp. 17–24, 14–15 July 2005

Adaptive Pipelined Neural Network Structure in Self-aware Internet of Things

Dhiya Al-Jumeily, Mohamad Al-Zawi, Abir Jaafar Hussain
and Ciprian Dobre

Abstract Self-Managing systems are a significant feature in Autonomic Computing which is required for system reliability and performance in a changing environment. The work described in this book chapter is concerned with self-healing systems; systems that can detect and analyse issues with their behavior and performance, and fixe or reconfigure as appropriate. These processes should occur in real-time to restore the desired functionality as soon as possible. The system should ideally maintain functionality during the healing process which occurs at runtime. Adaptive neural networks are proposed as a solution to some of these challenges; monitoring the system and environment, mapping a suitable solution and adapting the system accordingly. A novel application of a modified Pipelined Recurrent Neural Network is proposed in this chapter with experiments aimed to assess its applicability to online.

D. Al-Jumeily · M. Al-Zawi · A. J. Hussain (✉)
Applied Computing Research Group, Liverpool John Moores University,
Byrom Street, Liverpool L3 3AF, UK
e-mail: a.hussain@ljmu.ac.uk

D. Al-Jumeily
e-mail: d.aljumeily@ljmu.ac.uk

M. Al-Zawi
Institute of Applied Technology, Abu Dhabi, UAE
e-mail: mohamed.mousa@iat.ac.ae

C. Dobre
University Politehnica of Bucharest, 313, Splaiul Independentei, Office EG403,
Sector 6, 060042 Bucharest, Romania
e-mail: ciprian.dobre@cs.pub.ro

N. Bessis and C. Dobre (eds.), *Big Data and Internet of Things:*
A Roadmap for Smart Environments, Studies in Computational Intelligence 546,
DOI: 10.1007/978-3-319-05029-4_5, © Springer International Publishing Switzerland 2014

1 Introduction

Remarkable advances in technology have introduced increasingly complex and large-scale computer and communication systems. Autonomic computing has been projected as a remarkable challenge that will allow systems to self-manage this complexity, using sophisticated objectives and policies defined by humans. Internet of things (IoT) has exponentially increased the scale and the complexity of existing computing and communication systems; the autonomy is thus an imperative property for IoT systems. Self-healing is one of the most important components of autonomic computing. It has the ability to modify its own behavior in response to changes in the environment (in real-time), by repairing the detected faults. Hence, the system is capable of performing a reconfiguration action in order to recover from current faults [18].

Self-healing is used and desired in several fields such as distributed systems, defense software, and pervasive computing, robotics and control systems, programming language design, fault-tolerant computing, and middleware infrastructures [6, 12, 19].

A well-defined architectural style and requirements for self-healing systems contains the following features [17]

- Adaptability: The style should enable modification of a system's static (i.e., structural and topological) and dynamic (i.e., behavioral and interaction) aspects. The challenge occurs if the adaptation is applied during runtime.
- Dynamicity: Encapsulates system adaptability concerns during runtime (e.g., communication integrity and internal state consistency).
- Awareness: The style should support monitoring of the system's performance (state, behavior, correctness, reliability, and so forth) and recognition of anomalies in that performance.
- Observability: The style should enable monitoring of a resulting self-healing system's execution environment. Note that the system may not be able to influence changes in its environment (e.g., reestablishing failed network links), but it may plan changes within itself in response to the environment (e.g., performing in a degraded mode until the network link is reestablished).
- Autonomy: The style should provide the ability to address the anomalies (discovered through awareness and observability) in the performance of a resulting system and/or its execution environment. Autonomy is achieved by planning, deploying, and enacting the necessary changes.
- Robustness: The style should provide the ability for a resulting system to effectively respond to unforeseen operating conditions. Such conditions may be imposed by the system's external environment (e.g., malicious attacks, unpredictable behavior of the system's runtime substrate, unintended system usage), as well as errors, faults, and failures within the system itself. Note that this definition of robustness subsumes fault-tolerance.
- Distributability: The style should support effective performance of a resulting system in the face of different distribution/deployment profiles.

- Mobility: The style should provide the ability to dynamically change the (physical or logical) locations of a system's constituent elements.
- Traceability: The style should clearly relate a system's architectural elements to the system's execution-level modules in order to enable change enactment in support of the above requirements.

The above requirements are likely to be relevant to most self-healing systems, but additional requirements can be added as the system is evolving.

This work makes a number of contributions towards a better understanding of self-healing software requirements for autonomic distributed software engineering. Such contributions are summarized by using of a novel technique based on adaptive learning approach rather than software system engineering. Recursive neural networks have been proposed and used to solve the main challenges of self-healing, such as monitoring, interpretation, resolution, and adaptation. A modified pipelined neural network is introduced to fulfill the requirements in this field. This technique is examined and tested with different fields. Client server experiment illustrated promising results when compared to the outcomes of feedforward neural network. Also with the overcurrent relay experiment in the field of power system, good results are achieved using the pipelined neural network structure.

The reminder of this book chapter is organized as follows. Section 2 will be concerned with self-healing approaches, while Sect. 3 will shows the use of neural networks for self-healing. Section 4 will shows the experiments design. The conclusion of this chapter is illustrated in Sect. 5.

2 Self-healing Approaches

There are two approaches used to detect and recover the system from errors, integrated and external methods.

The integrated mechanism is described as a per-system ad hoc methodology [6] where it is based on the traditional method that assuming the system will be modified off-line. In this method, the self-healing is implemented internally within the system itself, which make it difficult to be separated and modified from the original system. The resulting adaptation knowledge is difficult to reuse in other systems, therefore the technique will not be able to determine the true source error [6]. Systems are required to be Non-Stop systems to meet the requirements and demands of developed system, since the system resources or needs might change frequently. Therefore, this traditional method needs to be replaced by more dynamic and flexible mechanism as described in [7].

In the externalized adaptation mechanism method the system behavior is monitored by external components of the running system; these components determine when the system behavior is within acceptable levels, and when it is out of the desired range and to accomplish this, the externalized mechanisms maintain one or

Fig. 1 Model based adaption
[6]

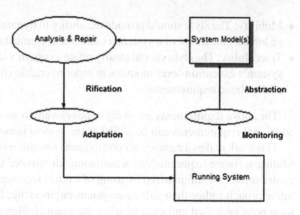

more system models, which provides an abstract, global view of the running system,
and support reasoning about system problems and repairs.

Externalize adaptation (some time called closed loop control) is a system model
based architecture as shown in Fig. 1, has the following features [6, 7, 9]:

- Different models can be used depending on the system quality of interest.
- Mechanism can be reused since it is not built in with the system.
- Easily to be changed, studied and reasoned.
- Monitoring and adaptation infrastructure can be shared.
- Enable the validity of any change.

Self-healing systems have the ability to modify their own behavior in response
to changes in their environment, such as resource variability, changing user needs,
mobility, and system faults, but there are several challenges needs to be addressed
such as monitoring, interpretation, resolution, and adaptation [2, 6–9].

- **Monitoring:** The monitoring capabilities should be added to the running system in
  non-intrusive way and it is essential to know what parameters need to be monitored.
- **Interpretation:** The challenges here are how to determine if there is a problem
  that requires system adaptation, and how to know the cause of that problem, also
  what are the optimum models to be used for a specific case?
- **Resolution:** Once the problem is identified, it is important to have a plan to select
  the best repair action and is it possible to guarantee that the suggested repair plan
  will work. In addition to that is it possible to improve the system even when no
  specific error has arisen.
- **Adaptation:** This is the last phase and it is important to implement the adaptation
  in real time (on the fly). The design of system components should respond to online
  adaptation, also there should be a backup plan if the operation fails or even if the
  outcomes of the applied resolution plan are not working properly.

Fig. 2 Block diagram of PRNN [15]

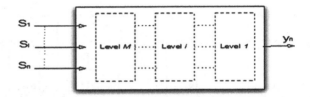

3 Neural Network and Self-healing

Neural network can be used in several fields; therefore this section will focus on the technique used in this research for the implementation of self-healing using supervised and unsupervised neural network. The main challenge was how to implement self-healing with all of its aspects of monitoring, interpretation, resolution, and adaptation in real time scenario/s. That was achieved using a modified Pipelined Recurrent Neural Network (PRNN).

3.1 Modified Pipelined Recurrent Neural Network (PRNN)

PRNN is designed by [11] and it consists of M separated modules. The modules are identical, each designed as a fully recurrent neural network with a single output neuron. Information flows into and out of the modules proceeds in a synchronized way. [11] used PRNN as a nonlinear adaptive predictor for non-stationary signals whose generation is governed by a nonlinear dynamical mechanism. The dynamic behavior of the predictor is demonstrated for the case of a speech signal, where the inputs are for a time series of one signal and the system is predicting the next value of the same signal. In general PRNN can be described in the block diagram of Fig. 2 as follows:

- PRNN is composed of M modules, each of which is designed to perform nonlinear adaptive filtering on an appropriately delayed version of the input signal vector.
- The M modules of the PRNN are identical, and each is designed as a fully connected recurrent network with a single output neuron.
- Information flow in and out of the modules proceeds in a synchronized fashion.

The PRNN used the well-known engineering concept of divide and conquer [11], meaning that breaking complex problem into a set of simpler problems in order to be easily solved. This is implemented in the PRNN as follows:

- The PRNN is composed of M separate modules, each of which is designed to perform a one-step nonlinear prediction on an appropriately delayed version of the input signal vector.
- The modules are identical, each designed as a recurrent neural network with a single output neuron, and this will be clear in the diagram of Fig. 3.

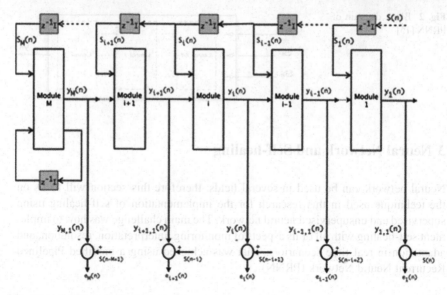

Fig. 3 A simplified pipelined neural network [11]

- Information flow into and out of the modules proceeds in a synchronized fashion.

The pipelined structure as mentioned earlier involves a total of M levels of processing. For every level there are a module and a comparator. The module consists of recurrent neural network. Figure 3 shows the details of level i. In every module, there are N neurons and a single output from each neuron. In addition to the external inputs, there are N feedback inputs and a constant input of $+1$ value. Therefore, there will be three different inputs; the $(p + N)$ inputs, and a bias input value of $+1$. Each module is fed back with $(N - 1)$ from the previous outputs, and the remaining output is applied to the next module.

The Module $i (1 \leq i \leq M - 1)$ is partially recurrent neural network, and module M, a delayed version of the output is also fed back to the input.

All the modules operate similarly in that all have the same number of external inputs and feedback nodes and also the same weight matrix.

The transfer function used for every neuron is a sigmoid function:

$$y_{k,i} \left(n + \Delta n \right) = \emptyset \left(v_{i,k}(n) \right) = \frac{1}{1 + e^{-v_{i,k}(n)}} \tag{1}$$

$$i = 1, \ldots, M; k = 1, \ldots, N$$

where $v_{i,k}(n)$ is the internal transfer function of the kth neurons, and $y_{i,k}$ is the output of the kth neuron of the ith module at the nth time point. The Δn is the processing time of a network.

Assuming that W is the weight matrix for each module $(N - by - (p + N + 1))$.

The element $w_{i,k}$ represents the weight of the connection to the kth neuron node from the lth input unit. The weight matrix W can be written as:

$$w = [w_1, \ldots, w_k, \ldots, w_N]^T \tag{2}$$

where w_k is a vector defined as:

$$w_k = \left[w_{k1}, w_{k2}, \ldots, w_{k(p+N+1)} \right] \tag{3}$$

The inputs of Module i is the $p - by - 1$ vector

$$s_i = [s(n - i), s(n - (i + 1)), \ldots, s(n - (i + p - 1)]^T \tag{4}$$

where p is the prediction order.

The feedback input is $(N - by - 1)$

$$r_i(n) = [r_{i,1}(n), r_{i,2}(n) \ldots, r_{i,N}(n)]^T \tag{5}$$

The final fixed input for each neuron is $+1$

At the nth time point, the outputs of neurons in Module i are denoted as:

$$y_i(n) = [y_{i,1}(n), \ldots, y_{i,k}(n) \ldots, y_{i,N}(n)]^T \tag{6}$$

where $y_{i,k}(n)$ is the output of neuron k of Module i.

The output $y_{i,1}(n)$ of Module i is limited in amplitude to the range $(0,1)$, therefore a normalization should be applied.

The output of the first neuron $y_{i,1}(n)$ of module i is fed directly to the following comparator and Module $i-1$. Each level computes an error signal using the following equation:

$$e_i(n) = \left[s(n - i + 1) - y_{i,1}(n) \right], \quad i = 1, 2, \ldots, M \tag{7}$$

where the sample $s(n - i + 1)$ of the input signal $s(t)$ is the desired response of Module i.

For illustration, let us assume the following:

$$M \text{ (Modules)} = 4$$

$$P \text{ (external inputs)} = 4$$

Number of Neurons per Module is $N = 2$, then the structure is depicted in Fig. 4.

Haykin and Li [11] was interested in one signal and they were looking for the prediction of one input signal with different time series, but the change in the algorithm for the current research will be based on the following:

• The error function is replaced here by:

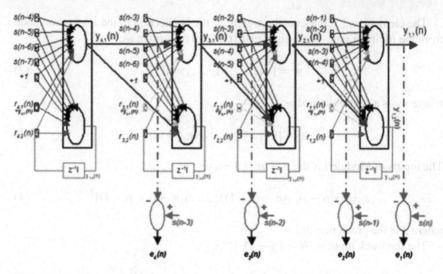

Fig. 4 Simple illustration of PRNN [15]

$$e_i(n) = \left[y_{desired}(n) - y_{i,1}(n) \right], \quad i = 1, 2, \ldots, M \tag{8}$$

- Two learning phases are applied; starting with supervised method (not more than 25 %) of the desired requirements then;
- Unsupervised algorithm of PRNN is used based on the initialization of the previous phase, so the learning and tuning will be online (on the fly) at real time scenario.
- The first learning phase started with real actual desired output.
- The online learning phases will be after a suggested period and that is achieved by creating a buffer to store the outputs, when that buffer reaches 100 % of its values, the algorithm will take the last 25 % (the same percentage of the first phase) as a new desired values for learning, if a healing plan is required by the system; those new inputs can be the sub outputs of each module, or the average of the sub outputs, to get good results.
- Unlike the original algorithm, different input signals are applied.
- The same inputs are applied at the same time for all the modules.

The overall cost function for the PRNN is still calculated by the following equation:

$$\varepsilon(n) = \sum_{i=1}^{M} \lambda^{i-1} e_i^2(n) \tag{9}$$

where λ is an exponential weighting factor that lies in the range $(0 < \lambda \leq 1)$.

The modified structure is described in Fig. 5.

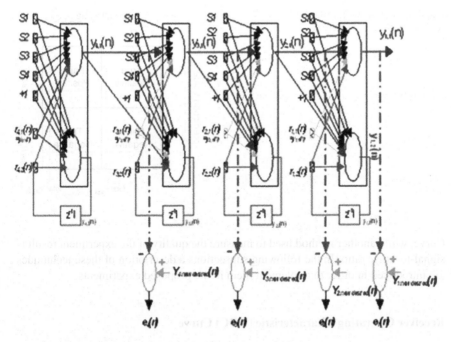

Fig. 5 Modified PRNN (the input signals and the comparator) [15]

4 Experiment Design

In this section several practical experiments from different fields are implemented to examine the use of neural networks in self-healing systems. The client-server experiment used by David et al. [9] is implemented in this research work with two methods based on supervised and unsupervised learning of artificial neural networks [13]. Another experiment from the field of electrical grid power system is introduced to test the neural network capabilities [14]. More than one technique is used to evaluate the results; such as Receiver Operating Characteristic (ROC) curve, analysis of variance techniques is implemented using the mean squared error (MSE), and the sum of squared errors (SSE) which are extremely powerful and commonly used procedures [3].

4.1 Experiment Evaluation Measurements

Different measures are used in this chapter to evaluate the obtained results. Some of these techniques are called analysis of variance such as the mean squared errors and the sum of squared errors, while the other calculate the area under its curve to be classified from poor test to a perfect or excellent experiment in Receiver Operating

Fig. 6 Confusion matrix and
common performance metrics
[15]

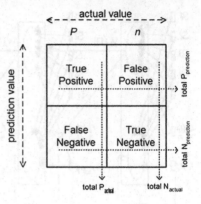

Curve, while another method used to measure the quality of the experiment result is
signal-to-noise ratio. In the following subsections a description of these techniques
are introduced in order to implement them in the proposed experiments.

Receiver Operating Characteristic (ROC) Curve

Receiver Operating Characteristics (ROC) curve is a useful technique for organizing
classifiers and visualizing their performance [5]. ROC curve is commonly used in
medical decision making, and in recent years have been increasingly adopted in the
machine learning and data mining research communities. It was originally developed
during World War II to analyze classification accuracy in differentiating signal from
noise in radar detection [22].

To show the basics of the curve, assume a binary classification of a two-class
prediction problem, in which the outcomes are labeled either as positive (P) or negative (N). The possible outcomes of the binary classification are shown in Fig. 6 and
defined as:

- True positive (TP): the prediction is P, and the actual value is P
- False positive (FP): the prediction is P, and the actual value is N
- True negative (TN): the prediction is N, and the actual value is N
- False negative (FN): the prediction is N, and the actual value is P

Several common metrics [5] can be calculated from the confusion matrix in
Fig. 6 Such as:

$$\text{Tp rate (true positive rate)} = \frac{\text{positives correctly classified}}{\text{total positives}} = \frac{\text{TP}}{\text{P}} \tag{10}$$

$$\text{Fp rate (false positive rate)} = \frac{\text{negatives incorrectly classified}}{\text{total negatives}} = \frac{\text{TP}}{\text{P}} \tag{11}$$

Fig. 7 ROC graph [15]

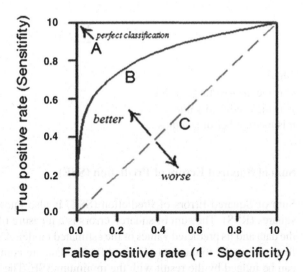

$$\text{Specifity} = \frac{\text{True negatives}}{\text{false positives} + \text{true negatives}} = (1 - fp \ rate)$$

(12)

$$\text{sensitivity} = \frac{\text{TP}}{\text{P}}$$

The true positive rate defines how many correct positive results occur among all the existing positive samples. Alternatively, false positive rate defines how many incorrect positive results occur among all negative samples available during the test.

The ROC curve is also called the sensitivity versus (1 − specificity) plot. In Fig. 7, the upper left corner or the (0, 1) point is called a perfect classification. The major diagonal (line of no-discrimination or random classifier) divides the ROC space. The numbers along the major diagonal represent the correct decisions made or good classification, and the numbers below this diagonal represent the errors between the various classes or poor classification.

Mean Squared Error (MSE)

Mean Squared Error (MSE) is a well-established tool for assessing the closeness to a target value [3]. MSE measures the average for the squares of the errors. The error is the amount by which the predicted value differs from the desired value. The mean squared error is important quantity in a number of practical situation such as the control of measurements processes [3]. From Eq. 13, the MSE index ranges from 0 to infinity, with 0 corresponding to the ideal case.

$$MSE = \frac{1}{n}\sum_{i=1}^{n}(y_i - y_i')^2 \tag{13}$$

where:
y_i is the desired/target value
y_i' is the predicted value
n is the number of samples

Sum of Squared Errors of Prediction (SSE)

Sum of Squared Errors of Prediction (SSE) is also known as the residual sum of squares (RSS). The sum of squared errors is a measure of the discrepancy between the data and its predicted values or the estimated model. A small SSE indicates a tight fit of the model to the data [4]. So different data are compared, the goodness-of-fit can be judged by the result with the minimum SSE. The following equation shows how to calculate the sum of squared errors:

$$SSE = \sum_{i=1}^{n}(y_i - y_i')^2 \tag{14}$$

where:
y_i is the desired/target value
y_i' is the predicted value
n is the number of samples

Signal-to-Noise Ratio (SNR)

Signal-to-Noise ratio (SNR) or the peak signal-to-noise ratio (PSNR) is measured in decibels. It is the ration between two signals. This ratio is often used as a quality measurement between the desired and the predicted signals. The higher the PSNR, the better the quality of the predicted signal in this measure, it describes the 'signal', which is represented by the desired vector, and the 'noise' is the error occurred in the prediction phase. So, if the experiment has a lower MSE and a high PSNR, therefore it is a better one. The following equation is used to calculate SNR as:

$$SNR = 10\log_{10}\left(\frac{R^2 \times n}{\sum_{i=1}^{n}(y_i - y_i')^2}\right) \tag{15}$$

where:
y_i is the desired/target value
y_i' is the predicted value

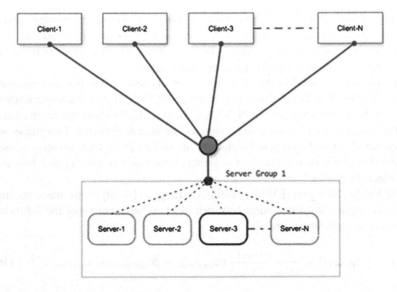

Fig. 8 General architecture of client-server experiment [15]

n is the number of samples

R is the maximum number in the desired vector

4.2 Client-Server Experiment

The client-server experiment was implemented by Garlan [9] using a class of web-based client server applications that are based on an architecture in which web clients access web resources by making requests to one of several geographically distributed server groups as illustrated in Fig. 8. Each server group consists of a set of replicated servers, and maintains a queue of requests, which are handled in FIFO order by the servers in the server group. Each server sends the results back directly to the requesting client.

David assumed that the organization managing the web service infrastructure wants to make sure that the quality of service for the customer is guaranteed, so the request-response latency for clients must be under a certain threshold and the active servers should be kept as loaded as possible to reduce costs.

Also it has been assumed that the model based adaptation system has two built-in adaptation mechanisms. First, it is possible to activate a new server in a server group or deactivate an existing server. Second, the client can change from one server group to another.

In this section, the same experiment concept has been implemented to test our technique, and the objective here is to maintain an optimum bandwidth for clients based on PRNN output in a way that it will switch between groups of running servers to make sure that the quality of service for the customer is guaranteed and the active servers are kept as loaded as possible. Also a Feedforward neural network is used to compare the results with PRNN (unsupervised learning).

The PRNN was used in the client-server experiment in which it will maintain an optimum bandwidth for clients in a way that PRNN will switch automatically to select an alternative active server. PRNN has been modified to meet our requirements and the inputs of the network contain various component elements. Two phases are implemented to test the proposed technique, in the first phase three servers are used in the server group, then the number of servers is increased to five to check how that will affect the results.

The bandwidth signal (BW) lies within the range of [0–10] Mbps. Basic normalization is applied to meet Neural Network input requirement using the following Min–Max normalization formula.

$$v_{norm}(i) = \frac{(v_i - v_{min})}{(v_{max} - v_{min})} (\alpha_{new\ max} - \alpha_{new\ min}) + \alpha_{new\ min} \tag{16}$$

PRNN should select another active server in case of server failure or when the BW is below an accepted level. The servers status signals (Srv1, 2, ….) are used to indicate if the servers are active or not; the status signal then has a binary value as 0 for non-active server while 1 value for the working server.

The output will direct the clients to the optimum server or group of servers in case of communication failure or a decrease of the BW signal level received by the neural network.

The experiment is designed based on the following:

- The total number of samples to be used is 200; it should be noted that any number of samples can be used and this number is used to test the proposed method.
- After invoking PRNN in the system, it should work for unlimited random samples; that is in a real time environment with online learning and adaptation.
- The first phase is done based on supervised learning of 50 samples (25 % of the overall samples, and it can be any percentage that initiate the PRNN weights and to build up the initial knowledge about the running system), so for the first experiment of 3 servers; 4 inputs as three servers and the bandwidth signals with its respective desired output value are fed to the PRNN, patch by patch and the weights are updated based on the modified cost function. Also for the 5 servers experiment, 6 inputs with the respective desired out are applied for the initial configuration.
- Once the first phase is finished, the second phase of unsupervised learning is started with the remaining 150 unseen samples with only 4 inputs epoch (and also 5 inputs for the second experiment of more servers) without its desired output since the system is supposed to predict that corresponding value which is the actual output. The outcomes are used for the unsupervised learning phase or the learning on the fly cycle.
- Different values of learning rate (λ) and forgetting factor (η) are used in order to enhance the output results, as described for PRNN structure. It should be noted

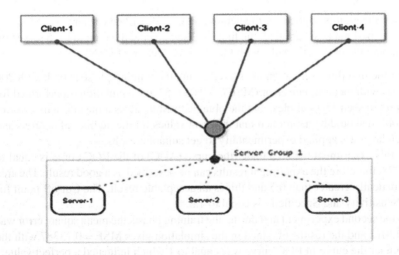

Fig. 9 Client-server experiment for 3 servers [15]

Fig. 10 PRNN I/O in client
server (3 server's example)
[15]

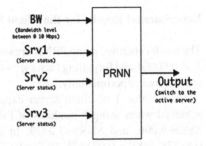

that several values were tested until the system achieved stability with the values
shown in this experiment.

• Number of cascaded modules *M* are changed between 2 and 10.

As shown in Fig. 9, the server group contains 3 servers to provide the clients with
the optimum bandwidth. Figure 10 shows the inputs and outputs of PRNN used in
this experiment.

In the next set of experiments, the same scenario is applied to another type of
network using only off-line learning or supervised learning. Feedforward neural
network (FFNN) is used as an external monitoring device.

The experiment is designed based on the following:

• The simulated network is implemented using multilayer perceptron (MLP) feedforward neural network. The neural was trained using the Levenberg-Marquart
learning (TRANLM algorithm) and two hidden layers with (15) neurons per layer
are used.
• The training phase is built with 200 samples, while the simulation is performed
for different unseen 150 samples.

- Normalization is applied for the neural inputs as described earlier.
- Two experiments are applied; the first for 3 servers and the second for 5 servers to compare the obtained results with the previous section of PRNN.

For the first three servers in the server group, the training phase is built with 200 samples with mean square error (MSE) = 0.034257, the simulation is performed for different unseen 150 samples, and the obtained results gives a mean square error of 0.07148. It should be mentioned that different values for the number of neurons and hidden layer are applied experimentally to get suitable results.

In this experiment the area under the curve (AUC) of the ROC curve is equal to 0.81778; therefore the experiment results can be described as a good result. The area is statistically greater than 0.5 and this is an acceptable result. The Cut-off point for best Sensitivity and Specificity is equal to 1.0.

In the second experiment and during the training phase, the mean square error was 0.0013012 and the results obtained in the simulation gives MSE = 0.13387 with the area under the curve in ROC curve was equal to 1 which indicated a perfect value.

Experimental Results for the Client Server

The results obtained using PRNN are summarized in Tables 1 and 2. The learning rate $\lambda = 0.00001$ and forgetting factor $\eta = 0.00001$. As mentioned before the parameters are selected experimentally.

From Table 1 of client server experiment of 3 servers group, the best results achieved when using 5 modules of PRNN with AUC = 0.73656, MSE = 0.06064, SSE = 9.096, and SNR = 7.0329. In contrast, the FFNN obtained with AUC = 0.81778, MSE = 0.071475, SSE = 10.7213, and SNR = 11.4585.

The results obtained from PRNN is acceptable since as discussed earlier it is unsupervised learning approach, while in FFNN was trained for 200 samples to get the mentioned results using the off-line learning. Both of the experiments can be described as very good tests since the area under the curve (AUC) is greater than 0.5 as discussed previously. Also in PRNN results, the mean squared errors and squared sum of errors was low compared to FFNN results, but better SNR is achieved with it.

Table 2 of 5 servers group shows that the FFNN obtained better results compared to PRNN. The best results that could be achieved by PRNN was when it was at two modules. The FFNN experiment shown here is described as a perfect test since the area under the curve of ROC is equal to 1, but even though the mean square value was higher than that of the previous experiment. It was clear that neither of the two methods obtained good results when the number of servers is increased in the group.

Table 1 Client server experiment for 3 server group [15]

	M	AUC	MSE	SSE	SNR
PRNN results	2	0.61818	0.063	9.4499	7.4227
	3	0.54906	0.0837	12.555	8.2874
	4	0.59118	0.080213	12.032	7.477
	5	0.73656	0.06064	9.096	7.0329
	6	0.75457	0.064683	9.7024	6.2627
	7	0.75108	0.069126	10.369	6.1184
	8	0.80188	0.11962	17.9425	6.6938
	9	0.76063	0.12308	18.4627	6.7615
	10	0.74485	0.1267	19.0051	6.9367
FFNN results	N/A	0.81778	0.071475	10.72131	11.4585

Table 2 Client server experiment for 5 servers group [15]

	M	AUC	MSE	SSE	SNR
PRNN results	2	0.7979	0.22587	33.88	8.6817
	3	0.55357	0.29914	44.8704	8.8407
	4	0.55345	0.2974	44.6103	8.8621
	5	0.54516	0.29544	44.3165	8.8789
	6	0.5568	0.2994	44.9096	8.9417
	7	0.52163	0.28576	42.8635	8.9806
	8	0.53185	0.28867	43.3	8.9851
	9	0.52624	0.28582	42.8733	8.9946
	10	0.52205	0.28578	42.8669	9.0088
FFNN results	N/A	1	0.13887	20.0805	8.7333

4.3 Numerical Power System Relay Experiment

Maintaining reliability and stability is an important subject in the power system
research area. Recently many IT applications have been implemented successfully in
power systems; such as numerical relays, digital Distribution Management Systems
(DMS), and Energy Management System (EMS). This section presents an implementation of an integrated external monitoring system to identify and recover from
failure in numerical over current relay using neural network. Two neural networks
are implemented to measure and predict the availability of substation numerical
protection in order to enhance their availability.

Overcurrent Relays

A fault in electrical power systems is any abnormal flow of current such as a short
circuit where the current flow bypasses the normal load. In three phase systems, a

fault may involve one or more phases and the ground, or may occur only between phases. In an earth fault, current flows into the earth. In power systems, protective devices detect fault conditions and operate circuit breakers and other devices to limit the loss of service due to a failure.

The protection relay is a device that is designed to initiate the disconnection of a part of an electrical installation or to operate an alarm signal whenever a fault or other abnormal condition is detected. In common with all relays a protection relay monitors one or more input signals and, when these satisfy predetermined conditions, the relay operates closing a pair of output electrical contacts. When used for protection, the closing of these output contracts causes the associated circuit breaker (*CB*) to trip and open the circuit that needs to be completed.

Overcurrent relays monitor the secondary current obtained from a current transformer whose primary winding is connected in to the high voltage current-carrying circuit. Typical they are rated at a nominal current of either 1 or 5 Amps. The current transformer therefore transforms the primary current to a more manageable level and also provides isolation between the high voltage of the power system and the low voltage of the relay and the protection equipment. The basic types of overcurrent relays are:

- The instantaneous overcurrent relay.
- The definite time overcurrent relay.
- The inverse characteristic overcurrent relay.

Although all three types are used for protecting distribution feeders, together with suitable combinations of them, the inverse characteristic overcurrent relays are the most common and it will be used in the experiment. Since the tripping times of the inverse characteristic relays become shorter as the fault current increases, they provide advantages over both instantaneous and definite time relays. The inverse characteristic relays are generally referred to as inverse definite minimum time overcurrent, or IDMT relays.

The time/current characteristics of IDMT relays enable them to be graded with other similar relays positioned at other points in the network so that they provide discrimination for faults on adjacent plant.

Instantaneous overcurrent relays are used in conjunction with IDMT relays and auto- reclose schemes to protect feeder circuits. In typical auto-reclose operation, the circuit breaker makes one attempt at reclosure after tripping to clear the fault and a successful reclosure or unsuccessful reclosure followed by lock-out of the circuit breaker may occur. There are two auto-reclosing schemes; a single and multi-shot scheme [1] or high speed and low speed scheme. Overcurrent relays are also used as high set elements for a great variety of applications where their setting corresponds to an excessive fault current. Definite time delayed overcurrent relays are used in conjunction with IDMT relays and auto-reclose schemes for feeder protection, as well as being used alone for earth fault protection.

Table 3 Relay characteristics of IEC 60255 [140, 141] [15]

Relay characteristic/curve type	Equation
Standard inverse (SI) ($\alpha = 0.02, \beta = 0.14$)	$t = TMS \times \frac{0.14}{i_r^{0.02}-1}$
Very inverse (VI) ($\alpha = 1, \beta = 13.5$)	$t = TMS \times \frac{13.5}{i_r-1}$
Extremely inverse (EI) ($\alpha = 2, \beta = 80$)	$t = TMS \times \frac{80}{i_r^2-1}$
Long time (LT) ($\alpha = 2, \beta = 120$)	$t = TMS \times \frac{120}{i_r-1}$

The main IDMT equation used here is:

$$t = \frac{\beta \times TMS}{\left(\frac{I}{I_o}\right)^\alpha - 1} \tag{17}$$

where:
$t = $ *The relay operating time*
$TMS = $ *The relay time multiplier*
$\alpha = $ *Curve constant (per IEC 60255)*
$\beta = $ *Curve constant (per IEC 60255)*
$I = $ *Faulty current*
$I_o = $ *Relay normal current*

IEC 60255 defines a number of standard characteristics as shown in Table 3.

where:
t is the relay operating time.
TMS is the relay time multiplier.
i_r is the multiple of the relay current setting.

Experiments Initial Setup

The prospective short circuit current of a fault can be calculated for power systems. The standard Inverse (SI) relay curve is used here in the following experiments to generate the data that will be used in the test. Also a single shot scheme is implemented in the experiment for low speed re-closing of the circuit breaker. The values of the Standard Inverse Relay are based on the characteristic equation in Table 3 with the following settings [1]:

- $TMS = 0.1$

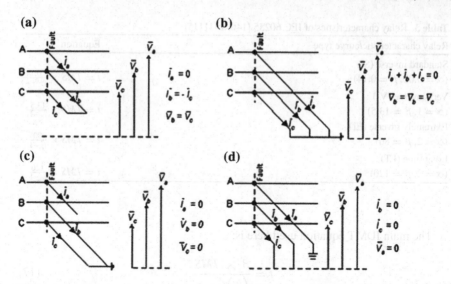

Fig. 11 Types of overcurrent fault. **a** Double-phase fault (B–C), **b** Three-phase-ground fault (A–B–C), **c** Double-phase-ground fault (B–C–E), **d** Single-phase-ground fault [15]

- $i_r = 1$ and 0.2 (for the overcurrent, and earth fault consequently)

The above values are the typical settings used in the distribution systems in some power system companies, and the experiment was simulated to real relays that work with the same sittings to overcome the faults at these sittings.

The data consists of ten parameters; the 3 phase currents or (A, B, C) or (R, Y, B), the neutral current (N), overcurrent time (t_{ph}), earth fault time (t_n), Relay status (R_f), and Trip Signal (T_s). Hence eight input signals are used in the following experiments. The remaining data represents the desired outputs of the Circuit Breaker (CB) control signal, and the dead time (T_d) which is the minimum time delay that is required before activating the CB output signal.

The generated data includes a number of different types of overcurrent system faults as shown in Fig. 11 which are implemented in the following experiments such as:

(a) Phase to Phase
(b) Three Phase to Ground
(c) Single Phase to Ground
(d) Phase to Phase with Ground

In Fig. 12, the PRNN is used as an external monitoring relay to work as a backup for an Inverse overcurrent relay which was described in the previous section. The existing relay is used to overcome the overcurrent problems, however if a failure occurs and there is no suitable action is applied, then PRNN will work as a digital relay to perform the desired action. The neural network (NN) inputs are described as follows:

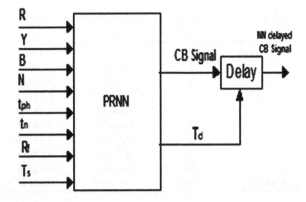

Fig. 12 PRNN I/O in over-current relay experiment [15]

Inputs from simulated power lines

- 3 phase currents (R, Y, B) or (A, B, C);
- Neutral current (N).

Inputs from simulated existing relay

- Overcurrent time (t_{ph}),
- Earth fault time (t_n),
- Relay status (R_f),
- Trip signal (T_s).

The main NN output is the Circuit Breaker (CB) control signal. The dead time (T_d) signal is another output which is needed as delay to activate the CB output signal.

The experiment is designed based on the following:

- The total number of samples to be used is 200.
- Only 25 % of the overall samples are used for the phase of supervised learning.
- The other phase of Nonsupervised learning is started with the remaining 150 unseen samples with only 8 inputs without its desired output since it is expected that PRNN will predict that corresponding value.
- Different values of learning rate (λ) and forgetting factor (η) are used in order to enhance the output results.
- Number of modules is $M=5$; Number of neurons per module is $N=8$.

The Feedforward Neural Network used in this experiment has the following properties:

- Two layer with 15 nodes per each, trained by Levenberg-Marquart learning algorithm.
- Initially, the neural network was trained with 250 samples; that was for the circuit breaker control signal (CB) and the Dead time (T_d) signals.

Table 4 The control signal (CB) overcurrent relay experiment [15]

	M	AUC	MSE	SSE	SNR
PRNN results	2	0.63575	0.56667	85	7.9862
	3	0.63575	0.56667	85	8.0666
	4	0.63663	0.57333	86	8.001
	5	0.63738	0.58	87	7.9006
	6	0.63738	0.58	87	7.8685
	7	0.63801	0.58667	88	7.8837
	8	0.63336	0.59333	89	7.7385
	9	0.63336	0.59333	89	7.723
	10	0.63336	0.59333	89	7.7067
FFNN results	N/A	0.85655	0.00138	0.207	28.6012

- Then the simulation was applied for 150 new unseen samples. The simulated results of CB control signal for the desired output versus the NN output and the mean square error is 0.00138, the results of Dead time (T_d) signals with MSE = 0.4825 has also been shown.

For dead time output (T_d) which will activate CB signal; the simulated result with mean square error performance equal to 0.4825; this shows a slight difference between the desired output and the Neural Network output; however this result is acceptable since the IDTM has $+/-$ 5 % accuracy limit as mentioned in Network Protection and Automation Guide and other researcher [16].

In this experiment looking at the CB signal output, the obtained result is very good since the AUC is 0.85655, while for Td signal a perfect experiment is obtained since AUC is equal to 1. The exceptional results are obtained here since FFNN was trained as a supervised learning method; therefore the neural network weights are tuned optimally for this experiment, while in unsupervised learning using PRNN the performance was less than the achieved results with supervised learning.

The Overcurrent Relay Experimental Results

In this experiment the learning rate $\lambda = 0.00005$ and forgetting factor $\eta = 0.00005$ that were noticed and identified by several test of PRNN. Tables 4 and 5 show that the optimum results were at the number of modules in PRNN is equal to five. Firstly, the control signal of the circuit breaker (CB) obtained here with AUC = 0.63738 while with FFNN it is 0.85655 which is better. Also the mean squared error with FFNN is 0.00138 which is better than what PRNN achieved and even in terms of SSE and SNR.

In contrast, for the Delay Signal (T_d), PRNN shows better results compared to FFNN in terms of MSE, SSE and SNR taking into account that AUC is accepted for the both techniques.

Table 5 The delay signal overcurrent relay experiment [15]

	M	AUC	MSE	SSE	SNR
PRNN results	2	0.85479	0.026986	4.048	8.7131
	3	0.85156	0.027002	4.0503	8.7178
	4	0.85293	0.026957	4.0435	8.6879
	5	0.85275	0.026958	4.0437	8.7138
	6	0.8575	0.026955	4.0432	8.7389
	7	0.85321	0.026904	4.0356	6.746
	8	0.85211	0.027722	4.1584	8.7241
	9	0.851811	0.02834	4.251	8.7113
	10	0.85216	0.028331	4.2497	8.7166
FFNN results	N/A	1	0.4825	72.375	3.165

4.4 Summary of the Experiments

Self-healing system simulation is implemented to check the system ability to investigate, diagnose and react to the failure. The experiments shows that the system automatically discovers, diagnose, and correct faults. Three experiments are implemented; the web-based client-server experiment where the clients access web resources by making requests to one of several geographically distributed server groups. The server group examined with three and five severs. It is required to make sure that the quality of service for the clients is guaranteed, so the request-response latency for clients must be under a certain threshold and the active servers should be kept as loaded as possible. The other experiment of overcurrent relays is implemented using neural networks to prevent and recover from system failure in power system field. In order to achieve all the experiments, PRNN and FFNN are used as unsupervised and supervised learning methods.

The main challenges of self-healing systems of monitoring, interpretation, resolution, and adaptation is solved with the use of a neural network. These challenges have been solved using an adaptive learning approach specifically by a novel technique of pipelined neural network [15] rather than software system engineering approach of

Different performance metrics are used to evaluate the results. The mean squared errors and the sum of squared errors as an analysis of variance technique is used, also ROC analysis tool is used for evaluating the performance of the experiments using the area under the curve (AUC) measurement. The obtained results were promising even though the results of supervised neural network in some experiments were better than the unsupervised technique but the main objective is achieved to implement self-healing in real time system.

5 Conclusions

Self-Management systems are a promising research area that has many interesting and important research problems that need further investigation. Different challenges related to self-configuring, self-healing, self-protection, and self-optimizing are investigated. Self-management systems or in other words 'autonomic computing' is needed in several fields such as distributed systems, electrical power systems, defense software, eSystems and more. Autonomic systems research started around 2001 for the purpose of managing the system complexity [20], and different companies such as IBM, HP, Microsoft, and sun microsystems started their research to overcome the challenges in this area [21]. The research needs to investigate complex systems' issues without the need of humans; these issues involves, knowledge for system stabilization, when and where an error state occurs, analyze the problem, and healing plan for recovering from failure [20].

The challenges in autonomic computing could be the way of implementation so that the recovery plan and system complexity should be hidden from the user. In addition to that the automatic adaptation and repairing have to be performed in the run time mode with a back-up plan in case of failure occurs during the run time adaptation.

The aim of the work described here is to investigate the fundamentals of self-healing systems then to apply a proposed solution that deals with the main challenges in this field specifically. The self-healing challenges [6, 7, 9]: of how to implement monitoring for the running system, which repair solution to be proposed, how to determine when there is a problem to be adapted, and finally the adaptation mechanism are solved using artificial neural network approach. The adaptive recursive neural network is proposed in this context to overcome the challenges during the running time. Moreover, this work focused on the external adaptation approach rather than internal technique. Different experiments were introduced to test the ability of the online learning neural network in the field of self-healing and bench marked with the off-line neural network.

This research makes a contribution towards better understanding of self-healing systems of the field of autonomic computing.

The chapter contribution is described in different position here, where a novel approach and method is used to build up a self-healing system that can respond to the adaptation in the running time. This method is achieved based on adaptive learning rather than software system engineering approach. Recurrent neural networks were the proposed method using a modified pipelined recurrent neural network algorithm. The four issues of self-healing are solved also in a way that the neural network supported monitoring, resolution, interpretation, and adaptation on the fly. This research explored self-healing in two different areas; in computer networks systems implemented by the client-server experiment and with distributed power systems by replacing the numerical relays automatically by the neural network to recover from failure. We believed that after the number of tests and experiments that took place during this research using online and off-line approaches, we have contributed to the field of autonomic computing by helping to pioneer a method that was proposed to overcome the challenges and problems in this area of research.

The challenging subject of self-healing systems manifests the system ability to investigate, diagnose and react to system malfunction or the system automatically discovers, diagnoses, and correct faults, which is required these days in several areas of distributed systems, the new era of internet (web3.0), computer networks, and complex software. After completing this research work, we can emphasize that the main issues in self-healing is how to monitor the running system and what are the important system behaviors to be monitored. Also the adaptation or reconfiguration is an important phase where a backup plan has to be ready if the failure occurred during implementing the modification of the system parameters or choosing the proposed solution. Several researchers used the software engineering approach to implement and overcome the self-healing issues while in this work using artificial intelligence method is used. The proposed modified pipelined recurrent neural network shows the ability to fulfill the requirements of self-healing. In general with some of the introduced experiments the continuous learning is achieved which leads to implementing self-healing during the running time using PRNN. Pipelined recurrent NN compares favorably with Feed forward NN, not only in terms of low mean square error with high signal-to-noise ratio, and low sum of squared error with AUC more than the threshold values, but also because of the real time learning and the adaptation phase occurs while the system is running. The proposed method in the client server experiment shows very good results for few servers per a group server while with increasing the servers poor performance is noticed.

Finally, the overall goals of this research work are achieved and we believed that a contribution towards better understanding and implementation of self-healing systems of the field of autonomic computing is performed.

References

1. Alstom, G.: T & D Protective Relays Application Guide, 3 edn. CEE relays Ltd (1987)
2. David Garlan, S.W.C., Schmerl, B.: Increasing system dependability through architecture-based self repair. In: Appears in Architecting Dependable Systems, 2003
3. Erik, H., Poul, T.: A statistical test for the mean squared error. J. Stat. Comput. Simul. **63**, 321–347 (1999)
4. Field, A.: Discovering Statistics Using SPSS (Introducing Statistical Methods S.), 2nd edn. SAGE Publication, Thousand Oaks (2005)
5. Fawcett, T.: ROC Graphs: Notes and Practical Considerations for Researchers. HP Laboratories, CA (2004)
6. Garlan, D.: Model-based adaptation for self-healing systems. Presented at the In ACM SIG-SOFT Workshop on Self-Healing Systems (WOSS'02). Charleston, SC, 2002a
7. Garlan, D.: Exploiting architectural design knowledge to support self-repairing systems. Presented at the The 14th International Conference on Software Engineering and Knowledge Engineering, Ischia, Italy, 2002b

8. Garlan, D., Chang, J.: Using Gauges forArchitecture-Based Monitoring and Adaptation. In: Proceedings of the Working Conference on Complex and Dynamic Systems Architecture, Brisbane, Australia http://repository.cmu.edu/compsci/690/ (2001)
9. Garlan, B.S.: Rainbow: Architecture-based self-adaptation with reusable infrastructure. IEEE Comput. Soc. **37**(10), (2004)
10. Hussain, A.J., Lisboa, P., El-Deredy, W., Al-Jumeily, D.: Polynomial pipelined neural network and its application to financial time series prediction. Lect. Notes Comput. Sci. **4304**, 597–606 (2006)
11. Haykin, S., Li, L.: Nonlinear adaptive prediction of nonstationary signals. IEEE Trans. Signal Process. **43**, 526–535 (1995)
12. Kon, F.: The case for reflective middleware. Presented at the Communications of the ACM, 2002
13. Mousa Al-Zawi, M., Hussain, A., Al-Jumeily, D., Taleb-Bendiab, A.: Using adaptive neural networks in self-healing systems. In: Proceedings of the 2nd International Conference on Developments in eSystems Engineering in Information Technology (DeSE'09), Abu Dhabi, UAE, pp. 227–232. 14–16 Dec 2009
14. Mousa Al-Zawi, M., Hussain, A., Taleb-Bendiab, A., Symons, A.: A survey: autonomic computing. In: 1st International Conference on Digital Communications and Computer Applications (DCCA2007), Jordan, pp. 973–979 (2007)
15. Mousa Al-Zawi, M.: Autonomic computing: using adaptive neural network in self-healing systems. PhD Thesis, Liverpool John Moores University (2012)
16. Mason, C.R.: The art and science of protective relaying, 1 edn. Wiley. Ariva, Network Protection & Automation Guide Barcelona, Spain (2002)
17. Mikic-Rakic, N.M., Medvidovic, N.: Architectural style requirements for self-healing systems. Presented at the Wass'02. Charleston, South Carolina, USA, 2002
18. Pereiraa, E., Pereirab, R., Taleb-Bendiabb, A.: Performance evaluation for self-healing distributed services and fault detection mechanisms. J. Comput. Syst. Sci. **72**, 1172–1182 (2006)
19. Peyman Oreizy, G., Taylor, R.N. et al.: An architecture-based approach to self-adaptive software. IEEE Intell. Syst. Appl. **14**, 54–62 (1999)
20. Sterritt, R.: Autonomic computing-a means of achieving dependability. In: Proceedings of IEEE International Conference on the Engineering of Computer Based Systems (ECBS'03), Huntsville, Alabama, USA, pp. 247–251 (2003)
21. Tosi, D.: Research Perspectives in Self-healing Systems. University of Milano, Bicocca (2004)
22. Van Erkel, R., Pattynama, P.M.T.: Receiver operating characteristic (ROC) analysis: basic principles and applications in radiology. Eur. J. Radiol. **27**, 88–94 (1998)

Spatial Dimensions of Big Data: Application of Geographical Concepts and Spatial Technology to the Internet of Things

Erik van der Zee and Henk Scholten

Abstract Geography can be considered an important binding principle in the Internet of Things, as all physical objects and the sensor data they produce have a position, dimension, and orientation in space and time, and spatial relationships exist between them. By applying spatial relationships, functions, and models to the spatial characteristics of smart objects and the sensor data, the flows and behaviour of objects and people in Smart Cities can be more efficiently monitored and orchestrated. In the near future, billions of devices with location—and other sensors and actuators become internet connected, and Spatial Big Data will be created. This will pose a challenge to real-time spatial data management and analysis, but technology is progressing fast, and integration of spatial concepts and technology in the Internet of Things will become a reality.

Abbreviations

ADE	Application Domain Extensions
ANPR	Automatic Number Plate Recognition
API	Application Programming Interface
AR	Augmented Reality
CEP	Complex Event Processing
EDA	Event Driven Architecture
ESB	Enterprise Service Bus

E. van der Zee (✉) · H. Scholten
VU University, FEWEB-RE, attn. SPINlab, De Boelelaan 1105, 1081 HV
Amsterdam, The Netherlands
e-mail: e2.vander.zee@vu.nl

E. van der Zee · H. Scholten
Geodan, President Kennedylaan, 1079 MB Amsterdam, The Netherlands
e-mail: h.j.scholten@vu.nl; info@geodan.nl

N. Bessis and C. Dobre (eds.), *Big Data and Internet of Things:*
A Roadmap for Smart Environments, Studies in Computational Intelligence 546,
DOI: 10.1007/978-3-319-05029-4_6, © Springer International Publishing Switzerland 2014

ESP	Event Stream Processing
ETL	Extract Transform Load
GGIM	Global Geospatial Information Management
GIS	Geographical Information System
GIS&T	Geographical Information Science and Technology
GML	Geography Markup Language
GPS	Global Positioning System
IDW	Inverse Distance Weighted
IoT	Internet of Things
JSON	Java Script Object Notation
LBS	Location Based Services
LoD	Level Of Detail
NFC	Near Field Communication
OGC	Open Geospatial Consortium
QR	Quick Response
RFID	Radio Frequency IDentification
SOA	Service Oriented Architecture
SEP	Simple Event Processing
SES	Sensor Event Service
SIR	Sensor Instance Registry
SOR	Sensor Observable Registry
SOS	Sensor Observation Service
SPS	Sensor Planning Service
SQL	Structured Query Language
SWE	Sensor Web Enablement
UN	United Nations
UWB	Ultra Wide Band

1 Introduction

1.1 Smart Cities

The world is faced with challenges in all three dimensions of sustainable development—economic, social, and environmental. The United Nations predicts that the world population will grow to 8.92 billion by 2050 and peak at 9.22 billion in 2075 [42]. At the same time, the population living in urban areas is projected to rise by 2.6 billion, increasing from 3.6 billion in 2011 to 6.3 billion in 2050 [43]. Furthermore, population growth becomes largely an urban phenomenon concentrated in the developing world [36].

Population growth and rapid urbanization, especially in developing countries, creates many economic, environmental and social problems, and calls for major changes in the way urban development is designed and managed. The concept of a

Smart City (or *Smart Environment*) can support these changes by using Information Technology and (Spatial) Big Data to monitor, steer and optimize processes in our environment in real-time.

A Smart City is defined by GSMA [16] as: "A city that makes extensive use of information and communications technologies, including mobile networks, to improve the quality of life of its citizens in a sustainable way. A Smart City combines and shares disparate data sets captured by intelligently-connected infrastructure, people and things, to generate new insights and provide ubiquitous services that enable citizens to access information about city services, move around easily, improve the efficiency of city operations and enhance security, fuel economic activity and increase resilience to natural disasters".

1.2 The Internet of Things and Big Data

New technological developments continue to penetrate countries in all regions of the world, as more and more people and objects are getting connected to the internet. More countries are reaching a critical mass in terms of ICT access and use, driven by the spread of mobile Internet [21]. This accelerates ICT diffusion and enables the development of Smart Environments. As network availability and speed are improving at a steady rate and computers become smaller, more energy efficient and lower priced, Internet connected devices with sensors and actuators are deployed on ever larger scales in our environment, creating an *Internet of Things (IoT)*.

Currently, the number of internet connected devices is rising exponentially. Cisco's Internet Business Solutions Group (IBSG) predicts some 25 billion devices will be connected by 2015, and 50 billion by 2020 [13]. The total amount of information in the world is estimated to have grown from 2.6 optimally compressed exabytes in 1986 to 15.8 in 1993, over 54.5 in 2000, and to 295 optimally compressed exabytes in 2007 [17]. An increasing part of this *Big Data* is created by internet connected devices, and consists of device states (properties), data collected by its embedded sensors and by humans using applications running on these devices (e.g. smart phones).

1.3 The Role of Geography in the Internet of Things

Geography is a holistic and broad interdisciplinary research field. Scholten et al. [38] describe the use of geospatial concepts and technology in a variety of study areas. The research field of Geographic Information Science and Technology (GIS&T) is well described by Dibiase et al. [9] in the GIS&T Body of Knowledge.

Geography can be considered an important binding principle in the Internet of Things, as all physical objects and the sensor data they produce have a position, dimension, and orientation in *space and time*, and spatial relationships exist between them. By applying spatial relationships, functions, and models to the spatial characteristics of smart objects and the sensor data, the flows and behaviour of objects and people in Smart Cities can be more efficiently monitored and orchestrated.

To be able to spatially analyze Big Data from internet connected devices for real-time spatial decision making, the Big Data has to be geo-referenced or "spatialized" (i.e. spatial coordinates have to be assigned to the data elements), to create *Spatial Big Data*.

1.4 Chapter Outline

This chapter has the following structure: Sect. 2 describes the spatial modeling of internet connected devices. In Sect. 3, the determination of the spatial properties (e.g. location) of internet connected devices is discussed. Section 4 shows how sensor data from internet connected devices can be georeferenced. Section 5 delineates the spatial context and spatial relationships between smart objects. In Sect. 6 real-time analysis of Spatial Big Data is discussed. Section 7 elaborates on the available spatial technology and standards for real-time spatial data management and spatial analysis of Big Data. Finally, Sect. 8 presents some smart city case studies and examples of the efficient use of spatial concepts and technology related to the Internet of Things.

2 Spatial Modelling of Things

2.1 Introduction

This section describes the spatial modeling of internet connected devices. Real-world objects ("things") have to be spatially modelled to be able to spatially analyse their properties and relationships. For most IoT use cases, modelling real-world objects with simple 2D or 2,5D point, line or polygon geometries is sufficient, but some use cases require a full 3D model of an object. For this, the CityGML information model can be used, which will be explained in Sect. 2.3.

2.2 Definition of a Thing

The Internet of Things is about connected things. But what do we actually consider a Thing? In a philosophical context, a Thing is an object, being, or entity [47]. The term 'object' is often used in contrast to the term 'subject'. The pragmatist Charles S. Peirce defines the broad notion of an object as anything that we can think or talk about [48]. Smart objects (or smart things) are defined by Serbanati et al.

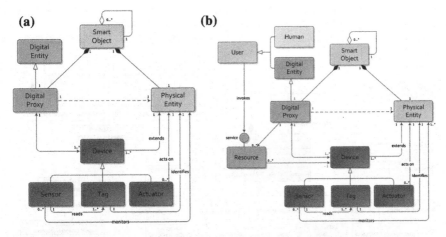

Fig. 1 **a** Conceptual model of a smart object; **b** Proposed internet of things reference model (*Source* Serbanati et al. [39])

[39] and Magerkurth [25] as objects which are directly or indirectly connected to the Internet, that can interact with their environment, and can describe their own possible interactions. Smart objects have a physical and digital representation and have a unique identity on the web. The conceptual models of a smart thing and a proposed IoT reference are shown in Fig. 1.

2.3 Spatial Modelling

Our environment consists of physical objects (or things), the Earth's natural objects (trees, rocks, etc.) and man-made artificial objects, some of which are smart objects. All these objects have spatial properties. According to Huisman and De By [18], spatial properties of objects are: (1) location ("where"); (2) shape ("what form"); (3) size ("how big"); and (4) orientation ("facing which direction"). The sphere of influence or effective range of an object (e.g. surveillance camera or siren) can be considered a fifth spatial property of an object. The spatial properties of an object can change in time.

Digital spatial representations (models) of real-world objects and beings have to be developed in order for computer systems to apply spatial algorithms. Use cases determine which of these spatial parameters are required to represent the object, and how the object will have to be spatially modelled. In most cases, only the location is needed, but in specific cases also shape, size and orientation matter.

Objects can be modelled using a vector or raster method. In the vector method, points, lines, regions, or solids can be used to represent an object. The raster method uses pixel or voxels. The vector method is often used for discrete data (i.e. with clearly defined borders) while raster representations are used for continuous data

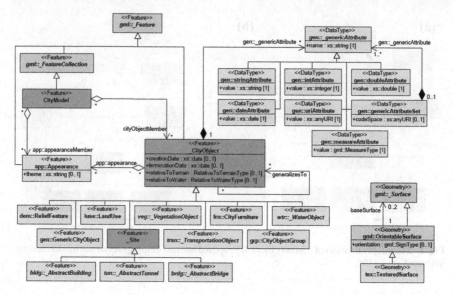

Fig. 2 CityGML's top level class hierarchy (*Source* Gröger et al. [15])

(e.g. distributions of height, pollution, temperature, etc.). However, it is possible to use and mix both models.

CityGML [15] is an information model that can be used for the spatial representation of (sets of) urban objects. CityGML provides common definitions of the basic entities, attributes, and relationships. The model contains 13 modules, i.e. Core; Appearance; Bridge; Building; CityFurniture; CityObjectGroup; Generics; LandUse; Relief; Transportation; Tunnel; Vegetation; and WaterBody. Figure 2 shows the top level class hierarchy of the CityGML information model.

CityGML supports different Levels of Detail (LoD) for various application requirements, e.g. for spatial analysis and modelling, less detail is required or needed than in the case of data visualization. Therefore, the same object can be represented in different LoDs simultaneously, enabling the analysis and visualization of the same object at different degrees of resolution.

As CityGML is a generic model, in most cases this model has to be tailored for specific situations. The GenericCityObject and GenericAttribute classes (defined within the Generics module) can be used for modelling objects that are not covered by the thematic classes or which require attributes not represented in CityGML.

Objects are often derived from, or have relationships to, objects in other databases or data sets. CityGML allows for making external references links to corresponding objects in external information systems (Fig. 3) using unique identifiers (URIs). In this way, external references can be made between the spatial representations of smart things in the CityGML model and their descriptions in external asset management systems.

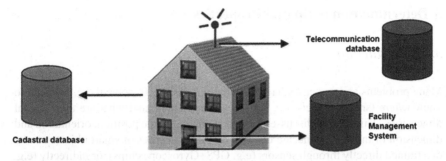

Fig. 3 External references (*Source* Gröger et al. [15])

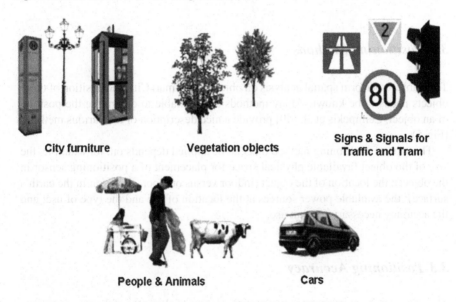

Fig. 4 Examples of prototypic shapes (*Source* Gröger et al. [15])

Most objects in CityGML are spatially modelled in real-world coordinates (e.g. buildings, bridges). In other cases (e.g. city furniture), objects are modelled as prototypes (see Fig. 4) of which the shape, size and orientation can be adapted using a transformation matrix that facilitates scaling, rotation, and translation of the prototype. The prototypes are spatially modelled using an internal coordinate system and positioned in real-world coordinates using a 2D or 3D base-point.

The CityObjectGroup class in CityGML can be used to group spatial objects. This is valuable, as objects often consist of a collection of smaller objects, some of which are smart objects. As the definition of a thing is quite broad, grouping gives the flexibility to spatially constitute things from other things.

3 Determination of Spatial Properties

3.1 Introduction

Many problems and use cases have a spatial dimension. To solve them, we need to know *where* (smart) objects are, how they are oriented, and what are their spatial dimensions. This section discusses the determination of the position, orientation and dimension of internet connected devices. Spatial properties of smart objects can be determined directly through sensors (e.g. GPS+Gyroscope chips), or indirectly (e.g. through the known position of the RFID scanner).

3.2 Positioning Methods

To be able to perform spatial analysis on objects in a Smart City, the position of these objects needs to be known. Many methods are available to determine the position of an object. Zeimpekis et al. [49] provide a nice description of the various methods (Fig. 5).

The type of positioning method that can be applied depends on, for example, the size of the object (available physical space for placement of a positioning sensor in the object); the location of the object (indoor versus outdoor, above/on/in the earth's surface); the available power sources at the location of use and the type of use; and the accuracy necessary for that use.

3.3 Positioning Accuracy

The accuracy of a measurement system is the degree of closeness of the measurements of a quantity to that quantity's actual (true) value. The precision of a measurement system is the degree to which repeated measurements under unchanged conditions show the same results. Dias [7] presents an overview of common positioning technologies related to accuracy and operation scales (Fig. 6).

For each IoT use case, the positional accuracy has to be evaluated. For this, metadata related to spatial properties is needed (e.g. which positioning technology or specific chip was used to determine coordinates).

3.4 Spatial Orientation

In some use cases, the orientation (pitch, roll, yaw) of an object becomes important. For example, in the case of a surveillance camera, it is useful to know in which

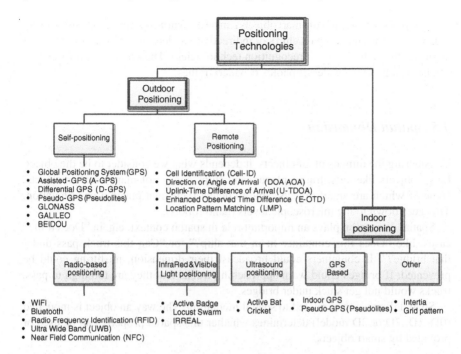

Fig. 5 Outdoor and indoor positioning methods (*Source* based on Zeimpekis et al. [49])

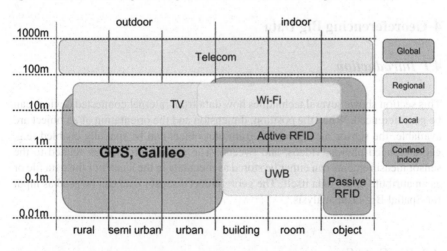

Fig. 6 Positioning technologies: accuracy and operation scales (*Source* Dias [7])

direction a camera is pointing and which area the camera is covering. Or, in the case of pictures taken with smartphones, when the orientations and positions of photos are stored, the positions of objects or people in that photo can be determined through, for example, Photosynth technology.

The spatial orientation of smart objects can be determined directly through sensors (e.g. Gyroscope and compass chip), or indirectly (e.g. determining the orientation of an object through 3D object recognition technologies). The orientation information can be stored in the header of photos or video frames.

3.5 Spatial Dimension

Considering the dimension of objects, it depends what we consider to be the object. Large objects like cars, trains, planes, and ships consist of many smaller objects, some of which are smart objects, able to detect their own position or orientation. This goes down to the microscopic levels (nano robots).

Spatial dimension plays an important role in spatial context, e.g. in "Does it fit?" cases, like "Does this container fit in that ship?", or "Can this truck pass under that bridge?". If containers could broadcast their dimension, misfitting could be prevented. If bridges could warn approaching trucks that they are too high to pass, trucks would not get stuck under bridges.

Spatial dimension is related to spatial modelling. The way an object is modelled (0D, 1D, 2D or 3D model) determines whether this spatial property can be used or provided by smart objects.

4 Georeferencing Big Data

4.1 Introduction

This section shows several techniques how data from internet connected devices can be georeferenced. When the position, dimension and the orientation of an object are available, the sensor measurements from that object can be spatially enabled (e.g. georeferenced photos, videos, and tweets). The spatial information related to the sensor measurements can either be stored as metadata in the header of the data file or as an attribute of the data itself. The georeferenced sensor data can be used as input for Spatial Big Data analysis.

4.2 Big Data Sources

Big Data can be produced by intelligent agents or by humans. Intelligent agents autonomously observe the environment through sensors (e.g. camera, microphone, chemical). People can also observe the environment using their natural sensors (eyes, ears, nose, tongue, tactile nerves) and brains. They can publish their observations through applications (e.g. social media applications like Twitter, YouTube, Flickr,

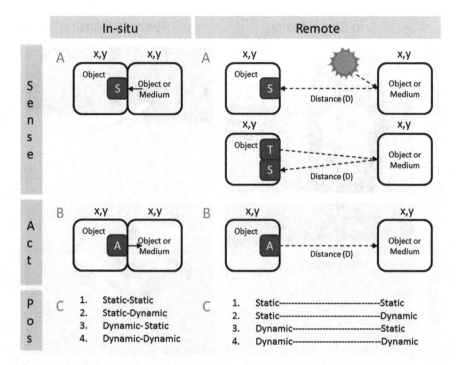

Fig. 7 In-situ versus remote sensing (**a**) and acting (**b**). **c** Indicates that while sensing or acting, the position of objects can be static or dynamic (*Source* based on Zeimpekis et al. [49])

Blogs) that run on fixed or mobile devices, e.g. smartphones [35]. Information on the status of smart objects (e.g. battery status) also adds to the data stream.

Sensors in smart objects can measure in-situ (i.e. in direct contact with an object or medium, e.g. a water temperature sensor in water) or remotely (i.e. in indirect contact with an object or medium, observing or interacting with an object or medium indirectly, e.g. a surveillance camera detecting cars' number plates from a distance). The same is true for in-situ and remote acting. Figure 7a, b show both situations.

During sensing, the position of objects can be static or dynamic. Figure 8 shows examples of sensors mounted on static or dynamic objects, (a) a moving smartphone measuring the air quality at a certain location in-situ (b) a moving satellite observing the earth remotely from space (c) fixed sensors in the asphalt measuring traffic speed and number of cars in-situ (d) fixed cameras remotely detecting number plates.

4.3 Geo-enabling Observations

To be able to perform spatial analysis on observations (events), the observations have to be geo-enabled. Figure 9 depicts five methods for geo-enabling events, depending on the capabilities of the object. In cases (a) and (b), the object has a positioning chip

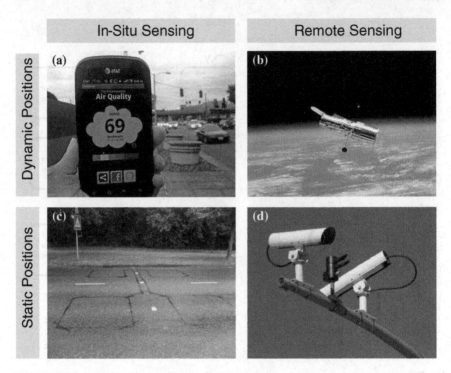

Fig. 8 Examples of in-situ and remote sensing objects (*Source* based on Zeimpekis et al. [49])

on board, the events can be spatialized by the object itself. In cases (c), (d), (e), and (f), the object does not have a positioning chip on board. In such cases, spatializing events can take place client- or server-side (depending on the capabilities of the object), using external geocoding or geotagging services. In the examples of Fig. 9, only server-side geo-enabling is elaborated. The spatial information can be stored either in the metadata (e.g. header) of a file (e.g. GeoTIFF), or in the event message itself (e.g. GeoJSON, GeoRSS, GeoSMS, KML, GML).

In case (a), the internal positioning chip (e.g. GPS) can be used to spatialize the events from sensors or apps installed on the object. The resulting georeferenced events can be sent to an event stream for further real-time spatial analysis. Also the position(s) of the object itself (with ID0) can be sent to the event stream. An example is a smartphone with a health app connected to external sensors (e.g. blood pressure, heart rate).

In Case (b), RFID tags on object ID1 are read by a mobile (GPS-enabled) mobile RFID reader ID2. The resulting RFID reading events can be spatialized using the location chip, thus creating georeferenced events that can be sent to an event stream for subsequent real-time spatial analysis. Also the position(s) of the object itself can be sent to the event stream.

In case (c), the object does not have an internal positioning chip, but the spatial position of the object and ID (ID3) are known (stored in a spatial object database).

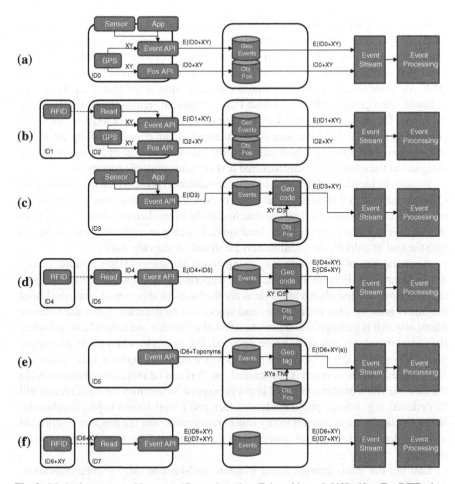

Fig. 9 Methods to geo-enable events (*Source* based on Zeimpekis et al. [49]). *Note* For RFID also Bluetooth, NFC, or similar methods can be read

Using a geocoding service (which returns an XY position based on an ID), events can be spatialized based on ID. Examples of this are smart assets with a fixed position e.g. surveillance cameras, environmental sensor devices, and mobile network antennas.

In case (d), an RFID chip (ID4) is read by an RFID reader (ID5) without a positioning chip. An event containing the IDs (ID4 and ID5) is created (e.g. a bankcard with ID4 was scanned by ATM with ID5, time=yyyy:mm:dd hh:mm:ss). When the position of the RFID reader with ID5 is known (stored in a spatial object database), the position of the RFID reader can be assigned to ID4. In this way, two georeferenced events can be created. With a different technology (e.g. Bluetooth) the distance between objects ID4 and ID5 can become greater creating a positioning error. Examples illustrating this case are RFID readers with a fixed position and without a positioning chip, such as ATM machines, public transport gates, and readers in logistic centers.

In case (e), the object does not have a GPS chip, nor a known position related to an ID. In this case, if the event message contains one or more toponyms, the event can be geotagged using the known positions of the toponyms. Examples are RSS feeds that contain toponyms such as the name of an address or a Point Of Interest (POI), or other references that are stored with a location in the spatial database. For example, the toponym "O'Leary's Irish Pub" + "Utrecht" returns lat 52.099124 long 5.115681. Other examples include a RSS feed from a blog page or a QR tag message containing toponyms. As the data is unstructured, it can contain spelling errors, so in some cases no match will be found. In other cases, more than one location will be assigned to the event, if the unstructured text contains multiple toponyms.

In case (f), location information (spatial coordinates) is encrypted in a Bar code or QR code, or is stored on board in an RFID or NFC chip. When scanned, the position of the object is revealed, and can be attached to the event. In cases where it is certain that an object will stay in the same location (e.g. a chip in a concrete wall or buried into the soil of a dike), the location may be stored on the chip itself.

A special way of positioning is by using object and subject (face, gesture) recognition algorithms. Footage of surveillance cameras (with known positions) or geotagged crowdsourced photos and videos are the basis for this type of analysis. Using the algorithms, number plates, faces and voices can be detected. Since the video or photo material is geotagged and timestamped, the location and time where and when the object or subject was seen can be deduced. For example, when a new geotagged photo is posted on the web (e.g. Picasa, Flickr), a photo recognition scan action can be triggered. When a person is recognized, an alert can be initiated subsequently. In the case the video or photo material is not geotagged, sometimes the location can still be deduced, e.g. when a photo contains a face and a well-known object (landmark). When the landmark is recognized by object recognition and the position is stored in a spatial database (e.g. Eiffel tower), then the position of the face (person) can be deduced.

Last but not least, through direct Machine-to-Machine (M2M) communication, smart objects without positioning capabilities can also retrieve a position from another nearby smart object that has a positioning chip, inferring thus its own position from the other object.

5 Spatial Context

5.1 Introduction

This section discusses the spatial relationships between a smart object with other (smart) objects. Once its spatial context is known, a smart object can efficiently interact with other smart objects in its vicinity. Spatial algorithms (relations, functions, models) can be used to determine the spatial context of smart objects.

5.2 Spatial Context of Things

Spatial context is an important aspect in the Internet of Things, as all physical and virtual objects have spatial relationships with other objects. This characteristic can be used for the effective deployment of smart objects. Spatial algorithms (relationships, functions and models) can be applied to geo-enabled objects, events, and their effect areas, in order to determine and analyse the spatial context. The definitions of spatial relationships and functions are standardized by ISO [20].

The spatial context of smart objects and events relates to: (1) the effective area and range of sensors and actuators of these objects or events; and (2) the spatial relationships between smart objects and events and other (smart) objects or events. Spatial context is applicable to both physical and virtual objects. Physical objects are objects tangible and visible in reality. Virtual spatial objects with real-world positions are objects that do not exist in reality, but which do have an influence in the environment, e.g. virtual zones (permit areas, administrative areas, danger zones). Alternatively, they can be virtual 3D objects (virtual sculptures), which can be made visible through, for example, augmented reality techniques. Smart virtual objects (with virtual sensors and actuators) can even create virtual events.

5.3 Effective Area

The first aspect of spatial context is the effective area. The sensors and actuators of smart objects have an effective area or sphere of influence. This can be a sensing range (e.g. the view area of a surveillance camera, the measuring range of a smoke detector), or an actuating range (e.g. the audible range of an air alarm, wifi transmitter range, light beam of a lighthouse). These ranges can be modelled as 2D or 3D spatial objects and can then be used in spatial context algorithms.

As spatial properties (location, shape, size, orientation) and non-spatial properties of smart objects and events can change over time, the spatial properties of effect areas can be static or dynamic in time as well. For example, when the focal length or tilt angle of a surveillance camera lens changes, the shape and size of the view area of the camera changes, and when the bearing of the camera is changed, the orientation of the view area changes as well.

5.4 Spatial Relationships

The second aspect of spatial context is the spatial relationship. A spatial relationship can be used to geographically select (smart) objects that match a certain spatial relationship condition. Figure 10 presents an overview of four commonly-used spatial relationships between (smart) objects that are spatially modelled as points, lines, or regions (polygons).

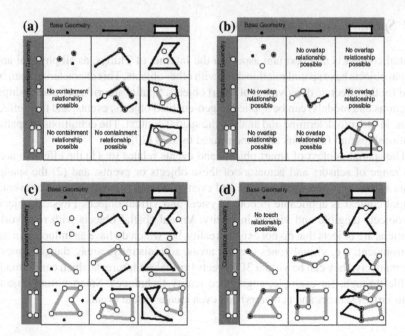

Fig. 10 Overview of common spatial relationships. **a** Contains. **b** Overlap. **c** Disjoint. **d** Touch. (*Source* Esri Inc. [10])

A commonly used spatial relationship in Location Based Services is the "contains" relationship. For example, when an tourist with a smartphone is entering a virtual zone (e.g. a municipality area), the "contains" spatial relation is true, and based on this, some action can be taken, e.g. sending a message with touristic information to that phone. Or, when a criminal with an electronic bracelet is entering a no-go zone, an alarm can be sent. Or when a smartphone or car is within 500 m from home, the carport opens and the coffee machine switches itself on. The possibilities of using spatial relationships in real-world use cases are virtually endless.

5.5 Spatial Functions

A spatial function creates one or more new spatial features based on the spatial properties of the input objects. The newly created spatial features can be associated or assigned to objects.

A commonly used spatial function is the "buffer" function (Fig. 11a). For example, based on the range property of an air alarm (e.g. 900 m) and the position of that object (x, y), a geographic buffer function can be applied, creating a new spatial object (region) that spatially represents the effective range of that air alarm. It can be associated with the air alarm object and subsequently be used in other spatial

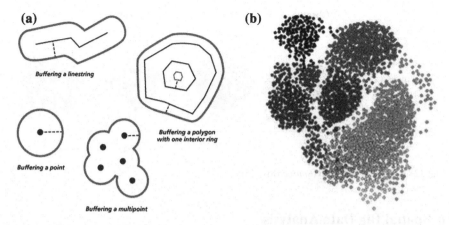

Fig. 11 Overview of common spatial operations (*Source* Esri Inc. [11])

algorithms and spatial analysis. Another common function is the spatial cluster function (Fig. 11b) which groups closely related positions of objects or events.

Other spatial functions include intersect, union, difference, convex and concave hull, spatial joins, routing, and geocoding. A geocoder/reversed geocoder returns x, y coordinates based on address information, and vice versa. A routing function calculates a route based on 'from', 'to', and 'via' locations. A spatial join makes it possible to transfer attributes from one object to another based on spatial relationships. Further, spatial interpolation functions can be used to predict sensor values at locations where no physical sensors are present. For this, input from nearby sensors is used. There are many spatial interpolation methods, e.g. splines, IDW, and Kriging.

5.6 Spatial Models

On the basis of historical and current data, a spatial model can predict future situations. Model output (predictions) can be used to initiate precautionary actions on things. For example, when a weather model predicts severe rainfall for a certain area, a flood model can calculate the expected excess rainwater. Subsequently, water drainage can be preventively intensified by stepping up the pumping activity (actuators) at certain locations in the effect area. An example of a flood forcasting model based on real-time sensor data is described by Berger [3].

Using weather sensors and spatial prediction models, weather patterns can be predicted and weather alerts can be sent to people's smartphones or to things, e.g. sun screens or other weather-dependent things. This can be combined with route planning, predicting routes that keep you dry.

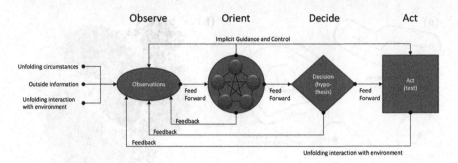

Fig. 12 Boyd's OODA loop (*Source* Boyd [5])

6 Spatial Big Data Analysis

6.1 Introduction

In the previous section it was shown how Big Data from smart objects can be geo-referenced. This section discusses real-time spatial analysis of this Spatial Big Data. Spatial characteristics can be used in real-time Spatial Big Data analysis, providing the necessary information, knowledge and wisdom to optimize processes and to solve problems in Smart Cities efficiently.

Once the positions and other spatial properties of objects and events are clear, real-time spatial analysis can be performed, and based on this analysis, decisions can be made. This process is described by the OODA loop. Real-time spatial analysis can be performed by integrating spatial concepts and technology in an Event Driven Architecture (EDA). The event stream engine in an EDA filters raw data events and outputs meaningful events. These can in their turn be used as input for (semi)automated processes, that steer actuators in smart objects. A *catalog of things* is an important component in such architecture, as this provides information on status and capabilities of smart objects. With current "smartness" of objects, the IoT has to be choreographed centrally. Once smart objects become more intelligent, decentral autonomous acting will become possible.

6.2 The Ooda Loop

In a smart city, large numbers of smart objects are connected to the Internet. These smart objects can be used either to monitor their environment (i.e. other objects, events or subjects) through their sensors or to act on the environment by using their actuators. To use the capabilities of smart objects efficiently, a continuous process of orchestration and choreography of smart objects is needed. This process is described by Boyd [5], see Fig. 12.

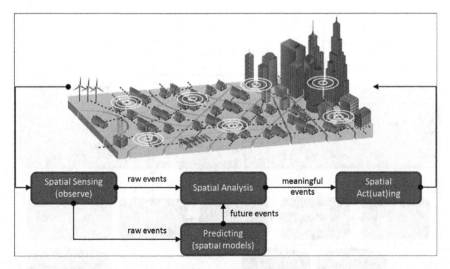

Fig. 13 Continuous loop of sensing, analysing, predicting, and act(uat)ing in a Smart City (*Source* based on Zeimpekis et al. [49])

The Observe-Orient-Decide-Act (OODA) model captures what happens between the onset of a stimulus and the onset of a reaction to that stimulus.

In all the phases of the OODA loop, spatial concepts and technology can be integrated and used to improve the efficiency of processes in a Smart City (Fig. 13).

Based on location, appropriate sensors and sensing ranges can be selected or activated. Spatial decisions can be made in real-time based on spatial characteristics of observations, and by applying spatial algorithms (relationships, functions, models) to positions of object and events. Furthermore, based on detected spatial patterns or exceeded spatial thresholds, appropriate actuators and actuating ranges can be selected or activated spatially.

6.3 Real-Time Spatial Analysis

Real-time spatial analysis of event streams can be realized by integrating spatial algorithms in a SOA-EDA (Service Oriented Architecture-Event Drive Architecture) configuration, as shown in Fig. 14.

In such architecture, raw events from smart objects are collected through feed adapters of an Enterprise Service Bus (ESB). From there, these events are sent to the EDA module. The first step in the EDA module is *event pre-processing*, which can be (geo)filtering, (geo)transformation, (geo)routing, or (geo)enrichment of events. Geocoding and geotagging of events can take place in this step, turning Big Data into Spatial Big Data. The result of the pre-processing step is filtered and enriched (georeferenced) events. When a smart object contains positioning capabilities (e.g.

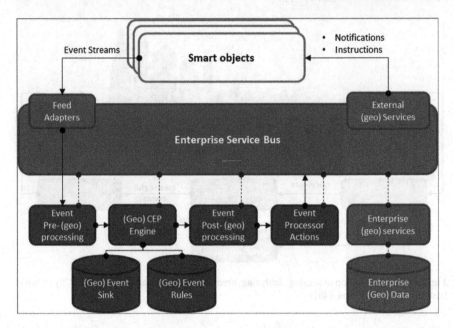

Fig. 14 Integration of enterprise (geo)services and spatial big data analysis in a SOA-EDA archi-tecture. EDA components are shown in *dark blue* (*Source* based on Zeimpekis et al. [49])

GPS), then events can already be georeferenced client-side and spatial pre-processing on the server-side is not needed.

The next step is real-time spatial event stream analysis by a *Geo-enabled Complex Event Processing (CEP) Engine*. In this step, raw filtered geo-events are analyzed against predefined geo event rules (spatial patterns). After processing, events can be stored in a spatial event sink for further causal processing or for spatial modeling purposes that need time series. The output of the CEP Engine is meaningful events.

An example of real-time spatial event stream analysis is credit card fraude detec-tion. When multiple credit card transactions are made from spatially remote places within a short period of time, it means that the credit card has been compromised and will be automatically blocked. In the same way, the behavior of smart objects and their sensors and actuators can be monitored and controlled based on their spatial characteristics and their spatial relationships with other smart objects.

After the CEP step, *post processing* can be performed when needed. Finally, *event processor actions* can be initiated based on the meaningful events from the CEP Engine, e.g. publish (geo) event, start business process, notify client, invoke (geo) service, capture (geo) event, or generate (geo) event. These actions will be processed by the ESB.

In all steps of the EDA module, spatial services from the Enterprise GIS system (e.g. geocoding, geotagging, routing, geoprocessing, spatial modeling, geostatistics, catalogs) can be used, and combined with other non-spatial business services through the ESB.

6.4 Catalogue of Things

In the Internet of Things, with its billions of sensors and actuators, it is vital to know where they are, and what they can measure or do. To effectively orchestrate and choreograph things with sensors and actuators in the IoT, a standardized catalogue service is required. This catalogue can be queried on location by objects or by operators, e.g. "Give me all air quality sensors in a range of 1.5 km from location x, y", or "Give me all cameras with Automatic Number Plate Recognition(ANPR) capabilities in an area of 500 m around abc street". A good "things" catalogue with spatial capabilities is vital and can be used to select the appropriate sensors or actuators in patterns or action scenarios. A catalogue service standard for geospatial data sets and webservices is available (OGC CS-W 2.0) Nebert et al. [28]. A candidate standard for a sensor catalogue service is proposed by Jirka and Nüst [22] as part of the OGC SWE standards. The application of semantic web concepts to sensor discovery has been described by Pschorr et al. [34].

6.5 Acting on Smart Objects

The last step in the OODA loop is acting. Based on sensor input and pattern analysis, notable events are generated by the Event Stream Processing (ESP) engine, and certain actions are initated. These actions or scenarios can be defined using business process models or business rules engines in an Enterprise Service Bus (ESB). This can include for example sending alerts, activating actuators in the right place, etc. Sensors and actuators can also react directly to each other through M2M communication, e.g. an irrigation system with smart soil moisture sensors and smart valves. When certain sensors detect drought, they can ask the valves to orientate themselves such that the water flows towards that sensor.

Physical actuators can be speakers, lights, motors, and electronic switches. Figure 15 shows examples of in-situ and remote acting objects, (a) a moving grass mower mowing the grass in-situ (b) a moving laser gun hitting a target remotely (c) a fixed automatic bollard that acts in-situ (d) a fixed air alarm acting remotely on ears through sound waves.

In most cases, smart objects contain both sensors and actuators (e.g. a surveillance dome camera contains an image sensor, a microphone sensor, and actuators like electro-motors to change the direction and zoom of a camera).

6.6 Central Versus Decentral Spatial Processing

The process of spatial orchestration and choreography of smart objects has to be managed either centrally or decentrally depending on the "smartness" of the object. In this respect, the different levels of complexity of the use case and the actors, related

Fig. 15 Examples of in-situ and remote acting objects (*Source* based on Zeimpekis et al. [49])

to the five phases of evolution of the IoT described by Pschorr et al. [6] have to be considered.

When smart objects become autonomous intelligent agents, they can operate independently and make certain (spatial) decisions autonomously (client-side), based on their spatial capabilities (e.g. calculating the shortest route). Additionaly, agents can acquire (pull) additional contextual spatial information (e.g. "Which smart objects with certain capabilities are closeby?") or receive (push) instructions ("Go to location x, y using route r and look there for a person with a red coat and blue pants.") from central systems. Intelligent systems can also retrieve contextual spatial information directly from other smart objects nearby (e.g. asking information from nearby sensors using M2M communication).

Depending on the intelligence of a smart object, their events are either simple measurement values or the outcome of a complex internal analysis, performed client-side by the smart object, signalling a problem or an impending problem, an opportunity, a threshold, or a deviation. For example, when a surveillance camera has built-in face and number plate recognition capabilities, it could act as an intelligent agent, looking out for a wanted person or car by comparing a photo of that person or car number plate with its own direct observations. It will only send a georeferenced event (alert) when that person or car is detected, instead of sending unnecessary continuous streams of raw camera data to a central server for processing.

Especially for moving autonomous agents like drones and robots, real-time spatial situational awareness (Where am I? Where am I going? Where are other (smart) objects and subjects? And, where are they going?) will be vital for efficient operation in the Internet of Things.

6.7 A Sensing and Acting Scenario

Sensors at a large chemical complex detect smoke or high temperatures, sent as georeferenced events. ESP Engine pattern analysis reveals "fire". As the position of the sensors is known, the fire can be pinpointed to a location. A smoke-plume model can calculate the direction of the smoke, using as input the location of the fire and using data from the closest weather sensors (using the "nearest" spatial algorithm). The model predicts a danger zone, people in the smoke-plume area are alerted by triggering the GSM towers in that area (using the "inside" spatial algorithm) to transmit a cell broadcast. The appropriate air alarms in the direction of the smoke are signalled (the properties of the air alarms are used to make the selection, range of air alarms = 900 m, inside algorithm with smoke area). The right actuators (sprinklers) in the fire area are activated, and the appropriate fire doors are closed. To select the right actuators and sensors, a sensor catalogue service is used. Then the closest fire department is warned (nearest algorithm), the position of the fire is sent to the fire truck and to the control rooms, the shortest route is calculated (routing algorithm, from position = fire station, to position = location of fire). The fire truck drives off, following the route. The firetruck's position if determined by its GPS sensor, and based on this position and the predicted route, the traffic light gives green light to the fire truck, open bridges on the route are closed, and automatic bollards are lowered.

7 Spatial Technology

7.1 Introduction

This section elaborates on the available spatial technology and standards for real-time spatial data management and spatial analysis of Big Data. Event Driven Architecture tools become spatially enabled and allow for real-time spatial analysis.

7.2 Geospatial Technology Stacks

In the past, geospatial technology used to be specialized 'island' technology, a niche using its own dedicated protocols and programming languages. But, recently, this situation has changed. Modern spatial technology components are based on W3C,

ISO, and OGC standards, and use regular programming languages like C++, C#, Java, and Python, making it easier to integrate geographical technology in service oriented (SO) or event driven (ED) architectures (A).

A typical geospatial SOA stack consists of a spatial database, a GIS server, spatial data loading (spatial-ETL) tools, and webservice APIs integration, for example through an Enterprise Service Bus (ESB). Nowadays, geospatial technology is progressing fast. Traditional relational databases all contain spatial capabilities (e.g. Oracle, Microsoft SQL Server, Postgres, MySQL), and also non-traditional NoSQL databases are also becoming spatially enabled (e.g. Neo4J, MongoDB, CouchDB).

Recently, various vendors have extended their products with EDA real-time spatial capabilities. Ali et al. [1] describes the spatio-temporal stream processing possibilities of Microsoft Streaminsight. Oracle provides real-time geostreaming and geofencing possibilities by combining the Oracle Spatial Database with the Oracle Complex Event Processing engine [40]. Esri Inc. released in 2013 their GeoEvent Processor module (Fig. 16) as an extension of the ArcGIS Server environment. This module can perform real-time geospatial analysis on geospatial events [27].

These developments enable the integration and leverage of spatial technology in the Internet of Things.

7.3 Geospatial Standards

In the geospatial domain, standardization efforts are coordinated by the Open Geospatial Consortium. "The Open Geospatial Consortium (OGC) is an international industry consortium of 484 companies, government agencies and universities participating in a consensus process to develop publicly available interface standards. The OGC® Standards support interoperable solutions that 'geo-enable' the web, wireless and location-based services and mainstream IT. The standards empower technology developers to make complex spatial information and services accessible and useful with all kinds of applications" [31]. The OGC has a broad user community and alliance partnerships with more than 30 standards organizations, amongst others, ISO, W3C, and OASIS.

The consortium has currently adopted 35 standards which are implemented worldwide in a variety of Geo-IT products and solutions. The standards can be grouped in catalogue services, data services, portrayal services, processing services, encodings and others. For the sensor and actuator networks and the IoT, the OGC has developed the Sensor Web Enablement (SWE) 2.0 standard suite [4]. Klopfer et al. [23] described real-world use cases with the OGC SWE standards. The SWE protocols have been tested in what are called OGC Web Services (OWS) Testbeds phase 4 (SOS), 5 (SPS), 6 (SES) and 7 (integration).

The OGC has identified the need for standardized interfaces for sensors and actuators in the Internet of Things. Therfore, the OGC formed a 'Sensor Web for IoT' Standards Working Group [19]. The IoT REST API SWG aims to develop such a standard based on existing IoT portals with consideration of the existing OGC

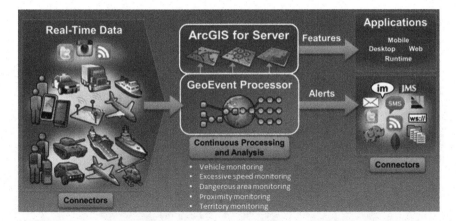

Fig. 16 Esri GeoEvent processor (*Source* Mollenkopf [27])

Sensor Web Enablement (SWE) standards. The OGC PUCK protocol [30] facilitates the publication of vendor-specific information of things, which makes it easier to "plug and play" things into sensor networks and the Internet of Things.

8 Case Studies and Examples

8.1 Introduction

Internet connected devices can be applied everywhere in Smart Cities, and can be used and shared in many use cases. Libelium [24] has defined an overview of 57 Internet of Things use cases in twelve categories, i.e. Cities, Environment, Water, Metering, Security & Emergencies, Retail, Logistics, Industrial Control, Agriculture, Animal Farming, Home Automation and eHealth, see Fig. 17.

An alternative research by Beecham Research [2] has defined nine categories of use cases, which are partly overlapping with the Libelium categories.

This section presents some Smart City case studies and examples of the efficient use of spatial concepts and technology related to the Internet of Things.

8.2 Smart Recycling Bins

Smart recycling bins with sensors can send an alert when the waste in the recycling bin reaches a certain level. As the positions of the recycling bins are known, spatial route-planning enables waste collection companies to manage (i.e. empty) their recycling bins in the most efficient manner, collecting waste on a "as-needed" basis

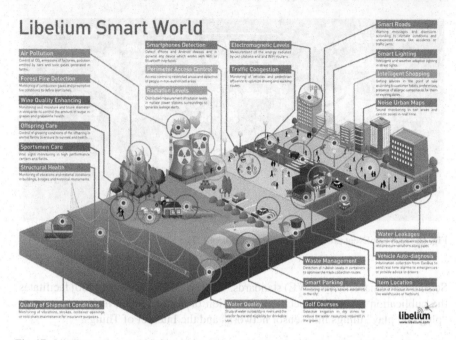

Fig. 17 Libelium smart world infographic—sensors for smart cities, internet of things and beyond (*Source* Libelium [24])

instead of collecting on scheduled regular basis (e.g. once per week). This contributes to substantial savings in fleet management costs. More information at http://www.smartbin.com (Fig. 18).

8.3 Smart Dikes (IJkdijk)

IJkdijk [32, 33] is an experimental dike which contains sensors that enable the continuous in-situ monitoring of the condition of the dike. When saturation occurs, the location of weaknesses and possible breaches can be detected at an early stage, and the necessary precautions can be initaited. More information at http://www.ijkdijk.nl.

8.4 Smart Roads (NDW)

The Nationaal Datawarehouse Wegverkeersgegevens [46] is a datawarehouse that receives real-time data from the highway. It is a system of sensors (ANPR cameras, inductive-loop vehicle classifier and speed sensors) and actuators (matrix signs, traffic

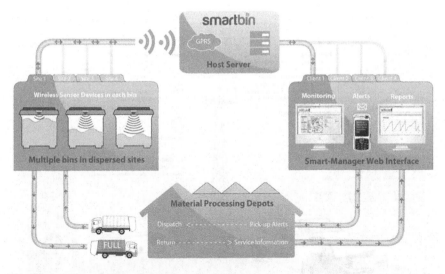

Fig. 18 Smartbin workflow (*Source* Smartbin [41])

lights) working closely together in semi-automated or automated ways to manage traffic jams, accidents. When the positions of cars are known, traffic jams can be pinpointed, and subsequently the traffic flow can be diverted or regulated by smart traffic lights. More information at http://www.ndw.nu.

8.5 Smart Cars (eCall)

A European Parliament adopted resolution, issued in June 2012, states that new vehicles have to be equipped with the eCall system by 2015. The eCall system in a car automatically calls 112 when an accident has happened. The warning message contains impact sensor information, as well as GPS coordinates. The information is sent to the nearest emergency response authority, which can use the GPS coordinates to calculate the shortest route to the accident. More information at https://ec.europa.eu/digital-agenda (Fig. 19).

8.6 Smart (ANPR) Cameras

In the detection of crime or the enforcement of the law, location plays an important role. As cameras become smarter, they are able to detect number plates and faces. ANPR cameras are used in The Netherlands for the enforcement of environmental zones to improve air quality. For this reason, trucks are not allowed in the city centre at certain hours of the day. When a truck is detected by ANPR cameras in

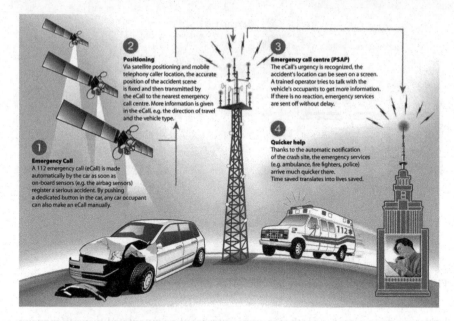

Fig. 19 E-Call, how it works (*Source* European Commission digital agenda [12])

the environmental zone at the wrong time, the number plate can be checked in the central car register and, based on this information, an automated fine procedure is initiated. More information at http://www.milieuzones.nl.

8.7 Smart Billboards

The latest development in location based marketing is smart billboards in public spaces, that can recognize gender and age of a person looking at the billboard using camera's and face recognition techniques. The ads are then focused on the characteristics of the person. Ng et al. [29] performed a survey on the state of the art in human gender recognition. As billboards can have fixed and dynamic positions (e.g. mounted on vehicles), ads can also be provided based on geofences (e.g. ice cream ads near beaches).

8.8 Smart Phones

Smart phones are the most known smart objects and are used in many use cases depending on the installed applications. Location based profile matching is a valuable application in smart cities, connecting people with the same hobbies, ideas or status (e.g. single) based on (smartphone) location. When two people with a matching profile are within a certain distance, they get an alert on their phones. Examples of location based mobile dating services include Brightkite, Grindr, Meetmoi, Okcupid, Sonar, and Skout. The same principle could be applied to "dating" of internet connected devices instead of people.

8.9 Smart Bracelets

More than 100,000 parolees and sex offenders are wearing ankle bracelets in the US. GPS ankle bracelets can be used to track and trace people geographically. When they enter a forbidden neighbourhood (geofence), an automated alarm with the position of the person can be sent to the nearest police station or officer in the field. GPS bracelets can also be attached to people, pets, or things to find them and bring them back more easily if they are missing or stolen. Micheal et al. [26] describe intriguing emerging ethics of humancentric GPS tracking and monitoring.

9 Epilogue

As the Internet of Things is a holistic concept, collaborative, cross-sectoral, and interdisciplinary research and thinking is needed to design it and to develop it further. It is important to realize that space (location, orientation, size and shape) and time play an indispensable role in the Internet of Things, as all objects and data they create have spatial properties and are spatially related to each other. Space and time can be the "glue" to connect smart physical devices in an efficient way. Geospatial concepts and technology should therefore be an integral part of an IoT architecture. With ever-increasing computing power and technological advancement in geospatial technology, real-time spatial data management and analysis of Spatial Big Data becomes possible.

The United Nations GGIM report on future trends in geospatial information management [45] mentions the importance of geospatial technology in relation to IoT. Also, the US Department of Labor designated geotechnology as one of the three most important emerging and evolving fields, along with nanotechnology and biotechnology [14]. Large enterprises like Google, Apple and Microsoft are increasingly using geospatial technology to create location based services. There is an increasing need for skilled people with knowledge on spatial concepts and technology, but unfortunately spatial concepts and technology are only briefly touched on, or even absent in the general IT curricula. A growing awareness is needed, so that general IT programmers and IT architects can become more familiar with geospatial concepts and technology. To achieve this, crossovers between geography and general IT curricula should be made.

Acknowledgments Writing this chapter would not have been possible without the support of our fellow researchers at the SPINlab of the VU University and our collegues at Geodan. We would like to thank them for their inspiring conversations, discussions, and thoughts on this subject. Further, we are grateful to the copyright owners for the use of figures and texts in this chapter. Finally, we are obliged to Laura Till and Patricia Ellman for taking a careful eye and a sharp pencil to review our material.

Author Biographies

Erik van der Zee MSc. has a background in Physical Geography and Business Economics. He works as senior Geo-IT consultant and researcher at Geodan (www. geodan.nl) and is a member of the Geodan Innovation Board. His expertise is in designing and implementing innovative geospatial IT architectures. At SPINlab, he is a researcher on sensor networks, the Internet of Things and Smart Cities. He also supervises PhD students within the EU funded MULTI-POS (http://www.multi-pos. eu) framework, an international initiative with 17 research institutes and associated commercial partners that addresses challenging research topics in the field of location services and technologies.

Prof. Dr. Henk Scholten is head of the SPatial INformation Laboratory (SPINlab, http://www.feweb.vu.nl/gis/spinlab) of the VU University Amsterdam. The SPINlab is a world-leading research centre for Geographical Information Science and Technology at the Department of Spatial Economics of the VU University Amsterdam. He is also CEO and founder of Geodan (www.geodan.nl), one of the largest European companies specialized in geospatial technology and system integration.

Henk Scholten and Erik van der Zee have recently contributed to the UN Global Geospatial Information Management (GGIM) report on the five- to ten-year vision of future trends in geospatial information management.

References

1. Ali, M., Chandramouli, B., Sethu, B., Katibah, R.: Spatio-temporal stream processing in microsoft streaminsight. Bulletin of the IEEE Computer Society Technical Committee on Data Engineering (2010)
2. Beecham Research: Sector map showing segmentation of the M2M Market. Available at: http://www.beechamresearch.com/article.aspx?id=4 (2013). Accessed 1 Aug 2013
3. Berger, H.E.J.: Flood forecasting for the river Meuse, hydrology for the water management of large river boons. In: Proceedings of the Vienna Symposium. IAHS Publication, Vienna (1991)
4. Botts, M., Percivall, G., Reed, C., Davidson, J.: OGC Sensor Web Enablement: Overview and High Level Architecture, 3rd edn. Open Geospatial Consortium Inc. (2007)
5. Boyd, J: A Discourse on Winning and Losing (1987)
6. Casaleggio Associati: Available at: http://www.slideshare.net/casaleggioassociati/the-evolution-of-internet-of-things (2011). Accessed 1 Aug 2013
7. Dias, E.S.: The Added Value of Contextual Information in Natural Areas: Measuring Impacts of Mobile Environmental Information. Vrije Universiteit, Amsterdam (2007)

8. Dias, E.S., et al.: Adding Value and Improving Processes Using Location-Based Services in Protected Areas. 11, pp. 291–302 (2004)
9. Dibiase, D., et al.: Geographic Information Science & Technology Body of Knowledge. Association of American Geographers (AAG), Washington, DC. ISBN- 10: 0–89291-267-7, ISBN-13: 978-0-89291-267-4 (2006)
10. Esri Inc.: Available at: https://developers.arcgis.com (2013a). Accessed Nov 2013
11. Esri Inc.: Available at: http://webhelp.esri.com (2013b). Accessed Nov 2013
12. European Commission Digital Agenda: Available at: http://ec.europa.eu/digital-agenda/en/ecall-time-saved-lives-saved (2013). Accessed Nov 2013
13. Evans, D.: The Internet of things—how the next evolution of the Internet is changing everything. CISCO Internet Business Solutions Group (IBSG) (2011)
14. Gewin, V.: Mapping opportunities. Nature **427**, 376–77 (2004)
15. Gröger, G., Kolbe, T.H., Nagel, C., Häfele, K.H.: OGC 12-019 City Geography Markup Language (CityGML) Encoding Standard Version: 2.0.0. Open Geospatial Consortium (OGC) (2012)
16. GSMA: Guide to Smart Cities—The Opportunity for Mobile Operators. GSMA, London (2013)
17. Hilbert, M., López, P.: The world's technological capacity to store, communicate, and compute information. Science **332**(6025), 60–65 (2011)
18. Huisman, O., De By, R.A.: Principles of Geographic Information Systems—An Introductory Textbook, Educational Textbook Series. International Institute for Geo-Information Science and Earth Observation (ITC), Enschede, The Netherlands. (2009)
19. IoT SWG: The OGC Forms "Sensor Web for IoT" Standards Working Group. Available at: http://www.opengeospatial.org/node/1650 (2013). Accessed Nov 2013
20. ISO: Reference ISO/IEC FDIS 13249-3:2011(E) Information Technology—Database languages—SQL Multimedia and Application Packages—Part 3: Spatial. ISO, Geneva, Switzerland (2011)
21. ITU: ISBN 978-92-61-14071-7 Measuring the Information Society. International Telecommunication Union, Geneva, Switzerland (2012)
22. Jirka, S., Nüst, D: OGC 10-171 OGC Sensor Instance Registry Discussion Paper. Open Geospatial Consortium Inc. (2010)
23. Klopfer, M., Simonis, I.: SANY—An Open Architecture for Sensor Networks. SANY Consortium (2009)
24. Libelium: 50 Sensor Applications for a Smarter World. Available at: http://www.libelium.com/top_50_iot_sensor_applications_ranking (2013). Accessed 1 Aug 2013
25. Magerkurth, C.: IoT-A (257521) Internet of Things—Architecture IoT-A Deliverable D1.4—Converged Architectural Reference Model for the IoT v2.0. European Commission. Deliverable within the Seventh Framework Programme (2007–2013), (2012)
26. Michael, K., McNamee, A., Michael, M.G.: The Emerging Ethics of Humancentric GPS Tracking and Monitoring. IEEE Xplore, Copenhagen (2006)
27. Mollenkopf, A.: Using ArcGIS GeoEvent Processor for Server to Power Real-Time Applications. Esri International Developer Summit. Esri Inc. (2013)
28. Nebert, D., Whiteside, A., Vretanos, P.: OpenGIS Catalogue Services Specification, vol. 202, 2nd edn. Open Geospatial Consortium Inc. (2007)
29. Ng, C.B., Tay, Y.H., Goi, B.-M.: Vision-based human gender recognition: a survey. In: Anthony, P., Ishizuka, M., Lukose, D. (eds.) PRICAI 2012: Trends in Artificial Intelligence (2012)
30. OGC: OGC PUCK Protocol Standard Version 1.4. Open Geospatial Consortium Inc. (2012)
31. OGC: Homepage. Available at: http://www.opengeospatial.org (2013). Accessed 1 Aug 2013
32. Pals, N., De Vries, A., De Jong, A., Broertjes, E.: Remote and in situ Sensing for Dyke Monitoring—te IJKDIJK experience. GeoSpatial Visual Analytics, pp. 465–475. NATO Science for Peace and Security Series C: Environmental Security (2009)
33. Peters, E., Van der Vliet, P.: Digidijk & Alert Solutions: Real Time Monitoring van Civieltechnische Constructies. (Civiele techniek), pp. 14–16. (2010)
34. Pschorr, J., Henson, C., Patni, H., Sheth, A.: Sensor Discovery on Linked Data Technical Report. Dayton, OH, USA: Knoesis Center, Department of Computer Science and Engineering. http://knoesis.org/library/resource.php?id=851 (2010)

35. Roche, S., Kloeck, K., Ratti, C.: Are 'Smart Cities' Smart Enough? Chap. 12. pp. 215–236. GSDI Association Press, Needham (2012)
36. Satterthwaite, D.: The transition to a predominantly urban world and its underpinnings. Human Settlements Discussion Paper Series 2007, (September, 2007)
37. Schmeisser R.: Location Based Marketing (2011)
38. Scholten, H.J., Van de Velde, R., Van Manen, N.: Geospatial Technology and the Role of Location in Science, 96th edn. Springer, Berlin (2009)
39. Serbanati, A., Medaglia, C.M., Ceipidor, U.B.: Building Blocks of the Internet of Things: State of the Art and Beyond, Chap. 20. In: Turcu, C. (ed.) Deploying RFID—Challenges, Solutions, and Open Issues. INTECH. doi:10.5772/19997 (2011)
40. Sharma, J.: Complex Event Processing, Oracle Spatial and Fusion Middleware products OBIEE Suite, JDeveloper (2011)
41. Smartbin: Available at: http://www.smartbin.com (2013). Accessed Nov 2013
42. United Nations: ST/ESA/SER.A/236 World Population to 2300. United Nations Publication (2004)
43. United Nations: ESA/P/WP/224 World Urbanization Prospects: The 2011 Revision. United Nations Publication (2012)
44. United Nations: ST/ESA/344-E/2013/50/Rev. 1, Sustainable Development Challenges—World Economic and Social Survey 2013. United Nations Publication, ISBN 978-92-1-109167-0, eISBN 978-92-1-056082-5 (2013a)
45. United Nations: D10227 Future Trends in Geospatial Information Management: The Five to Ten Year Vision. United Nations—Global Geospatial Information Management (GGIM) (2013b)
46. Viti, F. et al.: National data warehouse: how the Netherlands is creating a reliable, widespread, accessible data bank for traffic information, monitoring, and road network control. Trans. Res. Record (TRR Journal) **2049**, 176–185 (2008)
47. Wikipedia. Thing. Available at: http://en.wikipedia.org/wiki/Thing (2013a). Accessed 1 Aug 2013
48. Wikipedia: Object. Available at: http://en.wikipedia.org/wiki/Object_(philosophy) (2013b). Accessed 1 Aug 2013
49. Zeimpekis, V., Kourouthanassis, P., Giaglis, G.M.: UNESCO-EOLSS. In: Telecommunication Systems and Technologies, vol. I. UNESCO-EOLSS (2006)

Fog Computing: A Platform for Internet of Things and Analytics

Flavio Bonomi, Rodolfo Milito, Preethi Natarajan and Jiang Zhu

Abstract Internet of Things (IoT) brings more than an explosive proliferation of endpoints. It is disruptive in several ways. In this chapter we examine those disruptions, and propose a hierarchical distributed architecture that extends from the edge of the network to the core nicknamed Fog Computing. In particular, we pay attention to a new dimension that IoT adds to Big Data and Analytics: a massively distributed number of sources at the edge.

1 Introduction

The "pay-as-you-go" Cloud Computing model is an efficient alternative to owning and managing private data centers (DCs) for customers facing Web applications and batch processing. Several factors contribute to the economy of scale of mega DCs: higher predictability of massive aggregation, which allows higher utilization without degrading performance; convenient location that takes advantage of inexpensive power; and lower OPEX achieved through the deployment of homogeneous compute, storage, and networking components.

Cloud computing frees the enterprise and the end user from the specification of many details. This bliss becomes a problem for latency-sensitive applications, which require nodes in the vicinity to meet their delay requirements. An emerging wave of Internet deployments, most notably the Internet of Things (IoTs), requires mobility support and geo-distribution in addition to location awareness and low latency. We argue that a new platform is needed to meet these requirements; a platform we call Fog Computing [1]. We also claim that rather than cannibalizing Cloud Computing,

F. Bonomi · R. Milito · P. Natarajan (✉) · J. Zhu (✉)
Enterprise Networking Labs, Cisco Systems Inc., San Jose, USA
e-mail: prenatar@cisco.com

J. Zhu
e-mail: jiangzhu@cisco.com

N. Bessis and C. Dobre (eds.), *Big Data and Internet of Things:*
A Roadmap for Smart Environments, Studies in Computational Intelligence 546,
DOI: 10.1007/978-3-319-05029-4_7, © Springer International Publishing Switzerland 2014

Fog Computing enables a new breed of applications and services, and that there is a fruitful interplay between the *Cloud* and the *Fog*, particularly when it comes to data management and analytics.

Fog Computing extends the Cloud Computing paradigm to the edge of the network. While Fog and Cloud use the same resources (networking, compute, and storage), and share many of the same mechanisms and attributes (virtualization, multi-tenancy) the extension is a non-trivial one in that there exist some fundamental differences that stem from the Fog raison d'etre. The Fog vision was conceived to address applications and services that do not fit well the paradigm of the Cloud. They include:

- Applications that require very low and predictable latency—the Cloud frees the user from many implementation details, including the precise knowledge of where the computation or storage takes place. This freedom from choice, welcome in many circumstances becomes a liability when latency is at premium (gaming, video conferencing).
- Geo-distributed applications (pipeline monitoring, sensor networks to monitor the environment).
- Fast mobile applications (smart connected vehicle, connected rail).
- Large-scale distributed control systems (smart grid, connected rail, smart traffic light systems).

The emergence of the Internet of Things (IoT) brings a number of use cases of interest that fall in the above categories. It also brings Big Data with a twist: rather than high volume, in IoT big is the number of data sources distributed geographically. For these reasons we place IoT at the center stage of our discussion. We will examine some IoT use cases of interest, and will take a close look at analytics at the edge.

At this point we want to stress that the Fog complements the Cloud, does not substitute it. For the large class of Cloud intended applications the economies of scale (OPEX in particular) cannot be beaten. These efficiencies result from locating mega data centers where energy is inexpensive, and running them with a small team of qualified individuals. For the users (an individual, a small enterprise or even a big business), the flexibility of the pay-as-you-go model is very attractive. Combining these considerations, we state firmly that the Cloud paradigm is here to stay. On the other hand we argue that the model is not universally applicable, and that emergent IoT applications demand a platform with novel characteristics.

The rest of the chapter is organized as follows. Section 2 addresses the question: "what are the disruptive aspects of IoT?" to motivate the need for a hierarchical platform that extends from the edge to the core of the network. Section 3 brings up two use cases, chosen primarily to illustrate some key architectural and system-wide points. IoT brings a new dimension to Big Data and Analytics: massively distributed data sources at the edge. This is the theme of Sect. 4. Section 5 crystallizes a number of observations made in the previous section in some high-level architectural requirements. Section 6 presents an overview of the Fog software architecture, highlighting the essential technology components, followed by conclusions in Sect. 7.

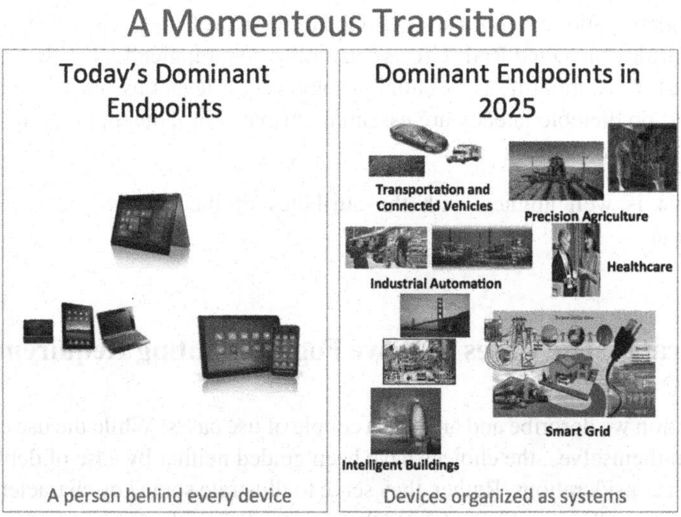

A Momentous Transition

Fig. 1 A momentous transition in IoT

2 What is Disruptive with IoT?

Current estimates place the number of endpoints (mostly smart phones, tablets, and laptops) around three or four billions. This number is expected to grow to a trillion (two to three orders of magnitude!) in a few years. This phenomenal explosion points to a major scalability problem assuming that nothing changes in the nature of the endpoints, the architecture of the platforms at the edge, and the communication model. Some scalability will have to be addressed, but the first critical issue to examine is the disruptive changes that IoT brings to the fore, changes that force to rethink the existing paradigm.

The first change refers the very nature of the endpoints. Today there is a person behind the vast majority of endpoints. The dominant landscape in IoT is different. As depicted in Fig. 1 most endpoints are organized into systems, with the system being a "macro endpoint". For instance, the many sensors and actuators in a Smart Connected Vehicle (SCV) communicate among them, but it is the vehicle as a unit that communicates with other vehicles, with the roadside units (RSUs) and with the Internet at large. The same is valid in the Smart Cities, Oil and Gas, Industrial Automation, etc. The first conclusion from this observation is that IoT forces us to adopt "system view", rather than an "individual view" of the endpoints. A number of consequences derive from this conclusion:

- A vibrant community of individual developers, working on the Android or the Apple platform, contributes today with endpoint applications. This community will not disappear, but as the IoT takes hold, domain expert companies will play an increasingly important role in the development of systems and applications

- A number of important IoT use cases, including Smart Cities, pipeline monitoring, Smart Grid, Connected Rail (CR) are naturally geographically distributed
- Support for fast mobility is essential in some very relevant use cases (SCV, CR)
- Low and predictable latency are essential attributes in industrial automation and other IoT use cases.

This list is well aligned with the attributes of the Fog we outlined in the Introduction.

3 Illustrative Use Cases to Drive Fog Computing Requirements

In this section we describe and analyze a couple of use cases. While the use cases are relevant in themselves, the choices have been guided neither by ease of deployment nor market considerations. Rather, they serve to illustrate some key characteristics of the architecture. We should emphasize that point solutions for each use case, or even for each major vertical do not cut it. Our intent is to abstract the major requirements to propose an architecture that addresses all, or at least the vast majority of the IoT requirements.

3.1 Use Case 1: A Smart Traffic Light System (STLS)

This use case considers a system of a system of traffic lights deployed in an urban scenario. The STLS is a small piece of the full-fledged system envisioned by Smart Connected Vehicle (SCV) and Advanced Transportation Systems, but rich enough to drive home some key requirements for the platform at the edge.

3.1.1 System Outline

- STLS calls for the deployment of a STL at each intersection. The STL is equipped with sensors that (a) measure the distance and speed of approaching vehicles from every direction; (b) detect the presence of pedestrians and cyclists crossing the street. The STL also issues "slow down" warnings to vehicles at risk to crossing in red, and even modifies its own cycle to prevent collisions.
- The STLS has three major goals: (a) accidents prevention; (b) maintenance of a steady flow of traffic (green waves along main roads); (c) collection of relevant data to evaluate and improve the system. Note that the global nature of (b) and (c), in contrast with the localized objective (a). Also note the wide difference in time scales: (a) requires real time (RT) reaction, (b) near-real time, and (c) relates to the collection and analysis of global data over long periods.

- To be specific, consider some numbers. Let us say that the green wave is set at 64 km/h (40 miles/h). A vehicle moving at 64 km/h travels 1.7 m in 100 ms. The policy requires sending an urgent alarm to approaching vehicles when collision with crossing pedestrians is anticipated. To be effective the local control loop subsystem must react within a few milliseconds—thus illustrating the role of the Fog in supporting low latency applications. Accident prevention trumps any other consideration. Hence, to prevent a collision the local subsystem may also modify its own cycle.[1] In doing so it introduces a perturbation in the green wave that affects the whole system. To dampen the effect of the perturbation a re-synchronization signal must be sent along all the traffic lights. This process takes place in a time scale of hundreds of milliseconds to a few seconds.

3.1.2 Key Requirements Driven by the STLS

This section discusses the requirements of a smart traffic light system; the requirements are also highlighted in (Table 1)

1. *Local subsystem latency*: In STLS a subsystem includes the traffic light, sensors and actuators in a local region such that the reaction time is on the order of <10 ms.
2. *Middleware orchestration platform*: The middleware handles a number of critical software components across the whole system, which is deployed across a wide geographical area. The components include:

   - The decision maker (DM), which creates the control policies and pushes them to the individual traffic lights. The DM can be implemented in a centralized, distributed or in a hierarchical way. In the latter, the most likely implementation, nodes with DM functionality of regional scope must coordinate their policies across the whole system. Whatever the implementation, the system should behave as if orchestrated by a single, all knowledgeable DM.
   - The federated message bus, which passes data from the traffic lights to the DM nodes, pushes policies from the DM nodes to the traffic lights, and exchanges information between the traffic lights.

3. *Networking infrastructure*: The Fog nodes belong to a family of modular compute and storage devices. However different the form factors, and the encasings (a ruggedized version is required for the traffic light), they offer common interfaces, and programming environment.
4. *Interplay with the Cloud*: In addition to the actionable real-time (RT) information generated by the sensors, and the near-RT data passed to the DM and exchanged among the set of traffic lights, there are volumes of valuable data collected by the system. This data must be ingested in a data center (DC)/Cloud for deep analytics that extends over time (days, months, even years) and over the covered territory. In a cursory review of the potential value of this data we list: (a) evaluation of

---

[1] The design of a good control policy for all scenarios is certainly a non-trivial task, beyond the scope of this discussion.

Table 1 Attributes of a smart traffic light system

Mobility	
Geo-distribution	Wide (across region) and dense (intersections and ramp accesses)
Low/predictable latency	Tight within the scope of the intersection
Fog-cloud interplay	Data at different time scales (sensors/vehicles at intersection, traffic info at diverse collection points)
Multi-agencies orchestration	Agencies that run the system must coordinate control law policies in real time
Consistency	Getting the traffic landscape demands a degree of consistency between collection points

the impact on traffic (and its consequences for the economy and the environment) of different policies; (b) monitoring of city pollutants; (c) trends and patterns in traffic. We emphasize here the interplay between the Fog and the Cloud, which must operate in mutual support.

5. *Consistency of a highly distributed system*: Visualize the STLS as a highly distributed collector of traffic data over an extended geographically data. Ensuring an acceptable degree of consistency between the different aggregator points is crucial for the implementation of efficient traffic policies.
6. *Multi-tenancy*: The Fog vision anticipates an integrated hardware infrastructure and software platform with the purpose of streamlining and making more efficient the deployment of new services and applications. To run efficiently, the Fog must support multi-tenancy. It must also provide strict service guarantees for mission critical systems such as the STLS, in contrast with softer guarantees for say, infotainment, even when run for the same provider.
7. *Multiplicity of providers*: The system of traffic lights may extend beyond the borders of a single controlling authority. The orchestration of consistent policies involving multiple agencies is a challenge unique to Fog Computing.

3.2 Use Case 2: Wind Farm

A wind farm offers a rich use case of Fog Computing, one that brings up characteristics and requirements shared by a number of Internet of Everything (IoE) deployments:

1. Interplay between real time analytics and batch analytics.
2. Tight interaction between sensors and actuators, in closed control loops.
3. Wide geographical deployment of a large system consistent of a number of autonomous yet coordinated modules—which gives rise to the need of an orchestrator.

3.2.1 System Outline

Modern utility-scale wind turbines are very large flexible structures equipped with several closed control loops aimed at improving wind power capture (measured as the ratio of actual to full capacity output for a given period) and power quality (affected by harmonic distortion) as well as reducing structural loading (hence extending lifetime and decreasing maintenance costs). A large wind farm may consist of hundreds of individual wind turbines, and cover an area of hundreds of square miles.

Power curves [2] depict power [MW] (both wind power and turbine power) versus wind speed [m/s]. There are four typical operating regions:

1. In region 1 the wind speed is so low (say, below 6 m/sec) that the available wind power is in the order of losses in the system, so it is not economically sound to run the turbine.
2. Region 2 (e.g. winds between 6 and 12 m/s) is the normal operating condition, in which blades are positioned to effect the maximum conversion of wind power into electrical power
3. The turbine power curve plateaus when the wind exceeds a certain threshold (say 12 m/s). In this, Region 3, power is limited to avoid exceeding safe electrical and mechanical load limits
4. Very high wind speeds (say, above 25 m/s) characterize Region 4. Here, the turbine is powered down and stopped to avoid excessive operating loads

In short, several controllers are used to tune the turbine (yaw and pitch) to the prevailing wind conditions in order to increase efficiency, and to stop it to minimize wear and prevent damage. These controllers operate in a semi-autonomous way at each turbine. Note, however, that a global coordination at the farm level is required for maximum efficiency. In fact, optimization at the individual level of each turbine may "wind starve" the turbines at the rear.

Central to the process of assessing a potential wind farm deployment is the study of atmospheric stability and wind patterns on a yearly and monthly basis, along with terrain characteristics. An operational wind farm requires fairly accurate wind forecasting at different time scales. We refer the reader to Botterud and Wang [3] for interesting details on the market operation. It suffices for our purposes to consider the following:

- Daily forecast used to submit bids to the independent system operator (ISO)
- Hourly forecast to adjust the commitment to changes in the operating conditions (forced outages, deviations from the forecast loads, etc.)
- Finer granularity (5 min interval) to dynamically optimize the wind farm operation.

3.2.2 Key Requirements Driven by the Wind Farm Use Case

However coarse, the above description touches on a few interesting characteristics of a wind farm operation. These characteristics are shared by a number of cases

Table 2 Attributes from the wind farm use case

Mobility	
Geo-distribution	Wide and dense but confined to the farm
Low/predictable latency	Critical at the turbine level
Fog-cloud interplay	Diverse time scales and sources (weather forecasting and local wind, market conditions and trends)
Multi-agencies orchestration	
Consistency	Applies to the actionable wind data within the farm

emerging under the IoE umbrella. We have a large-scale, geographically distributed system that includes thousands of sensors and actuators. The system consists of a large number of semi-autonomous modules or sub-systems (turbines). Each subsystem is a fairly complex system on its own, with a number of control loops. Established organizing principles of large-scale systems (safety, among others) recommend that each subsystem should be able to operate semi-autonomously, yet in a coordinated manner. This system requires a platform that includes (Table 2):

- *Networking Infrastructure*: An efficient communication network between the subsystems, and between the system and the Internet at large (cloud).
- *Global controller*: A controller with global scope, possibly implemented in a distributed way. The controller builds an overall picture from the information fed from the subsystems, determines a policy, and pushes the policy for each subsystem. The policy is global, but individualized for each subsystem depending on its individual state (location, wind incidence, conditions of the turbine). The continuous supervisory role of the global controller (gathering data, building the global state, determining the policy) requires low latency achievable locally in the edge type deployment we call the Fog.
- *Middleware orchestration platform*: A middleware that mediates between the subsystems and the cloud.
- *Data Analytics*: This system generates huge amounts of data. Much of this information is actionable in real time. It feeds the control loops of the subsystems, and it also used to renegotiate the bidding terms with the ISO when necessary. Beyond this real (network) time applications, the data can be used to run analytics over longer periods (months, years), over wider scenarios (including other wind farms, and other energy data). The cloud is the natural place to run this rich batch analytics.

3.3 Key Attributes of the Fog Platform

The use cases discussed above bring up a number of attributes that differentiate the Fog Computing Platform from the Cloud. Those attributes do not apply uniformly to every use case. Mobility, for instance, a critical attribute in Smart Connected Vehicle and Connected Rail, plays no role in the STLS and Wind Farm cases. Multi-agent orchestration, critical in STLS, does not apply in the Wind Farm case. This is another facet of the heterogeneity of the Fog. The idea is to deploy a common platform that supports a wide range of verticals, rather than point solutions for each vertical. However, different verticals may exercise different attributes of the platform.

Along these lines is worth analyzing the concept of multi-tenancy. Both Cloud and Fog support multi-tenancy, i.e., the support of a multiplicity of client-organizations without mutual interference. There are subtle differences, though, in the nature of those client-organizations in the Cloud and Fog environments.

Typical users of the Cloud include individuals, and enterprises. Some of the latter off-load completely their IT to the Cloud; many overflow during periods of peak demand. The pay-as-you-go, task-oriented model dominates. The Fog platform also supports task-oriented compute and storage requests, but the dominant mode is different. In the two use cases described the operations run 24×7 every day of the year, with possible periods of fluctuations in the demand. Similar models apply to many other verticals and use cases. In fact the IoT landscape brings a new set of actors and novel interactions among them: the owner of the infrastructure, the agency or agencies that run the service, the users of the service. Business models that determine the different levels of responsibility and the actual exchange of service and money among the participants will structure these interactions. Development of these business models goes beyond the scope of this chapter. It is worth, though, to revisit the STLS use case to get insight on the issues involved. One or more agencies run the system based on a platform owned by a service provider. Other services (a pollution monitoring system, for example), run by different agencies coexist on the same platform. Note that in this example the concept of the end user (ex: pedestrian crossing the street, approaching vehicles) is diluted, as is the way to pay for the service.

4 Geo-distribution: A New Dimension of Big Data

Big Data today is currently characterized along three dimensions: Volume, Velocity, and Variety. As argued in Sect. 2, Many IoT use cases, including STLS, Smart Cities, Smart Grids, Connected Rail (CR), and pipeline monitoring are naturally distributed. This observation suggests adding a fourth dimension to the characterization of Big Data, namely, *geo-distribution*. Consider, for instance, the monitoring of a pipeline, or the measurement of pollution level (air quality, pollen, and noise) throughout a city. The big N in these scenarios is neither the number of terabytes nor rate of data generated by any individual sensor, but rather the number of sensors (and actuators) that are *naturally distributed*, and that has to be managed as a coherent whole. The call for "moving the processing to the data" is getting louder. There is a need for *distributed intelligent platform at the Edge (Fog Computing)* that manages distributed compute, networking, and storage resources.

In short, IoT/IoE transforms the landscape more dramatically than the image of zillions of sensors suggests. Together with a number of challenges, IoT/IoE brings

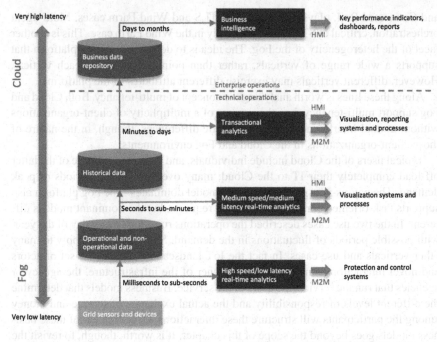

Note: HMI = Human to machine; M2M = Machine to machine

Fig. 2 Many uses of the same data [5]

new opportunities: this is a game that requires (a) a solid presence from the edge to the core that includes networking, compute and storage resources; and (b) the creation of an ecology of domain expert partners to attack these nascent market successfully.

In many use cases there is a rich interplay between the edge and the core of the network, because the data generated has different requirements and uses at different time scales. This topic is further discussed next.

4.1 Interplay Between the Edge (Fog) and the Core (Cloud)

Many use cases of interest actually demand an active cooperation between the Edge and the Core. The scope of the data processed, narrow in space and time at the edge, and wide at the core, is typical in many verticals of interest (Smart Connected Vehicle, Oil and Gas, Connected Rail, Smart Communities, etc.). It suggests a hierarchical organization of the networking, compute, and storage resources.

To illustrate the point consider an example taken from Smart Grid, as depicted in Fig. 2. The figure shows a whole range of time-scales, from milliseconds to months. The machine-to-machine (M2M) interactions at the millisecond-sub second

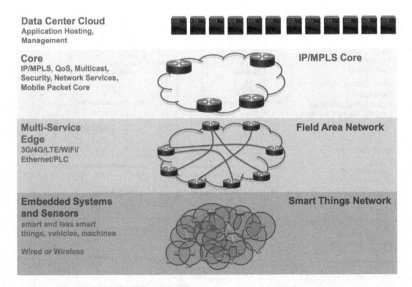

Fig. 3 Fog distributed infrastructure for IoT/IoE

level cover the real-time actionable analytics (breakers, etc.). The second level (second-to-sub minute) includes both M2M processes and human-to-machine (H2M) interactions (visualization, starting/stopping home appliances). Not reflected in the figure is the fact as we climb up the time scale the scope gets wider in coverage. For instance, the business intelligence level that operates on data collected over months may cover a wide geographical region, possibly a whole country and beyond. In contrast, the sub-second level refers to more localized loops aimed essentially to keep the grid stable.

5 High-Level Architectural and System-Wide View

Cloud Computing consists of mostly homogenous physical resources that are deployed and managed in a centralized fashion. Fog complements and extends the Cloud to the edge and endpoints; Fog's distributed infrastructure comprising of heterogeneous resources needs to be managed in a distributed fashion. Figure 3 shows the various players in the distributed Fog infrastructure, ranging from data centers, core of the network, edge of the network, and end points. The Fog architecture enables distributed deployment of applications requiring computing, storage and networking resources spanning across these different players.

Similar to Cloud, Fog architecture supports co-existence of applications belonging to different tenants. Each tenant perceives its resources as dedicated, and defines its own topology. Figure 4 shows two active tenants, A and B, with their respective applications. The distributed applications for A has one Cloud component, and two Fog components. The application for B has one cloud component, one component

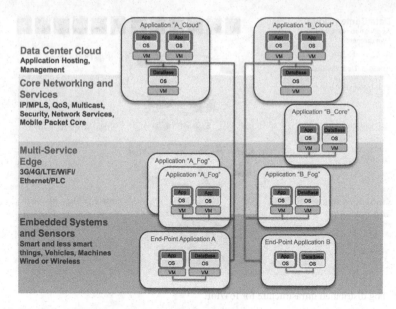

Fig. 4 Distributed IoT/IoE applications on the fog infrastructure

in the Core, and a Fog component. A virtual network topology is allocated for each tenant.

In order to fully manifest this distributed architecture, Fog relies on technology components for scalable virtualization of the key resource classes:

- Computing, requiring the selection of hypervisors in order to virtualize both the computing and I/O resources.
- Storage, requiring a Virtual File System and a Virtual Block and/or Object Store.
- Networking, requiring the appropriate Network Virtualization Infrastructure (e.g., Software Defined Networking technology).

Similar to Cloud, Fog leverages a policy-based orchestration and provisioning mechanism on top of the resource virtualization layer for scalable and automatic resource management. Finally, Fog architecture exposes APIs for application development and deployment.

6 Software Architecture

The use cases and the requirements discussed in previous sections help nail down the following key objectives of the Fog software architecture:

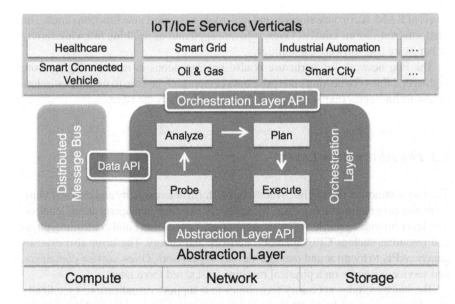

Fig. 5 Components in fog architecture

- Fog nodes are heterogeneous in nature and deployed in variety of environments including core, edge, access networks and endpoints. The Fog architecture should facilitate seamless resource management across the diverse set of platforms.
- The Fog platform hosts diverse set of applications belonging to various verticals— smart connected vehicles to smart cities, oil and gas, smart grid etc. Fog architecture should expose generic APIs that can be used by the diverse set of applications to leverage Fog platform.
- The Fog platform should provide necessary means for distributed policy-based orchestration, resulting in scalable management of individual subsystems and the overall service.

The rest of this section discusses the components of Fog architecture (Fig. 5), explaining in detail how the above objectives are met.

6.1 Heterogeneous Physical Resources

As discussed earlier, Fog nodes are heterogeneous in nature. They range from high-end servers, edge routers, access points, set-top boxes, and even end devices such as vehicles, sensors, mobile phones etc.[2] The different hardware platforms have varying

---

[2] Consistently with the "system view" of IoT we regard a Connected Vehicle, a manufacturing cell, etc. as end devices.

levels of RAM, secondary storage, and real estate to support new functionalities. The platforms run various kinds of OSes, software applications resulting in a wide variety of hardware and software capabilities.

The Fog network infrastructure is also heterogeneous in nature, ranging from high-speed links connecting enterprise data centers and the core to multiple wireless access technologies (ex: 3G/4G, LTE, WiFi etc.) towards the edge.

6.2 Fog Abstraction Layer

The Fog abstraction layer (Fig. 5) hides the platform heterogeneity and exposes a uniform and programmable interface for seamless resource management and control. The layer provides generic APIs for monitoring, provisioning and controlling physical resources such as CPU, memory, network and energy. The layer also exposes generic APIs to monitor and manage various hypervisors, OSes, service containers, and service instances on a physical machine (discussed more later).

The layer includes necessary techniques that support virtualization, specifically the ability to run multiple OSes or service containers on a physical machine to improve resource utilization. Virtualization enables the abstraction layer to support multi-tenancy. The layer exposes generic APIs to specify security, privacy and isolation policies for OSes or containers belonging to different tenants on the same physical machine. Specifically, the following multi-tenancy features are supported:

- Data and resource isolation guarantees for the different tenants on the same physical infrastructure
- The capabilities to inflict no collateral damage to the different parties at the minimum
- Expose a single, consistent model across physical machine to provide these isolation services
- The abstraction layer exposes both the physical and the logical (per-tenant) network to administrators, and the resource usage per-tenant.

6.3 Fog Service Orchestration Layer

The service orchestration layer provides dynamic, policy-based life-cycle management of Fog services. The orchestration functionality is as distributed as the underlying Fog infrastructure and services. Managing services on a large volume of Fog nodes with a wide range of capabilities is achieved with the following technology and components:

- A software agent, *Foglet*, with reasonably small footprint yet capable of bearing the orchestration functionality and performance requirements that could be embedded in various edge devices.

- A distributed, persistent storage to store policies and resource meta-data (capability, performance, etc) that support high transaction rate update and retrieval.
- A scalable messaging bus to carry control messages for service orchestration and resource management.
- A distributed policy engine with a single global view and local enforcement.

6.3.1 Foglet Software Agent

The distributed Fog orchestration framework consists of several Foglet software agents, one running on every node in the Fog platform. The Foglet agent uses abstraction layer APIs to monitor the health and state associated with the physical machine and services deployed on the machine. This information is both locally analyzed and also pushed to the distributed storage for global processing.

Foglet is also responsible for performing life-cycle management activities such as standing up/down guest OSes, service containers, and provisioning and tearing down service instances etc. Thus, Foglet's interactions on a Fog node span over a range of entities starting from the physical machine, hypervisor, guest OSes, service containers, and service instances. Each of these entities implements the necessary functions for programmatic management and control; Foglet invokes these functions via the abstraction layer APIs.

6.3.2 Distributed Database

A distributed database, while complex to implement is ideal for increasing Fog's scalability and fault-tolerance. The distributed database provides faster (than centralized) storage and retrieval of data. The database is used to store both application data and necessary meta-data to aid in Fog service orchestration. Sample meta-data include (discussed more in the next subsection):

- Fog node's hardware and software capabilities to enable service instantiation on a platform with matching capabilities.
- Health and other state information of Fog nodes and running service instances for load balancing, and generating performance reports.
- Business policies that should be enforced throughout a service's life cycle such as those related to security, configuration etc.

6.3.3 Policy-Based Service Orchestration

The orchestration framework provides policy-based service routing, i.e., routes an incoming service request to the appropriate service instance that confirms to the relevant business policies. The framework achieves this with the help of the policy

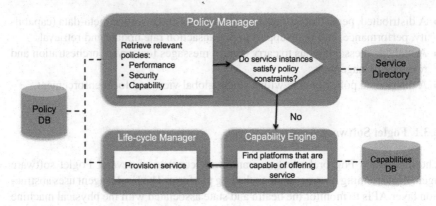

Fig. 6 Policy-based orchestration framework

manager (Fig. 6). This section discusses the workflow associated with policy-based orchestration.

Administrators interact with the orchestration framework via an intuitive dashboard-style user interface (UI). Admins enter business policies, manage, and monitor the Fog platform through this UI. The UI offers policy templates that admins can refine based on needs. The policy framework is extensible and supports a wide variety of policies. Few example policies include:

- Policies to specify thresholds for load balancing such as maximum number of users, connections, CPU load etc.
- Policies to specify QoS requirements (network, storage, compute) with a service such as minimum delay, maximum rate etc.
- Policies to configure device, service instance in a specific setting.
- Policies to associate power management capabilities with a tenant/Fog platform.
- Policies to specify security, isolation and privacy during multi-tenancy.
- Policies that specify how and what services must be chained before delivery, ex: firewall before video service.

Business policies specified via the UI are pushed to a distributed policy database (Fig. 6). The policy manager is triggered by an incoming service request. The policy manager gathers relevant policies i.e., those pertaining to the service, subscriber, tenant etc. from the policy repository. The policy manager also retrieves meta-data about currently active service instances from the services directory. With these two sets of data, the policy manager tries to find an active service instance that satisfies the policy constraints, and forwards the service request to that instance. If no such instance is available, then a new instance must be created. For that purpose, the policy manager invokes the capability engine whose job is to identify a ranked list of Fog nodes whose capabilities match the policy constraints for instantiating the new service. The capability engine hands over this ranked list to the life cycle manager that provisions the service on a Fog device. The life cycle manager may reach out to the policy repository to identify device, service, and network configuration policies while provisioning the new instance.

The orchestration functionality is distributed across the Fog deployment such that the logic in Fig. 6 is embedded in every Foglet. The distributed control provides better resiliency, scalability, and faster orchestration for geographically distributed deployments.

6.4 North-Bound APIs for Applications

The Fog software framework exposes northbound APIs that applications use to effectively leverage the Fog platform. These APIs are broadly classified into data and control APIs. Data APIs allow an application to leverage the Fog distributed data store. Control APIs allow an application to specify how the application should be deployed on the Fog platform.

Few example APIs:

- Put_data(): To store/update application-specific data and meta-data on the Fog distributed data store.
- Get_data(): To retrieve application-specific data meta-data from the Fog distributed data store.
- Request_service(): To request for a service instance that matches some criteria.
- Setup_service(): To setup a new service instance that matches some criteria.
- Install_policy(): To install specific set of policies for a provider, subscriber in the orchestration framework.
- Update_policy (): To configure/re-configure a policy with a specific set of parameters (ex: thresholds for a load balancing policy).
- Get_stats(): To generate reports of Fog node health and other status.

7 Conclusions

This chapter looked at Fog Computing, a hierarchical and distributed platform for service delivery consisting of compute, storage, and network resources. We examined key aspects of Fog computing, and how Fog complements and extends Cloud computing. We looked at use cases that motivated the need for Fog, emphasizing Fog's relevance to several verticals within IoT and Big Data space. We also provided a high-level description of Fog's software architecture, highlighting the different technology components necessary to achieve the Fog vision.

Acknowledgments This work would not have been possible without the support of our colleagues, mentioned here in alphabetical order: Hao Hu, Mythili S. Prabhu, and Rui Vaz.

References

1. Bonomi, F., Milito, R., Zhu, J., Addepalli, S.: Fog computing and its role in the internet of things. In: Proceedings of the 1st edn. of the MCC workshop on Mobile cloud computing (2012)
2. Pao, L., Johnson, K.: A tutorial on the dynamics and control of wind turbines and wind farms. In: American Control Conference (2009)
3. Botterud, A., Wang, J.: Wind power forecasting and electricity market operations. In: International Conference of 32nd International Association for Energy Economics (IAEE), San Francisco, CA (2009)
4. Cristea, V., Dobre, C., Pop, F.: Context-aware environ internet of things. Internet of Things and Inter-cooperative Computational Technologies for Collective Intelligence Studies in Computational Intelligence, vol. 460, pp. 25–49 (2013)
5. Haak, D.: Achieving high performance in smart grid data management. White paper from Accenture (2010)

Part II
Advanced Models and Architectures

Big Data Metadata Management in Smart Grids

Trinh Hoang Nguyen, Vimala Nunavath and Andreas Prinz

Abstract Smart home, smart grids, smart museum, smart cities, etc. are making the vision for living in smart environments come true. These smart environments are built based upon the Internet of Things paradigm where many devices and applications are involved. In these environments, data are collected from various sources in diverse formats. The data are then processed by different intelligent systems with the purpose of providing efficient system planning, power delivery, and customer operations. Even though there are known technologies for most of these smart environments, putting them together to make intelligent and context-aware systems is not an easy task. The reason is that there are semantic inconsistencies between applications and systems. These inconsistencies can be solved by using metadata. This chapter presents management of big data metadata in smart grids. Three important issues in managing and solutions to overcome them are discussed. As a part of future grids, some concrete examples from the offshore wind energy are used to demonstrate the solutions.

1 Introduction

Advanced technologies are making the vision for living in smart environments become realistic. Recently, several concepts within the smart environments have been introduced, such as smart home, smart transport, smart grids, smart museum,

T. H. Nguyen (✉) · V. Nunavath · A. Prinz
Department of Information and Communication Technology, University of Agder,
Grimstad, Norway
e-mail: trinh.h.nguyen@uia.no

V. Nunavath
e-mail: vimala.nunavath@uia.no

A. Prinz
e-mail: andreas.prinz@uia.no

N. Bessis and C. Dobre (eds.), *Big Data and Internet of Things:*
A Roadmap for Smart Environments, Studies in Computational Intelligence 546,
DOI: 10.1007/978-3-319-05029-4_8, © Springer International Publishing Switzerland 2014

and smart cities. These smart environments are built based upon the Internet of Things (IoT) paradigm where lots of devices, sensors, appliances are connected through the Internet. These devices produce vast amounts of data, thus making the management of data a highly challenging task. Another common feature and an important problem of these smart environments is that each of them involves data modeling, information analysis, integration and optimization of large amounts of data coming from various smart appliances in diverse formats. The data are then processed by different intelligent systems with the purpose of providing efficient system planning, power delivery, and customer operations. Even though there are known technologies for developing most of these smart environments, putting them together to make intelligent and context-aware systems is not an easy task. The reason is that there are semantic inconsistencies between applications and systems. These inconsistencies can be solved by using metadata.

Typically, data are a collection of raw and unorganized symbols that represent real-world states. The information is the processed, organized, and structured data according to a given context [1, 60]. The context of related data and processes will decide the role as information of the captured data. Principally, information is the structured data with semantics. For example, if data are used for documentation or analysis, the data become information. Without metadata, the data cannot easily become information and incomplete or inaccurate metadata or too much metadata can cause misinterpretation of data [55]. Metadata should be therefore managed in a way that data can be easily interpreted and transformed to information.

Metadata management is a key to make data integration successful [25]. It has to be taken into consideration in the development of systems since it helps in making the systems scalable. For formal metadata management, semantic technologies have been developed. Ontology, which is a part of semantic technologies, plays a significant role in managing metadata of a domain. Ontologies can be used to support data integration in terms of facilitating knowledge sharing and data exchange between participants in a domain. In ontologies, concepts, properties, relations, functions, constraints, and axioms of a particular domain are explicitly defined [18]. We use semantic technologies to exploit the semantics of data, and hence ease metadata handling in smart environments.

In this chapter, we discuss how to manage big data metadata in smart grids with a particular focus on (1) knowledge sharing and data exchange, (2) derived data from relations between concepts, and (3) data quality as metadata. We will present a developed ontology model for offshore wind energy metadata management as an example of domain concept descriptions. IEEE P2030 points out that ontology might be a good option to create formal representation of real-world systems or objects composing these systems within smart grids [24]. As the number of devices is increasing tremendously, and many of them will be used in smart environments, it is important to make sure that any future system is scalable enough to keep pace with the technologies. Metadata models, as a backbone of any system, also need to be considered thoughtfully. The models need to be developed so that the following requirements are fulfilled.

- The models need to be compatible with existing data resources and future applications.
- Minimum effort is used to modify the models when integrating new devices.
- New devices' metadata are described in a way that discovery and access to them are easy.
- It must provide a guide to structuring, sharing, storing, and representing the big data in smart grids.
- The semantics of data needs to be exploited and clearly defined.
- Since it is not feasible to attach metadata with individual data, the metadata models must be related to data sources.

The rest of the chapter is organized as follows. Section 2 gives some background information about the areas that we discuss in this work. Section 3 presents some challenges of big data metadata management that we attempt to tackle. Section 4 describes solutions and approaches to overcoming the challenges. Section 5 discusses our solutions and gives some remarks on future work. Finally, Sect. 6 concludes the chapter.

2 Background

This section describes the background of metadata, semantic technologies, IoT and smart grids. The relations between these areas are also highlighted.

2.1 Metadata

The term "metadata" was first introduced in 1968 by Philip R. Bagley to refer to descriptive data that provided information about other data in a database environment [51]. In different contexts, the term metadata is interpreted in different ways, for example, metadata are data about data; or metadata are machine-readable information about electronic resources or other things; or metadata are structured information that describes, explains, locates an information resource [54]. Basically, metadata are descriptors that describe a way of identifying information. Data without metadata result in blind decision making [55]. In other words, without metadata, data have no identifiable meaning. For instance, when a user searches for information, he will receive a list of search results from a search engine. The search engine looks up for requested information from huge amounts of data based on search terms, tagging content, and other metadata associated with data. Metadata provide the necessary documentation for users by answering who, what, when, where, why, and how questions upon the users' requests.

Metadata put data into a context so that the data can be understood by users and become information. Besides the general role as descriptors, metadata can be used for:

- information classification—information is classified into different categories based on content, purpose, location, area, etc;
- information discovery—a large amount of time is used to look for things, and many of them cannot be found due to the lack of descriptions. Metadata therefore enhance information discovery and knowledge sharing;
- information interpretation—a poor description of data may lead to wrong decision making or business loss due to wrong interpretation of the data;
- data integration—when we integrate data from various sources in different formats and platforms, metadata are the only option that can make a foundation for data integration [55];
- device discovery—based on metadata of devices such as location, type, and other features devices can be discovered either automatically or semi-automatically by a system.

2.2 Big Data Metadata Management

Big data is characterized with volume, variety and velocity [61]. Volume is considered as a huge amount of data which can hold terabytes to petabytes of data which come from different devices, applications, and systems. Velocity is the speed at which the data comes in, and variety means many data types and data formats. Structured, semi-structured and unstructured data are involved in big data [14]. Data often come from machines, sensors, social networks such as Facebook, Tweets, smart phones and other cell phones, GPS devices and other sources making it complex to manage [45]. According to a report from McKinsey Global Institute, every year, over 30 billion original documents with data are created. 85 % of the data will never be retrieved, 50 % of the data is duplicates, and 60 % of stored documents are obsolete. $1 and $10 are the costs to create a document and to manage it, respectively [31]. As the amount of data increases, the cost of management also increases. It is important to describe and manage metadata so that only important and necessary data are stored and provided to users when requested. Since data are used for making decisions by different applications and systems, the quality of data is one of concerns.

Not all of the data captured from sensors or devices are useful, only a part of the data is. Data are transformed to information only if the data are used for particular purposes, e.g., modeling, documentation. Part of the information will become knowledge in terms of abstraction and perception. Users are not interested in information (numbers), they are interested in knowledge, i.e., what can be derived from the information. For example, if a user wants to know about the temperature in a wind turbine hub, he will probably not expect to get a number or set of numbers as a response, but he will probably want to get either "Normal", "Cold", or "Hot".

Eventually, only part of the knowledge will be transformed to wisdom if the knowledge is used to serve some actionable intelligence [46]. Every step of the transformation involves management of data, information, and knowledge. Management of big data metadata concerns a way to manage big data metadata such that metadata are good enough to enable knowledge extraction from big data.

2.3 Smart Grids and Internet of Things

Smart grids are the future generation of power grids where the energy is managed in a way that both consumers and energy producers will get more benefits from the grid in terms of reduction of expenditure on energy and reduction of carbon emissions. Indeed, it enables consumers to utilize lower tariff charges during off-peak periods and energy producers to react efficiently during peak periods. Smart grids are also used to effectively response to the fluctuations of renewable sources such as wind and solar when they are integrated in a power grid.

A smart grid is an electricity network that efficiently delivers sustainable, economic, and secure electricity supplies by intelligently integrating the actions of all users connected to it, including generators, consumers and those that do both [15]. On the consumer side, smart grids involve many smart meters and smart appliances, for example, smart washing machines, and dishwashers. The number of smart appliances is increasing dramatically. These devices are connected directly to the Internet. A large amount of sensors are used in these devices to make sure that every single change can be detected, managed and controlled. On the energy provider side, intelligent applications are used to maintain balance between demand and supply. Smart grids will bring the decision making gradually from a centralized level to local and finally to automatic.

In order to make a grid become smart, different technologies and applications are involved, e.g., advanced metering infrastructure (AMI), distribution management system (DMS), geographic information system (GIS), outage management systems (OMSs), intelligent electronics devices (IEDs), wide-area measurement systems (WAMS), and energy management systems (EMSs) [11]. These systems are driven effectively by IoT [56].

In IoT, things are connected in such a way that machines and applications can understand our surrounding environments better and therefore make intelligent decisions and respond to the dynamics of the environments effectively [5]. These things communicate to each other over the Internet. Advantages of IoT will contribute a lot to the effort of making smart grids in terms of real-time monitoring and control. Smart grid applications require quick response time no matter how big the data are. One example of such a system is an energy trading system which allows energy consumers or third parties to bid for energy prices in advance [12].

Due to characteristics of smart grids, a number of challenges are encompassed with the development of smart grids such as support heterogeneous participants, flexible data schema (e.g., add new or remove old appliances), complex event processing,

Fig. 1 An example of conceptual model for smart grid communication

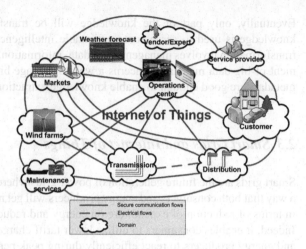

privacy and security [57]. Thus, data from IoT alone are not enough. The data must be used together with the domain knowledge, machine interpretable metadata, services, etc. to become useful.

Figure 1 illustrates a conceptual model for smart grid communication with a focus on offshore wind as an energy generator. The model is based on the *Smart Grid Interoperability Panel* promoted by the National Institute of Standards and Technology (NIST) [38].

Each domain is a high-level grouping of organizations, individuals, and systems of the offshore wind industry. Communication between stakeholders in the same domain may have similar characteristics and requirements. The communication flows are bidirectional. In this model, smart meters, smart appliances are installed at households, sensors are embedded on wind turbines, and intelligent programs are used at operations center.

Metadata are significant in the smart grid context. It is needed for organizing and interpreting data coming from energy market, service providers, customers, power grid, and power generators. Managing metadata in such a varied environment is a challenging task.

2.4 Semantic Technologies

Semantic technologies have been developed to make metadata understandable by a machine. Ontology is a part of semantic technologies that plays a significant role in managing metadata of a domain. There are several ontology languages such as SHOE, OIL, DAML-ONT, DAML+OIL, and OWL [20, 30]. Web Ontology Language (OWL), a language proposed by World Wide Web Consortium (W3C) Web Ontology Working Group, is being used intensively by research communities as well as industries. Ontologies can be represented by using Resource Description Framework (RDF)/RDFS (RDF Schema). However, a number of other features are missing in RDFS such as cardinality restrictions, logical combinations (intersections, unions or complements), and disjointness of classes. Let us examine some concrete cases within the offshore wind energy. The first case is that in RDF, we cannot state that *HydraulicSystem* and *HeatingSystem* are disjoint classes. The second case concerns the lack of cardinality restrictions, e.g., the fact that a wind power plant (WPP) can have more than one wind turbine converter component (WCNV) cannot be expressed in RDF, but it can be done in OWL using the following axiom $WPP \sqsubseteq (\geq 1\ hasWPP\text{-}Component.WCNV)$. OWL is an extension of RDFS, in the sense that OWL uses the RDF meaning of classes and properties [2, 7, 20]. The design of OWL was influenced by its predecessors DAML+OIL, the frames paradigm and RDF [20].

In OWL, Owl:Thing is a built-in most general class and is the class of all individuals. It is a superclass of all OWL classes. Classes are defined using owl:Class. A class defines a group of individuals that belong together. Individuals are also known as instances. Individuals can be referred to as being instances of classes. Note that the word concept is sometimes used in place of class. Classes are a concrete representation of concepts. Owl:Nothing is a built-in most specific class and is the class that has no instances. It is a subclass of all OWL classes. There are two types of properties in OWL ontology, they are object property and data type property. Properties in OWL are also known as roles in description logics and relations in Unified Modeling Language (UML). An object property relates individuals to other individuals (e.g., *hasWPP-Component* relates *WPP* to *WPP components*). An object property is defined as an instance of the built-in OWL class owl:ObjectProperty. A data type property relates individuals to data type values (e.g., *hasOilPressure*, *hasWindSpeed*). A datatype property is defined as an instance of the built-in OWL class owl:DatatypeProperty. A property in OWL can be transitive, functional, symmetric, or inverse.

OWL DL (DL stands for "Description Logic") is a variant of OWL. It was developed to support existing DL and to provide a possibility of working with reasoning systems. In this work, OWL DL is used to develop ontologies. The OWL DL semantics is very similar to the $\mathcal{SHOIN}^{(\mathcal{D})}$ Description Logic. It provides maximum expressiveness and it is decidable [20]. OWL DL abstract syntax and semantics can be found in [41].

2.5 Ontology Reasoning and Querying

A reasoner is a piece of software that is able to infer logical consequences from a set of asserted facts or axioms. It is used to ensure the quality of ontologies. It can be used to test whether concepts are non-contradictory and to derive implied relations. Reasoning with inconsistent ontologies may lead to erroneous conclusions [3]. There are some existing DL reasoners such as FaCT, FaCT++, RACER, DLP and Pellet. A reasoner has the following features: satisfiability, consistency, classification, and realization checking [49]. Given an assertional box $\mathcal{A}$ (ABox contains assertions about individuals), we can reason w.r.t a terminological box T(TBox contains axioms about classes) about the following:

- Consistency checking: ensures that an ontology does not contain any contradictory facts. An ABox $\mathcal{A}$ is consistent with respect to T if there is an interpretation I which is a model of both $\mathcal{A}$ and T.
- Concept satisfiability: checks if it is possible for a class to have any instances. Given a concept C and an instance a, check whether a belongs to C. $\mathcal{A} \models C(a)$ if every interpretation that satisfies $\mathcal{A}$ also satisfies $C(a)$.
- Classification: computes the subclass relations between all named classes to create the complete class hierarchy. Given a concept C, retrieve all the instances a which satisfy C.
- Realization: computes the direct types for each of the individuals. Given a set of concepts and an individual a, find the most specific concept(s) C (w.r.t. subsumption ordering) such that $\mathcal{A} \models C(a)$.

For relational database (RDB), Structured Query Language (SQL) is the query language of choice. But for ontologies, SPARQL and SQWRL (Semantic Query-Enhanced Web Rule Language) [39] are used to build queries. SPARQL is an RDF query language and SQWRL is a SWRL-based language for querying OWL ontologies. SPARQL extensions such as SPARQL-DL [48] and SPARQL-OWL [27] can be used as OWL query languages in many applications. But SPARQL cannot directly query entailments made using OWL constructs since it has no native understanding of OWL [39].

3 Challenges in Managing Big Data Metadata in Smart Grids

There are a number of challenges associated with management of big data metadata such as metadata quality, metadata provenance, semantics, and metadata alignment. In this section, we attempt to tackle three challenges in managing smart grids' big data metadata.

3.1 Knowledge Sharing and Information Exchange

In a diverse environment such as smart grids, meters, appliances, and applications are developed by different companies and vendors. Many of them use their own proprietary data formats, protocols, and platforms, thus data exchange is impeded. Using approved standards would contribute to solving such problems since they can make the data exchange unambiguous. The standards can be seen as a means of interoperability, a dictionary of data that can be used to manage, simplify, and optimize data models [9]. However, there are some problematic issues related to existing international standards for data exchange. For instance, it takes some years to approve a standard internationally, but it seems that new technologies are proposed every year. As a result, novel concepts and terms are introduced, but they are not immediately described in these international standards.

The lack of widely accepted standards prevents the interoperability between smart devices, applications, smart meters, and renewable sources [47]. The Institute of Electrical and Electronics Engineers (IEEE), and NIST have recommended a list of standards that should be considered while developing smart grids [24, 38]. These standards have been developed by different working groups, leading to a lack of harmonizations. Although these standards describe different parts of smart grids, they share a common feature, i.e., the smart grid concepts. The question here is how to structure the domain concepts such that semantics is exploited effectively, knowledge sharing and data exchange are eased, and new concepts are updated in knowledge bases timely.

3.2 Relations Between Concepts

Ontologies can be used to support data integration in terms of facilitating knowledge sharing and data exchange between participants in a domain. Ontologies describe the relations between concepts and their properties. These relations are metadata since relations can lead to computability of derived data. This opens several possible paths for calculation and gives users the possibility of selecting the most suitable one. However, there is a lack of a formal description of such relations in ontologies. One important question in managing metadata in ontologies is how to handle relations so that the selection of data (independent of type of data: base or derived data) can be done at runtime depending on the actual situation.

3.3 Data Quality

It is normal to use more than one sensor to measure, e.g., pressure or temperature at a particular point. The quality of each sensor is different from the others and depends on the conditions. In offshore wind energy, a couple of sensors are deployed on a windmill and they frequently measure and deliver the data to the users and applications by means of services. As sensors are prone to failures their results might be inaccurate, incomplete, and inconsistent [50]. Therefore, the data quality should be handled in such a way that users and applications can specify the desired quality level of the data. Only when the data source has the requested quality descriptions it would be used for further processing. One of the issues related to data quality is the handling of data quality at user level in enterprise applications where there is a potentially large number of data sources with quality information. Another issue is that sometimes none of the available data sources has the required quality information.

In this case, how a system should respond to such a request should be considered and a way to provide requested data to users should be investigated.

4 Solutions

This section presents three solutions and approaches to overcoming the challenges described in Sect. 3. Smart grids involve vast amounts of data from consumers, generators, billing, and management. Here, we use a case in which offshore wind energy plays a role as a renewable energy source generator to demonstrate our points.

4.1 Semantic-enhanced Concept Modeling

This section discusses the solution to the challenge described in Sect. 3.1. We look into how the semantic technologies can help us to solve the challenges with taking into consideration the requirements for the developed metadata models presented in Sect. 1.

4.1.1 The Information Model

An information model plays an important role in building a smart grid. It not only provides a common basis for understanding the general behavior of smart grid communication, but also facilitates the collaboration process between smart grid stakeholders due to shared concepts with a common semantics. An example of sharing common concepts between partners of offshore wind energy is illustrated in Fig. 2.

Availability and reliability of data are significant for any systems and partners. Offshore wind partners can efficiently perform their work using the available data. For example, wind speed information is the input to a wind speed prediction program. The output from the program can be used with the generator speed to predict the availability of wind power in the next few hours. In order to optimize wind farm efficiency, wind farm operations information regarding wind direction, active power, status of blades, etc. is needed. The weather forecast and energy market information is used to manage wind power production as well as maintenance for wind turbines (e.g., a wind turbine can be stopped when consumer demand is low). An information model is developed based on the IEC 61400-25 standard [23] to keep pace with the continual introduction of new technologies. More details about the information model can be found in [37].

Fig. 2 Data pie chart for the offshore wind industry [36]

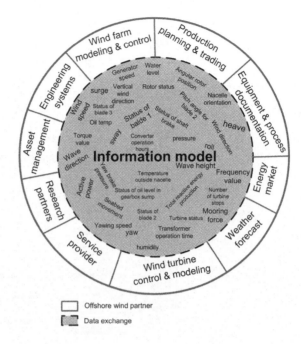

- ☐ Offshore wind partner
- ▨ Data exchange

4.1.2 An Offshore Wind Ontology

An information model represents the knowledge concerning specific domain communication. In particular, the purpose of creating an offshore wind information model is to facilitate the process of agreement on data exchange models as well as collaborations among offshore wind partners. We use the developed information model to build an offshore wind ontology (OWO) as depicted in Fig. 3. The idea of creating OWO from the terminologies is to share, reuse knowledge, and reason about behaviors across a domain and task. It is also a key instrument in developing the semantic web in which information is given well-defined meaning, better enabling computers and people to work in cooperation [8]. An ontology helps to make an abstract model of a phenomenon by identifying the relevant concepts of that phenomenon [53].

Suppose several different sources/data storages contain wind turbine information. If these sources share and publish the same underlying ontology of the terms they all use, then computer agents can extract and aggregate information from these different sources. The agents can use this aggregated information to answer user queries or to provide input data to other applications. For example, a SQWRL query over OWO that is used to get oil pressure and pitch angle set point of the wind power plant which has ID is "2300249", is expressed as follows:

$WF(?p)\,\hat{}\,hasID(?p, "2300249")\,\hat{}\,hasWPPComponent(?p, ?comp)\,\hat{}\,hasOilPressure(?comp, ?pres)$
$\hat{}\,hasPitchAngleSetPoint(?comp, ?pitchAngle) \to sqwrl : select(?p, ?pres, ?pitchAngle)$

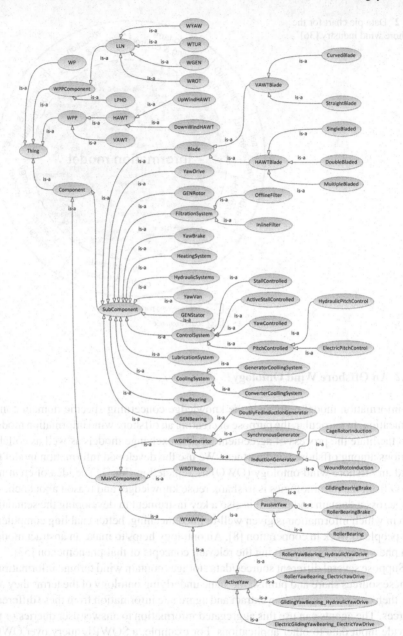

Fig. 3 OWO visualization

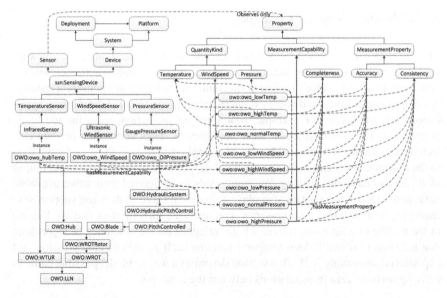

Fig. 4 An example of the alignment of the SSN ontology to OWO

4.1.3 Semantic Sensor Network Ontology

As the number of devices and appliances grows, the number of sensors embedded in such devices will also grow. Ontologies are an adequate way to model sensors and their capabilities [35]. Sensor metadata are used for selecting sensor sources and for integrating with other data sources [28]. Thus sensor metadata are important and needs to be exploited. However, sensor metadata alone cannot make a grid become smart. These metadata must be associated with metadata from devices and appliances that are participated in the grid.

The W3C semantic sensor network incubator group has introduced a semantic sensor network (SSN) ontology[1] to describe sensors, observations, and measurements. The ontology describes sensors and their properties such as accuracy, precision, resolution, measurement range, and capabilities. The ontology includes models for describing changes or states in an environment that a sensor can detect and the resulting observation [13]. An example of the alignment of the SSN ontology to the developed OWO is depicted in Fig. 4.

The developed OWO can be connected to SSN to share common information such as measurement values from sensors embedded on a wind power plant. At the same time, OWO can still guarantee complete description of a wind power plant data model. These two ontologies should be maintained separately since the number of concepts in these ontologies might grow as new technologies are introduced.

---

[1] http://www.w3.org/2005/Incubator/ssn/ssnx/ssn

4.2 Relations Between Concepts

Missing data can be caused by network disconnection, device faults, and software bugs. In some cases, where monitoring of devices or components is extremely important, a single missing value of a data point could lead to wrong predictions or damage of components. In the wind energy domain, many prediction and monitoring applications are employed, for example, power output prediction, wind turbine blade monitoring. The performance of these applications relies very much on data collected from the wind turbines. Missing of a single data item in the set of input data to these applications can make the applications produce wrong output or no output at all. In this case, the missing data item needs to be derived from other available data items. Derivation of data also plays a significant role in decision support systems [43]. For instance, in time-series data analysis, missing data that are located in the middle of a time-series have a high influence on the efficiency of algorithms that are used to reveal hidden temporal patterns such as vector autoregression and exponential smoothing [62]. This section describes a way to model possible paths to deriving missing data from relations between the concepts.

4.2.1 Derived Data Modeling

Data are classified into two categories: base data and derived data [19]. Base data are those data obtained from data sources. Derived data are those data obtained by combining or computing from base data. The combination and computation of base data are based on relations between domain concepts.

Derived data are described by derived classes and derived attributes. A derived attribute is an attribute that is derived from other attributes in the same class or from different classes that have relationships with the class that contains the attribute. If all attributes of a class are derived, the class is called derived class [4].

Derived data give an advantage of storing data since there is no need to store derived data in a database. Another advantage is that the structure of the data storage is undisclosed to users, derived attributes are accessed via user interface.

Guaranteeing the correctness of derived data is an important task because applications that use the data might produce wrong results if they receive insufficient input. Therefore, derived data need to be handled in such a way that its correctness is ensured. Formally modeling of derived data can help us to figure out different aspects of handling the data, and hence guaranteeing the correctness.

We use UML [17] to model the concepts in the wind domain. UML is based on object-oriented design concepts and is independent of any specific programming language. We also use Object Constraint Language (OCL) to express constraints in UML models [59]. OCL is a complement of UML. It makes models precise, consistent, and complete. In this work, we add OCL constraints to our models to tackle the derived issue mentioned in Sect. 3.2. We analyze two wind energy related

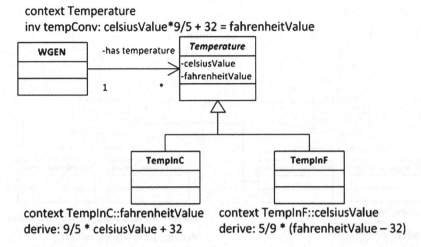

context Temperature
inv tempConv: celsiusValue*9/5 + 32 = fahrenheitValue

context TempInC::fahrenheitValue   context TempInF::celsiusValue
derive: 9/5 * celsiusValue + 32   derive: 5/9 * (fahrenheitValue – 32)

Fig. 5 Temperature conversion

cases where derived data play an significant role. We use the ontology introduced in Sect. 4.1.2 to demonstrate the cases.

4.2.2 Derived Data Within One Concept

Temperature measurement can be presented in different units such as Fahrenheit (F) or Celsius (C). The relation between F and C is as follows.

$$F = \frac{9}{5} * C + 32 \tag{1}$$

or

$$C = \frac{5}{9} * (F - 32) \tag{2}$$

The derivation can be obtained during execution time, for example, the authors of [10] use SWRL to define the transformation between temperature measurement units. However, such an approach will limit the possibility of expressing complex equations. A better approach is to attach formulas directly to properties in ontologies such as [22]. Let us consider a simple ontology describing the wind turbine generator (WGEN) concept and temperature as one of its properties. Figure 5 illustrates a formal model of temperature conversion using UML and OCL.

WGEN denotes the wind turbine generator class as described in [23]. *Temperature* is an abstract class that contains two attributes: the *celsiusValue* and *fahrenheitValue*. The two classes *TempInC* and *TempInF* contain rules to convert temperature unit from C to F and from F to C, respectively.

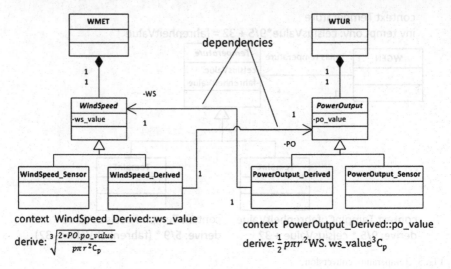

Fig. 6 Derivative relationships between two concepts

4.2.3 Derived Data Between Two Concepts

Let us consider an offshore wind farm scenario where many sensors are located on a wind turbine to capture information. What if one of them loses the connection? Information related to that one will be lost. How can we utilize other devices to derive that information so that the monitoring of the wind turbine is still ensured? Figure 6 shows how to make use of derived data from two parameters within the wind domain. The basic mathematical relation between wind speed and power output is expressed in Eq. (3) [33].

$$P_{avail} = \frac{1}{2} \rho \pi \, r^2 v^3 C_p \tag{3}$$

where P_{avail} denotes the available power output (W), ρ denotes air density (kg/m^3), r denotes blade length (m), v is the wind speed (m/s), and C_p denotes the power coefficient. Please note that the power coefficient is not constant; it depends on other factors such as rotational speed of the turbine, pitch angle, and angle of attack [34].

4.2.4 Derived Data with More than Two Concepts

What happens if one more parameter is added to the system? As an extension of the two concept model, we can have a model for three parameters as shown in Fig. 7.

Equation (3) can be rewritten as follows:

$$P_{avail} = \frac{1}{2} \rho \pi \, r^2 C_p (\frac{r}{TSR})^3 \omega^3 \tag{4}$$

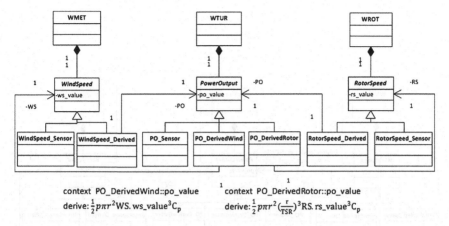

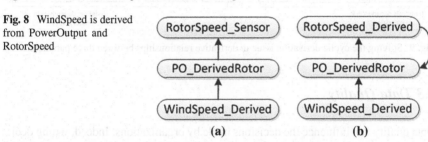

Fig. 7 Derivative relationships between three concepts

Fig. 8 WindSpeed is derived from PowerOutput and RotorSpeed

where TSR is tip speed ratio, ω (rpm) is the rotational speed of the blade. The TSR value can be obtained from the blade manufacturer, otherwise let TSR equal 7 since it is the most widely reported value in three bladed wind turbines [42]. We can then easily obtain *PO_DerivedRotor* as shown in Fig. 7.

A simple path, which is extracted from the model described in Fig. 7, is shown in Fig. 8a where *WindSpeed_Derived* can be derived from *PO_DerivedRotor* which can be derived from *RotorSpeed_Sensor*.

If we choose *RotorSpeed_Derived* instead of *RotorSpeed_Sensor*, this leads to a cyclic dependency as shown in Fig. 8b. Cyclic dependencies have to be avoided, as they cannot be computed.

Figure 9 depicts a model which is the extension of the model illustrated in Fig 7. In order to solve the derivation cycle issue, the transitive closure of the dependency *dependsOn* should not be reflexive.

The transitive closure of *dependsOn* is expressed in OCL as follows:

---

contextProperty
inv cycleRestriction : **not** *self .dependsOn.closure()− > include(self)*

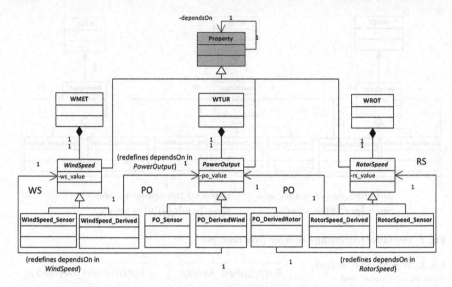

Fig. 9 Solving the cyclic derivation issue in derivative relationships between three parameters

4.3 Data Quality

Data quality can influence the decisions made by organizations. Indeed, wrong decisions can be made because of poor quality data [21, 52]. Data quality describes the characteristics of data and hence gives users a better view on data they want to request for. We consider data quality as metadata. Data quality has several dimensions which are criteria for selecting the most suitable data source according to users' requests. This section presents a solution to the challenge posed in Sect. 3.3.

4.3.1 Data Quality Dimensions

There are more than 17 data quality dimensions which have been mentioned in literature, e.g., accuracy, completeness, timeliness, consistency, access security, data volume, confidence, and understandability [6, 16, 29, 58]. The most commonly used quality dimensions are *accuracy*, *completeness*, and *timeliness* [44]. The other dimensions such as *confidence*, *value-added*, and *coverage* are only suggested by a couple of studies because these dimensions can be either derived from the other dimensions or applicable only in a few domains. There is no unique definition for each data quality dimension, so we describe the dimensions based on existing definitions and our understanding. Table 1 shows the notation that we use in our definitions.

Accuracy is defined as how close the observed data are to reality. According to the ISO 5725 standard [26], accuracy consists of precision and trueness.

Table 1 Table of notation

Symbol	Explanation
D	Data source
R	Reference data source (reality)
N_D	Total number of data points in D
N_R	Total number of data points in R
d_i	A single data point in D
r_i	Real value corresponding to d_i
x_i	$d_i - r_i$
$t(r_i)$	The moment when the data point i is due
$t(d_i)$	The moment when the data point i is available

We assume that the sensors are calibrated, meaning that the trueness is very close to zero. Therefore, we only consider precision as the accuracy in our system. A statistical measure of the precision for a series of repetitive measurements is the standard deviation. Let μ denote the trueness ($\mu = 0$). Thus, the accuracy of data source D can be obtained using Eq. (5).

$$Acc(D) = \sqrt{\frac{1}{N_D} \sum_{i=1}^{N_D} (x_i - \mu)^2} = \sqrt{\frac{1}{N_D} \sum_{i=1}^{N_D} (d_i - r_i)^2} \tag{5}$$

Completeness is defined as the ratio of the number of successful received data points to the number of expected data points. The completeness of the data source D can be calculated using Eq. (6).

$$Compl(D) = \frac{N_D}{N_R} \tag{6}$$

Timeliness is the average time difference between the moment a data point has been successfully received and the moment it is produced. The timeliness of data source D is calculated using Eq. (7).

$$Time(D) = \frac{\sum_{i=0}^{N_D}(t(d_i) - t(r_i))}{N_D} \tag{7}$$

4.3.2 Combination and Computation of Data Quality

By combining existing data sources, it is possible to improve the quality of data to meet the user defined requirement. The combination of data sources is defined as the process of constructing a data source from existing data sources. We present three simple methods to combine data quality: D1 (E) D2, D1 $\bigoplus$ D2, and D1 (A) D2.

- D1 (A) D2: taking a conventional average of the data sources D1 and D2.

Table 2 Combination results

Method	Completeness	Accuracy	Timeliness
D1 (A) D2	$\overline{P(D1)} \cdot \overline{P(D2)}$	$\sqrt{\frac{Acc(D1)^2 + Acc(D2)^2}{4}}$	$\approx \frac{3}{2} Time(D1)$
D1 $\oplus$ D2	$\overline{P(D1)} \cdot \overline{P(D2)}$	$\frac{P(D1) * Acc(D1) + \overline{P(D1)} * P(D2) * Acc(D2)}{P(D1) + \overline{P(D1)} * P(D2)}$	$\frac{Compl(D1) * Time(D1) + \overline{P(D1)} * P(D2) * Time(D2)}{P(D1) + \overline{P(D1)} * P(D2)}$
D1 (E) D2	$\overline{P(D1)} \cdot \overline{P(D2)}$	$\alpha Acc(D1) + \overline{\alpha} Acc(D2)$	$\frac{Time(D1) * Time(D2)}{Time(D1) + Time(D2)}$

Table 3 Quality combination relations

Combination method	Completeness	Accuracy	Timeliness
D1 (A) D2	✓	✓	×
D1 $\oplus$ D2	✓	–	–
D1 (E) D2	✓	–	✓

- D1 $\oplus$ D2: use data points from data source D1 if available, otherwise use D2.
- D1 (E) D2: pick up the earliest received data point from either D1 or D2.

Table 2 gives an overview of all combination methods with data quality dimensions. These methods are used to generate the virtual data source from the real data sources. P(D1) denotes the probability of the event *D1 having data available* and P(D2) denotes the probability of the event *D2 having data available*. Acc(D1) and Acc(D2) are the accuracy (precision) of D1 and D2, respectively. α is the probability of the event a data point $D1_i$ arrives before a data point $D2_i$.

The following assumptions are made in order to obtain Table 2. (1) Data sources D1 and D2 are independent and normally distributed; (2) timeliness *Time*(D1) and *Time*(D2) of D1 and D2 are two independent distributed exponential random variables.

The combination methods have different effects on the data quality dimensions. A quality dimension can increase or decrease depending on a combination method. Table 3 shows relation the between the combination operations and the data quality dimensions, where (✓) indicates that it can be better than both of D1 and D2, (−) means it varies from case to case, and (×) means it is worse than both of D1 and D2.

According to this table, all three methods can increase the completeness. By using the average method, the combined data source would have better accuracy. However, it makes the timeliness become worse. For the $\oplus$ method, both the accuracy and timeliness of the combined data source varies from case to case. The (E) method helps to increase the completeness and timeliness, but not the accuracy. If the timeliness is the critical choice, the (E) method is recommended to use.

4.3.3 A Data Quality-Based Framework for Data Source Selection

We have developed a framework for data source selection based on data quality dimensions. An overview of the framework is shown in Fig. 10. The framework offers ways to manage data sources, to insert a new data source, and to provide the

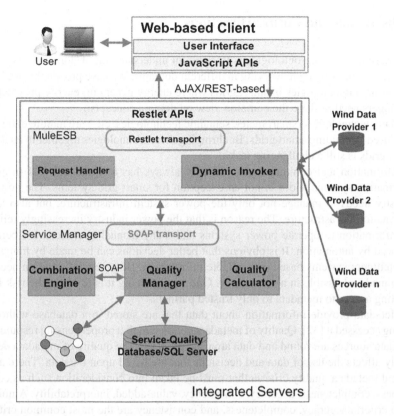

Fig. 10 An overview of a quality-based data source handling framework

best suited data source to users. Due to limitation of space, we cannot describe the prototype in detail. More information about the prototype implementation can be found in [44].

The prototype contains three main parts: web-based client application, integrated servers (IS), and data provider services. The web-based client application receives requests from users and forwards them to the IS. The client is in charge of data visualization in terms of graphs. The IS is responsible for data quality handling and communicating with data providers. The data providers store the data and provide addresses to access those data. The IS consists of an open source enterprise service bus, MuleESB and the *Service Manager* which contains the *Combination Engine*, the *Quality Manager*, the *Quality Calculator*, and the *Service-quality Database*.

5 Discussion and Future Directions

One reason of having ontologies is to share an understanding of domain concepts between partners who are working in different domains. We have proven the usefulness of having ontologies in smart grids where energy generator, energy providers, consumers need to share the common view on domain concepts.

Many technologies (smart meter, semantic technologies, etc.) are mature enough to be used in building smart grids. But bringing these technologies together to enable smart grids is still a challenging task.

Information and communication security always has a significant role in any information systems and it is not an exception for smart grid systems. The power industry needs to manage not only the power system infrastructure, but also the information infrastructure. The reason is that the power industry increasingly relies on information to operate power systems and many manual operations are being replaced by automation. It is obvious that better decisions can be made by humans or intelligent systems based on available information. However, information needs to be made accessible in a secure way. One way of doing it is to lower the risk by granting access to metadata to only trusted partners.

Metadata provide information about data that are stored in a database without having accessed it [32]. Quality of metadata guarantees that proper sensing resources and data sources are found and data are used properly. The quality of metadata definitely affects the use of data and decisions that are based upon the data. There are several metadata quality criteria that must be taken into consideration such as correctness, completeness, accuracy, consistency, value-added, interpretability. Among these criteria accuracy, completeness, and consistency are the most common criteria for measuring metadata quality in literature [40]. The challenge is among those metadata quality dimensions which ones are the most important and how to check their quality. Another challenge that has not been addressed in this work is tracking provenance of metadata when it comes to metadata combination and enhancement. Besides management of metadata, agreement on the definition of concepts is also an important task since without understanding the definitions, metadata may be misinterpreted or misused. We plan to tackle these challenges in future work.

6 Conclusions

Technologies bring us closer to our vision for living in smart environments. Even though there are available technologies for us, it is still not an easy task to bring all the technologies together. A smart grid is an example of a smart environment. In smart grids, a huge number of smart meters, sensors, smart appliances, and other smart devices are employed and connected to Internet. This leads to issues in handling and processing vast amounts of data, and integrating these devices in a network so that they can communicate with each other through intelligent systems and applications.

In this chapter, we have discussed issues related to management of big data metadata in smart grids. Three problems were addressed: concept modeling for knowledge sharing and data exchange, formal description of derived data from concept relations, and data quality handling. We have also proposed solutions and approaches to solving these problems. Some concrete examples within the offshore wind energy were used to demonstrate our points.

This work shows that the semantic technologies are mature enough to be used in the development of smart grids in particular and smart environments in general. The work also proves that data quality can be improved in some cases by combining different data sources that provide measurements about the same physical phenomenon. Relations between concepts not only describe real-world objects/phenomena, but also open several possible paths for calculation and give users the possibility of selecting the most suitable one.

Acknowledgments This work has been (partially) funded by the Norwegian Centre for Offshore Wind Energy (NORCOWE) under grant 193821/S60 from the Research Council of Norway (RCN). NORCOWE is a consortium with partners from industry and science, hosted by Christian Michelsen Research.

References

1. Ackoff, R.L.: From data to wisdom. J. Appl. Syst. Anal. **16**, 3–9 (2010)
2. Antoniou, G., Harmelen, F.v.: Web ontology language: OWL. Handbook on Ontologies. International Handbooks on Information Systems, pp. 91–110. Springer, Berlin Heidelberg (2009)
3. Baclawski, K., Kokar, M., Waldinger, R., Kogut, P.: Consistency checking of semantic web ontologies. The Semantic Web ISWC 2002, pp. 454–459 (2002)
4. Balsters, H.: Modelling database views with derived classes in the UML/OCL-framework. In: UML 2003-The Unified Modeling Language. Modeling Languages and Applications, pp. 295–309. Springer (2003)
5. Barnaghi, P., Wang, W., Henson, C., Taylor, K.: Semantics for the internet of things: early progress and back to the future. Int. J. Semant. Web Inf. Syst. (IJSWIS) **8**(1), 1–21 (2012)
6. Baumgartner, N., Gottesheim, W., Mitsch, S., Retschitzegger, W., Schwinger, W.: Improving situation awareness in traffic management. In: Proceedings of the International Conference on Very Large Data Bases (2010)
7. Bechhofer, S., Van Harmelen, F., Hendler, J., Horrocks, I., McGuinness, D., Patel-Schneider, P., Stein, L., et al.: OWL web ontology language reference. W3C Recommendation **10**, 10 (2004)
8. Berners-Lee, T., Fischetti, M.: Weaving the Web: The Original Design and Ultimate Destiny of the World Wide Web by Its Inventor. HarperInformation, 256 p. (2000)
9. Bredillet, P., Lambert, E., Schultz, E.: CIM, 61850, COSEM standards used in a model driven integration approach to build the smart grid service oriented architecture. In: First IEEE International Conference on Smart Grid Communications (SmartGridComm), 2010, pp. 467–471 (2010)
10. Bröring, A., Maué, P., Janowicz, K., Nüst, D., Malewski, C.: Semantically-enabled sensor plug & play for the sensor web. Sensors **11**(8), 7568–7605 (2011)
11. Camacho, E.F., Samad, T., Garcia-Sanz, M., Hiskens, I.: Control for renewable energy and smart grids. The Impact of Control Technology, Control Systems Society, pp. 69–88 (2011)
12. Chen, J., Chen, Y., Du, X., Li, C., Lu, J., Zhao, S., Zhou, X.: Big data challenge: a data management perspective. Front. Comput. Sci. **7**(2), 157–164 (2013)

13. Compton, M., Barnaghi, P., Bermudez, L., García-Castro, R., Corcho, O., Cox, S., Graybeal, J., Hauswirth, M., Henson, C., Herzog, A., et al.: The SSN ontology of the W3C semantic sensor network incubator group. Web Semant.: Sci. Serv. Agents World Wide Web **17**, 25–32 (2012)
14. Datastax Corporation: Big Data: Beyond the Hype. White paper (2013)
15. ETP: Smart Grids—Strategic Deployment Document for Europe's Electricity Networks of the Future (2010)
16. Geisler, S., Weber, S., Quix, C.: Onotology-based data quality framewrok for data stream applications. In: 16th International Conference on Information Quality, Nov 2011, Adelaide, AUS (2011)
17. Ghazel, M., Toguyéni, A., Bigand, M.: An UML approach for the metamodelling of automated production systems for monitoring purpose. Comput. Ind. **55**(3), 283–299 (2004)
18. Gruber, T.R., et al.: Toward principles for the design of ontologies used for knowledge sharing. Int. J. Hum. Comput. Stud. **43**(5), 907–928 (1995)
19. Hachem, N.I., Qiu, K., Serrao, N., Gennert, M.A.: GaeaPN: A Petri Net model for the management of data and metadata derivations in scientific experiments. Worcester Polytechnic Institute, Computer Science Department, Technical Report WPI-CS-TR-94 1 (1994)
20. Horrocks, I., Patel-Schneider, P., Van Harmelen, F.: From SHIQ and RDF to OWL: the making of a web ontology language. Web semant.: Sci. Serv. Agents World Wide Web **1**(1), 7–26 (2003)
21. Huang, K.T., Lee, Y.W., Wang, R.Y.: Quality Information and Knowledge. Prentice Hall PTR (1998)
22. Iannone, L., Rector, A.L.: Calculations in OWL. In: OWLED (2008)
23. IEC: IEC 61400 Wind Turbines—part 25: Communications for Monitoring and Control of Wind Power Plants (2006)
24. IEEE: Guide for Smart Grid Interoperability of Energy Technology and Information Technology Operation with the Electric Power System (EPS), End-Use Applications, and Loads. IEEE Std 2030-2011 pp. 1–126 (2011)
25. Informatica Corporation: Metadata Management for Holistic Data Governance. White paper (2013)
26. ISO: ISO 5725–2: 1994: Accuracy (Trueness and Precision) of Measurement Methods and Results—Part 2: Methods for the Determination of Repeatability and Reproductibility. International Organization for Standardization (1994)
27. Kollia, I., Glimm, B., Horrocks, I.: SPARQL query answering over OWL ontologies. In: The Semantic Web: Research and Applications, pp. 382–396. Springer (2011)
28. Le-Phuoc, D., Nguyen-Mau, H.Q., Parreira, J.X., Hauswirth, M.: A middleware framework for scalable management of linked streams. Web Semant.: Sci. Serv. Agents World Wide Web **16**, 42–51 (2012)
29. Lee, Y.W., Strong, D.M., Kahn, B.K., Wang, R.Y.: AIMQ: a methodology for information quality assessment. Inf. Manage. **40**(2), 133–146 (2002)
30. Lenzerini, M., Milano, D., Poggi, A.: Ontology representation and reasoning. Universit di Roma La Sapienza, Roma, Italy, Technical report NoE InterOp (IST-508011) (2004)
31. Manyika, J., Chui, M., Brown, B., Bughin, J., Dobbs, R., Roxburgh, C., Byers, A.H.: Big data: the next frontier for innovation, competition, and productivity. Technical report, McKinsey Global Institute (2011)
32. Margaritopoulos, T., Margaritopoulos, M., Mavridis, I., Manitsaris, A.: A conceptual framework for metadata quality assessment. Universitätsverlag Göttingen 104 (2008)
33. Muljadi, E., Pierce, K., Migliore, P.: Control strategy for variable-speed, stall-regulated wind turbines. In: Proceedings of the 1998 American Control Conference, vol. 3, pp. 1710–1714. IEEE (1998)
34. Muyeen, S., Tamura, J., Murata, T.: Wind turbine modeling. Stability Augmentation of a Grid-Connected Wind Farm, pp. 23–65 (2009)
35. Neuhaus, H., Compton, M.: The semantic sensor network ontology. AGILE workshop on challenges in geospatial data harmonisation, Hannover, Germany, pp. 1–33 (2009)

36. Nguyen, T.H., Prinz, A., Friiso, T., Nossum, R.: Smart grid for offshore wind farms: towards an information model based on the iec 61400-25 standard. In: IEEE PES Innovative Smart Grid Technologies (ISGT), 2012, pp. 1–6 (2012). doi:10.1109/ISGT.2012.6175686
37. Nguyen, T.H., Prinz, A., Friisø, T., Nossum, R., Tyapin, I.: A framework for data integration of offshore wind farms. Renew. Energy **60**, 150–161 (2013)
38. NIST: NIST Framework and Roadmap for Smart Grid Interoperability Standards, Release 2.0. NIST Special Publication 1108R2 edn. (2012)
39. O'Connor, M., Das, A.: SQWRL: a query language for OWL. In: Proceedings of 6th OWL: Experiences and Directions, Workshop (OWLED2009) (2009)
40. Park, J.R.: Metadata quality in digital repositories: a survey of the current state of the art. Cataloging Classif. Q. **47**(3–4), 213–228 (2009)
41. Patel-Schneider, P.F., et al., Hayes, P., Horrocks, I., et al.: OWL web ontology language semantics and abstract syntax. W3C Recommendation **10** (2004)
42. Ragheb, M., Ragheb, A.M.: Wind turbines theory-the betz equation and optimal rotor tip speed ratio. In: Carriveau, R. (ed.) Fundamental and Advanced Topics in Wind Power, pp. 19–37 (2011)
43. Ramirez, R.G., Kulkarni, U.R., Moser, K.A.: Derived data for decision support systems. Decis. Support Syst. **17**(2), 119–140 (1996)
44. Rasta, K., Nguyen, T.H., Prinz, A.: A framework for data quality handling in enterprise service bus. In: Third International Conference on Innovative Computing Technology (INTECH), 2013, pp. 491–497 (2013)
45. Rossouw, L., Re, G.: Big data-big opportunities. RISK **16**(2) (2012)
46. Sheth, A., Anantharam, P., Henson, C.: Physical-cyber-social computing: an early 21st century approach. IEEE Intell Syst **28**(1), 78–82 (2013). doi:10.1109/MIS.2013.20
47. Singh, A.: Standards for smart grid. Int. J. Eng. Res. Appl. (IJERA) (2012)
48. Sirin, E., Parsia, B.: SPARQL-DL: SPARQL query for OWL-DL. In: OWLED, vol. 258 (2007)
49. Sirin, E., Parsia, B., Grau, B., Kalyanpur, A., Katz, Y.: Pellet: a practical OWL-DL reasoner. Web Semant.: Sci. Serv. Agents World Wide Web **5**(2), 51–53 (2007)
50. Snyder, B., Kaiser, M.J.: Ecological and economic cost-benefit analysis of offshore wind energy. Renew. Energy **34**(6), 1567–1578 (2009)
51. Solntseff, N., Yezerski, A.: A survey of extensible programming languages. Ann. Rev. Autom. Prog. **7**, 267–307 (1974)
52. Strong, D.M., Lee, Y.W., Wang, R.Y.: Data duality in context. Commun. ACM **40**(5), 103–110 (1997)
53. Studer, R., Benjamins, V., Fensel, D.: Knowledge engineering: principles and methods. Data Knowl. Eng. **25**(1–2), 161–197 (1998)
54. Tambouris, E., Manouselis, N., Costopoulou, C.: Metadata for digital collections of e-government resources. Electron. Libr. **25**(2), 176–192 (2007)
55. Tannenbaum, A.: Metadata Solutions: Using Metamodels, Repositories, XML, and Enterprise Portals to Generate Information on Demand. Addison-Wesley Longman Publishing Co., Inc., Boston (2001)
56. Vermesan, O., Friess, P., Guillemin, P., Gusmeroli, S., Sundmaeker, H., Bassi, A., Jubert, I.S., Mazura, M., Harrison, M., Eisenhauer, M., et al.: Internet of things strategic research roadmap. In: Vermesan, O., Friess, P., Guillemin, P., Gusmeroli, S., Sundmaeker, H., Bassi, A., et al. (eds.) Internet of Things: Global Technological and Societal Trends, pp. 9–52 (2011)
57. Wagner, A., Speiser, S., Harth, A.: Semantic web technologies for a smart energy grid: Requirements and challenges. In: ISWC Posters and Demos (2010)
58. Wang, R.Y., Strong, D.M.: Beyond accuracy: what data quality means to data consumers. J. Manag. Inf. Syst. 5–33 (1996)
59. Warmer, J., Kleppe, A.: The object constraint language: getting your models ready for MDA. Addison-Wesley Longman Publishing Co., Inc., Boston (2003)
60. Xu, H.: Critical success factors for accounting information systems data quality. Ph.D. thesis, University of Southern Queensland (2009)

61. Zikopoulos, P.C., Eaton, C., DeRoos, D., Deutsch, T., Lapis, G.: Understanding Big Data. The McGraw-Hill Companies (2012)
62. Zubcoff, J., Pardillo, J., Trujillo, J.: A UML profile for the conceptual modelling of data-mining with time-series in data warehouses. Inf. Softw. Technol. **51**(6), 977–992 (2009)

Context-Aware Dynamic Discovery and Configuration of '*Things*' in Smart Environments

**Charith Perera, Prem Prakash Jayaraman, Arkady Zaslavsky,
Peter Christen and Dimitrios Georgakopoulos**

Abstract The Internet of Things (IoT) is a dynamic global information network consisting of Internet-connected objects, such as RFIDs, sensors, actuators, as well as other instruments and smart appliances that are becoming an integral component of the future Internet. Currently, such Internet-connected objects or '*things*' outnumber both people and computers connected to the Internet and their population is expected to grow to 50 billion in the next 5–10 years. To be able to develop IoT applications, such '*things*' must become dynamically integrated into emerging information networks supported by architecturally scalable and economically feasible Internet service delivery models, such as cloud computing. Achieving such integration through discovery and configuration of '*things*' is a challenging task. Towards this end, we propose a Context-Aware Dynamic Discovery of Things (CADDOT) model. We have developed a tool *SmartLink*, that is capable of discovering sensors deployed in a particular location despite their heterogeneity. *SmartLink* helps to establish the direct communication between sensor hardware and cloud-based IoT middleware platforms. We address the challenge of heterogeneity using a plug in architecture. Our prototype tool is developed on an Android platform. Further, we employ the Global Sensor Network (GSN) as the IoT middleware for the proof of

C. Perera (✉) · P. P. Jayaraman · A. Zaslavsky · D. Georgakopoulos
CSIRO Computational Informatics, Canberra, ACT 0200, Australia
e-mail: charith.perera@csiro.au; charith.perera@anu.edu.au

P. P. Jayaraman
e-mail: prem.jayaraman@csiro.au

A. Zaslavsky
e-mail: arkady.zaslavsky@csiro.au.au

D. Georgakopoulos
e-mail: dimitrios.georgakopoulos@csiro.au

C. Perera · P. Christen
Research School of Computer Science, The Australian National University,
Canberra, ACT 0200, Australia
e-mail: peter.christen@anu.edu.au

N. Bessis and C. Dobre (eds.), *Big Data and Internet of Things:*
A Roadmap for Smart Environments, Studies in Computational Intelligence 546,
DOI: 10.1007/978-3-319-05029-4_9, © Springer International Publishing Switzerland 2014

concept validation. The significance of the proposed solution is validated using a test-bed that comprises 52 Arduino-based Libelium sensors.

1 Introduction

The Internet of Things (IoT) [4] first received attention in the late 20th century. The term was firstly coined by Kevin Ashton [3] in 1999. *"The Internet of Things allows people and things[1] to be connected Anytime, Anyplace, with Anything and Anyone, ideally using Any path/ network and Any service"* [18]. As highlighted in the above definition, connectivity among devices is a critical functionality that is required to fulfil the vision of IoT. The following statistics highlight the magnitude of the challenge we need to address. Due to the increasing popularity of mobile devices over the past decade, it is estimated that there are about 1.5 billion Internet-enabled PCs and over 1 billion Internet-enabled mobile devices today. The number of *'things'* connected to the Internet exceeded the number of people on earth in 2008 [23]. By 2020, there will be 50–100 billion devices connected to the Internet [45]. Similarly, according to BCC Research, the global market for sensors was around \$56.3 billion in 2010. In 2011, it was around \$62.8 billion, and it is expected to increase to \$91.5 billion by 2016, at a compound annual growth rate (CAGR) of 7.8 % [5].

The above statistics allow us to conclude that the growth rate of sensors being deployed around us is increasing over time and will keep its pace over the coming decade. Over the last few years, we have witnessed many IoT solutions making their way into the market [40]. The IoT market has already been fragmented, with many parties competing with a variety of different solutions. Broadly, these IoT solutions can be divided into two segments: sensor hardware-based solutions [27] and cloud-based software solutions [14, 17, 31]. Some products specifically address one segment, while others address both. In this chapter, we propose a Context-Aware Dynamic Discovery of Things (CADDOT) model in order to support the integration of *'things'* into cloud-based IoT solutions via dynamic discovery and configuration by also addressing the challenge of heterogeneity. We reduce the complexity of the *'things' configuration process* and make it more user friendly and easier to use. One major objective is to support non-technical users by allowing them to configure smart environments without technical assistance.

This chapter makes the following contributions. We propose a model, CADDOT, that can be used to configure sensors autonomously without human intervention in highly dynamic smart environments in the Internet of things paradigm. To support this model, we developed a tool called *SmartLink*. *SmartLink* is enriched with context-aware capabilities so it can detect sensors using different protocols such as TCP, UDP,

---

[1] We use both terms, *'objects'* and *'things'* interchangeably to give the same meaning as they are frequently used in IoT related documentation. Some other terms used by the research community are 'smart objects', 'devices', 'nodes'. Each 'thing' may have one or more sensors attached to it.

Bluetooth and ZigBee. CADDOT is designed to deal with highly dynamic smart environments where sensors are appearing and disappearing at a high frequency. This chapter also presents the results of experimental evaluations performed using 52 sensors measuring different types of phenomenon and using different communication sequences.

We explain how our model can be used to enrich the existing solutions proposed in the research field. The chapter is organized as follows. We present background information and motivation in Sect. 2. In Sect. 3, we discuss the functional requirements of an ideal IoT configuration process. We discuss related work in Sect. 4. The proposed CADDOT model is introduced in Sect. 5. The design decisions we made are justified and compared with alternative options in Sect. 6. Implementation details and evaluations are presented in Sects. 7 and 8 respectively. The lessons learnt are discussed in Sect. 9. Open challenges are presented in Sect. 10 and we conclude the chapter in Sect. 11 with indications for future work.

2 Background and Motivation

This section briefly highlights the background details of the challenge we address in this chapter. Firstly, we explain the challenges in the smart environment from the perspective of dynamic discovery and configuration of '*things*'. Secondly, we discuss the concept of sensing as a service and its impact on the IoT. At the end, we present the importance of the configuration of '*things*' in the big data domain.

2.1 Smart Environment

A smart environment can be defined as "*a physical world that is richly and invisibly interwoven with sensors, actuators, displays, and computational elements, embedded seamlessly in the everyday objects of our lives, and connected through a continuous network*" [46]. Smart environments may be embedded with a variety of smart devices of different types including tags, sensors and controllers, and have different form factors ranging from nano to micro to macro sized. As also highlighted by Cook and Das [13], device communication using middleware and wireless communication is a significant part of forming a connected environment. Forming smart environments needs several activities to be performed, such as discovery (i.e. exploring and finding devices at a given location), identification (i.e. retrieving information about devices and recognizing them), connection establishment (i.e. initiating communication using a protocol that the device can understand), and configuration. Further, users may combine sensors and services to configure smart environments where actuators are automatically triggered based on conditions [25]. In smart home environments, Radio Frequency for Consumer Electronics (RF4CE) has been used to perform atuomated configuration of consumer devices [43]. However, such techniques cannot be used to configure low-level smart *'things'*.

2.2 Sensing as a Service

The sensing-as-a-service model [37] provides sensing capabilities as a service similar to other models such as infrastructure-as-a-service (IaaS), platform-as-a-service (PaaS), and software-as-a-service (SaaS). Mobile devices are widely used to collect data from inbuilt or external sensors [42].

It envisions that sensor descriptions and capabilities are posted on the Internet so the interested consumer can get access to the corresponding sensors by paying a fee [37]. The sensing as a service model is expected to drive the IoT from the business point of view by creating a whole new set of opportunities and values. It has been predicted that individuals as well as, private and public organizations will deploy sensors to achieve their primary objectives [8, 37]. Additionally, they will share their sensors with others so a collectively value-added solution can be built around them. Such sensor deployments and data collection allows the creation of real-time solutions to address tough challenges in Smart Cities [29, 37]. In order to support sensor deployments, easy-to-use *'things'* discovery and configuration tools need to be developed. Such a set of tools will stimulate the growth of sensor deployments in the IoT. They will help the non-technical community to become involved in building smart environments efficiently and effectively.

2.3 Big Data Challenge

Big Data [6] mainly comprises six categories of data, as illustrated in Fig. 1a transaction data, scientific data, sensor data, social media data, enterprise data, and public data. The sensor data category is expected to be generated by the growing number of sensors deployed in different domains, as illustrated in Fig. 1. The data streams coming from *'things'* will challenge the traditional approaches to data management and contribute to the emerging paradigm of big data. Collecting sensor data on a massive scale, which creates big data, requires easy-to-use sensor discovery and configuration tools that help to integrate the *'things'* into cloud-based IoT middleware platforms. Big data has been identified as a secondary phase of the IoT, where new sensors are cropping up and organizations are now starting to analyse data, that in some cases, they have been collecting for years.

This work is also motivated by our previous work which focused on utilising mobile phones and similar capacity devices to collect sensor data. In DAM4GSN [38], we proposed an application that can be used to collect data from sensors built into mobile phones. Later, we proposed MoSHub [33] that allows a variety of different external sensors to be connected to a mobile phone using an extensible plu-

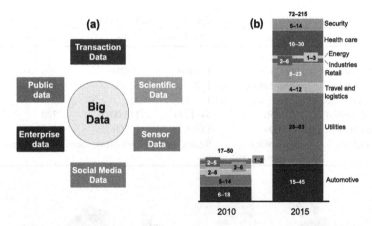

Fig. 1 a Big data comprises six categories of data **b** Data generated from the IoT will grow exponentially as the number of connected nodes increases. Estimated numbers of connected nodes based on different sectors are presented in millions [28]

gin architecture. MoSHub also configures the cloud middleware accordingly. Later in MOSDEN [34], we developed a complete middleware for resource-constrained mobile devices. MOSDEN is capable of collecting data from both internal and external sensors. It can also apply SQL-based fusing on data streams in real time. As we mentioned earlier, in order to collect data from sensors, first we need to discover and configure the sensors in such a way that the cloud can communicate with them. In our previous efforts, discovery and configuration steps were performed manually. In this chapter, we propose an approach that can be used to discover and configure sensors autonomously.

3 Functional Requirements

The '*things*' *configuration process* detects, identifies, and configures sensor hardware and cloud-based IoT platforms in such a way that software platforms can retrieve data from sensors when required. In this section, we identify the importance, major challenges, and factors that need to be considered during a configuration process. The process of sensor configuration in IoT is important for two main reasons. Firstly, it establishes the connectivity between sensor hardware and software systems wich makes it possible to retrieve data from the deployed sensor. Secondly, it allows us to optimize the sensing and data communication by considering several factors as discussed below. Let us discuss the following research problem: *Why is sensor configuration challenging in the IoT environment?*. The major factors that make sensor configuration challenging are (1) *the number of sensors*, (2) *heterogeneity*, (3) *scheduling, sampling rate, communication frequency*, (4) *data acquisition*, (5) *dynamicity*, and (6) *context* [36].

Table 1 Heterogeneity in term of Wireless Communication Technology

	ZigBee	GPRS-GSM	WiFi	Bluetooth
Standard	802.15.4		802.11b	802.15.1
System resources	4–32 KB	16 MB+	1 MB+	250 KM+
Batterylife (days)	100–1,000+	1–7	0.5–5	1–7
Network size (nodes)	2^{64}	1	32	7
Bandwidth (KB/s)	20–250	64–128+	11,000	720
Transmissionrange (m)	1–100+	1,000	1–100	1–10+
Success metrics	Reliability, power, cost	Reach, quality	Flexibility, speed	Convenience, cost

Gas Sensor Node

• Carbon Monoxide – CO
• Carbon Dioxide – CO2
• Oxygen – O2
• Methane – CH4
• Hydrogen – H2
• Ammonia – NH3
• Isobutane – C4H10
• Ethanol – CH3CH2OH
• Toluene – C6H5CH3
• Hydrogen Sulfide – H2S
• Nitrogen Dioxide – NO2
• Ozone – O3
• Hydrocarbons – VOC

Event Sensor Node

• Pressure/Weight
• Bend
• Vibration
• Impact
• Hall Effect
• Tilt
• Temperature (+/-)
• Liquid Presence
• Liquid Level
• Luminosity
• Presence (PIR)
• Stretch

Fig. 2 Heterogeneity in term of sensing/measurement capabilities of sensor nodes

1. Number of Sensors: When the number of sensors that need to be configured is limited, we can use manual or semi-autonomous techniques. However, when the numbers grow rapidly towards millions and billions, as illustrated in Fig. 1b, such methods become extremely inefficient, expensive, labour-intensive, and in most situations impossible. Therefore, large numbers have made sensor configuration challenging. An ideal sensor configuration approach should be able to configure sensors autonomously as well as within a very short time period.

2. Heterogeneity: This factor can be interpreted in different perspectives. (1) Heterogeneity in terms of the communication technologies used by the sensors, as presented in Table 1. (2) Heterogeneity in terms of measurement capabilities, as presented in Fig. 2 (e.g. temperature, humidity, motion, pressure). (3) The types of data (e.g. numerical (small in size), audio, video (large in size)) generated by the sensors are also heterogeneous. (4) The communication sequences and security mechanisms used by different sensors are also heterogeneous (e.g. exact messages/commands and the sequence that needs to be followed to successfully communicate with a given sensor). As illustrated in Fig. 3, some sensors may need only a few command passes and others may require more. Further, the messages/commands understood by each sensor may also vary. These differences make the sensor configuration process challenging. An ideal sensor configuration approach that is designed for the IoT paradigm should be able to handle such heterogeneity. It should also be scalable and should provide support for new sensors as they come to the market.

3. Scheduling, Sampling Rate, and Network Communication: The sampling rate defines the frequency with which sensors need to generate data (i.e. sense the

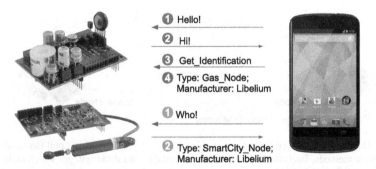

Fig. 3 Heterogeneity in term of communication and message/command passing sequences. Some sensors may need only a few message/command passes and others may require more. The messages/commands understood by each sensor may also vary

phenomenon) (e.g. sense temperature every 10 s). Deciding the ideal (e.g. balance between user requirement and energy consumption) sampling rate can be a very complex task and has a strong relationship with *(6) Context* (see below). The schedule defines the timetable for sensing and data transmission (e.g. sense the temperature only between 8 am and 5 pm on weekdays). Network communication defines the frequency of data transmission (e.g. send data to the cloud-based IoT platform every 60 s). Designing efficient sampling and scheduling strategies and configuring the sensors accordingly is challenging. Specifically, standards need to be developed in order to define schedules that can be used across different types of sensor devices.

4. Data Acquisition: Such methods can be divided into two categories: based on responsibility and based on frequency [36]. There are two methods that can be used to acquire data from a sensor based on responsibility as illustrated in Fig. 4: push (e.g. the cloud requests data from a sensor and the sensor responds with data) and pull (e.g. the sensor pushes data to the cloud without continuous explicit cloud requests). Further, based on frequency, there are two data acquisition methods: instant (e.g. send data to the cloud when a predefined event occurs) and interval (e.g. send data to the cloud periodically). Pros, cons, and applicabilities of these different approaches are discussed in [36]. Using the appropriate data acquisition method based on context information is essential to ensure efficiency.

5. Dynamicity: This means the frequency of changing positions/appearing/disappearing of the sensors at a given location. IoT envisions that most of the objects we use in everyday lives will have sensors attached to them in the future. Ideally, we need to connect and configure these sensors to software platforms in order to analyse the data they generate and so understand the environment better. We have observed several domains and broadly identified different levels of dynamicity based on mobility.[2] Sensors that move/appear/disappear at a higher frequency (e.g. RFID

---

[2] It is important to note that the same object can be classified at different levels depending on the context. Further, there is no clear definition to classify objects into different levels of dynamicity. However, our categorization allows us to understand the differences in dynamicity.

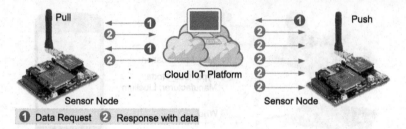

Pull

Push

Cloud IoT Platform

Sensor Node

Sensor Node

❶ Data Request ❷ Response with data

Fig. 4 Data can be retrieved from a sensor using both push (*right side*) and pull (*left side*) communication methods. Each method has its own advantages and disadvantages which make them suitable for different situations

and other low-level, low-quality, less reliable, cheap sensors that will be attached to consumables such as stationery, food packaging, etc.) can be classified as highly dynamic. Sensors embedded and fitted into permanent structures (such as buildings and air conditioning systems) can be classified as less dynamic. An ideal sensor configuration platform should be able to efficiently and continuously discover and re-configure sensors in order to cope with high dynamicity (Fig. 4).

6. Context: Context information plays a critical role in sensor configuration in the IoT. The objective of collecting sensor data is to understand the environment better by fusing and reasoning them. In order to accomplish this task, sensor data needs to be collected in a timely and location-sensitive manner. Each sensor needs to be configured by considering context information. Let us consider a scenario related to smart agriculture to understand why context matters in sensor configuration. *Severe frosts and heat events can have a devastating effect on crops. Flowering time is critical for cereal crops and a frost event could damage the flowering mechanism of the plant. However, the ideal sampling rate could vary depending on both the season of the year and the time of day. For example, a higher sampling rate is necessary during the winter and the night. In contrast, lower sampling would be sufficient during summer and daytime. On the other hand, some reasoning approaches may require multiple sensor data readings. For example, a frost event can be detected by fusing air temperature, soil temperature, and humidity data. However, if the air temperature sensor stops sensing due to a malfunction, there is no value in sensing humidity, because frost events cannot be detected without temperature. In such circumstances, configuring the humidity sensor to sleep is ideal until the temperature sensor is replaced and starts sensing again.* Such intelligent (re-)configuration can save energy by eliminating ineffectual sensing and network communication.

4 Related Work

In this section, we review some of the state-of-the-art solutions developed by the research community, as well as commercial business entities. Our review covers both mature and immature solutions proposed by start-up initiatives as well as large-scale projects. Our proposed CADDOT model as well as the *SmartLink* tool help to overcome some of the weaknesses in the existing solutions.

There are commercial solutions available in the market that have been developed by start-up IoT companies [40] and the research divisions of leading corporations. These solutions are either still under development or have completed only limited deployments in specialized environments (e.g. demos). We discuss some of the selected solutions based on their popularity. *Ninja Blocks* (ninjablocks.com), *Smart-Things* (smartthings.com), and *Twine* (supermechanical.com) are commercial products that aim at building smart environments [40]. They use their own standards and protocols (open or closed) to communicate between their own software systems and sensor hardware components. The hardware sensors they use in their solutions can only be discovered by their own software systems. In contrast, our pluggable architecture can accommodate virtually any sensor. Further, our proposed model can facilitate different domains (e.g. indoor, outdoor) using different communication protocols and sequences.

In addition, the CADDOT model can facilitate very high dynamicity and mobility. *HomeOS* [15] is a home automation operating system that simplifies the process of connecting devices together. Similar to our plugin architecture, *HomeOS* is based on applications and drivers which are expected to be distributed via an on-line store called *HomeStore* in the future. However, *HomeOS* does not perform additional configuration tasks (e.g. scheduling, sampling rate, communication frequency) depending on the user requirements and context information. Further, our objective is to develop a model that can accommodate a wider range of domains by providing multiple alternative mechanisms, as discussed in Sect. 6. Hu et al. [21] have proposed a sensor configuration mechanism that uses the information store in TEDS [22] and SensorML [7] specifications. Due to the unavailability and unpopularity of TEDS among sensor manufacturers, we simulate TEDS using standard communication message formats, as explained in Sect. 6.

Actinium [26] is a RESTful runtime container that provides Web-like scripting for low-end devices through a cloud. It encapsulates a given sensor device using a container that handles the communication between the sensor device and the software system by offering a set of standard interfaces for sensor configuration and life-cycle management. The Constrained Application Protocol (CoAP), a software protocol intended to be used in very simple electronics devices that allows them to communicate interactively over the Internet, has been used for communication. Pereira et al. [32] have also used CoAP and it provides a request/response interaction model between application end-points. It also supports built-in discovery of services and resources. However, for discovery to work, both the client (e.g. a sensor) and the server (e.g. the IoT platform) should support CoAP. However, most of the sensor manufacturers do not provide native support for such protocols. *Dynamix* [9] is a plug-and-play context framework for Android. *Dynamix* automatically discovers, downloads, and installs the plugins needed for a given context sensing task. *Dynamix* is a stand-alone application and it tries to understand new environments using pluggable context discovery and reasoning mechanisms. Context discovery is the main functionality in *Dynamix*. In contrast, our solution is focused on dynamic

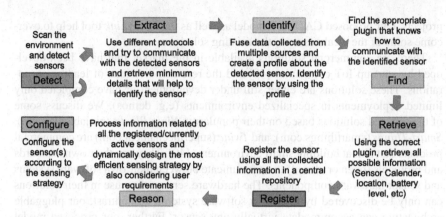

Fig. 5 Context-aware Dynamic Discovery of Things (*CADDOT*) model for configuration of things in the IoT paradigm consists of eight phases

discovery and configuration of '*things*' in order to support a sensing as a service model in the IoT domain. We employ a pluggable architecture which is similar to the approach used in *Dynamix*, in order to increase the scalability and rapid extension development by third party developers. The Electronic Product Code (EPC) [16] is designed as a universal identifier that provides a unique identity for every physical object anywhere in the world. EPC is supported by the CADDOT model as one way of identifying a given sensor. Sensor integration using IPv6 in building automation systems is discussed in [24]. Compton et al. [12] have used a Device Profile for Web Services[3] (DPWS) to encapsulate both devices and services. DPWS defines a minimal set of implementation constraints to enable secure web service messaging, discovery, description, and eventing on resource-constrained devices. However, discovery is only possible if both ends (client and server) are DPWS-enabled.

5 Overview of the CADDOT Model

Previously, we identified several major factors that need to be considered when developing an ideal sensor configuration model for the IoT. This section presents a detailed explanation of our proposed solution: Context-aware Dynamic Discovery of Things (CADDOT). Figure 5 illustrates the main phases of the proposed model.

Phases in CADDOT model: The proposed model consists of eight phases: *detect, extract, identify, find, retrieve, register, reason,* and *configure*. Some of the tasks mentioned in the model are performed by the *SmartLink* tool and other tasks are performed by the cloud middleware. Some tasks are performed collectively by both *SmartLink* and the cloud.

---

[3] http://docs.oasis-open.org/ws-dd/ns/dpws/2009/01

1. Detect: Sensors are configured to actively seek open wireless access points (WiFi or Bluetooth) to which they can be connected without any authorization, because in this phase sensors do not have any authentication details. Sensors will receive the authentication details in phase **phase 8**). As a result, in this phase sensors are unable to connect to an available secured network. The mobile device that *SmartLink* is installed in becomes an open wireless access point (hotspot) so the sensors can connect to it. However, it is important to note that there are different application strategies that *SmartLink* can use to execute the CADDOT model, as discussed in Sect. 6.

2. Extract: In this phase, *SmartLink* extracts information from the sensor detected in the previous phase. Each sensor may be designed to respond to different message-passing sequences, as illustrated in Fig. 3, depending on the sensor manufacturer and the sensor program developer. Even though the sensors and the *SmartLink* may use the same communication technology/protocol (e.g. TCP, UDP, Bluetooth), the exact communication sequence can vary from one sensor to another. Therefore, it is hard to find the specific message-passing sequence that each sensor follows. To address this challenge, we propose that every sensor will respond to a common message during the communication initiation process. Alternatively, CADDOT can support multiple initiation messages (extraction mechanisms). However, such alternative approaches will increase the time taken to extract a minimum set of information from a given sensor due to multiple communication attempts that need to be carried out until a sensor successfully responds. For example, *SmartLink* broadcasts a message [WHO], as illustrated in (*C1*) in Fig. 10, where the sensors are expected to respond by providing a minimum amount of information about themselves, such as a sensor's unique identification number, model number/name, and manufacturer. This is similar to the TEDS mechanism discussed in [21]. It is important to note that we propose this [WHO] constraint only for minimum information extraction. Once the sensor is identified, subsequent communications and heterogeneity of message-passing sequences are handled by matching plugins.

3. Identify: *SmartLink* sends all the information extracted from the newly detected sensor to the cloud. Cloud-based IoT middleware queries its data stores using the extracted information and identifies the complete profile of the sensor. The descriptions of the sensors are modelled in an ontology.[4]

4. Find: Once the cloud identifies the sensor uniquely, this information is used to find a matching plugin (also called drivers) which knows how to communicate with a compatible sensor at full capacity. The IoT middleware pushes the plugin to *SmartLink* where it is installed.[5]

5. Retrieve: Now, *SmartLink* knows how to communicate with the detected sensor at full capacity with the help of the newly downloaded plugin. Next, *SmartLink* retrieves the complete set of information that the sensor can provide

---

[4] This is an extended version of an SSN ontology (www.w3.org/2005/Incubator/ssn/ssnx/ssn). The detailed description of our extended ontology is out of the scope of this chapter.

[5] In practice, the IoT middleware sends a request to the application store (e.g. Google Play). The application store pushes the plugin to the *SmartLink* autonomously via the Internet.

(e.g. configuration details such as schedules, sampling rates, data structures/types generated by the sensor, etc.). Further, *SmartLink* may communicate with other available sources (e.g. databases, web services) to retrieve additional information related to the sensor.

6. Register: Once all the information about a given sensor has been collected, registration takes place in the cloud. The sensor descriptions are modelled according to the semantic sensor network ontology (SSNO) [12]. This allows semantic querying and reasoning at a later stage to perform operations such as sensor search [35]. Some of the performance evaluation related to the SSN ontology and semantic querying is presented in [39].

7. Reason: This phase plays a significant role in the sensor configuration process. It designs an efficient sensing strategy. Reasoning takes place in a distributed manner. The cloud IoT middleware retrieves data from a large number of sensors and identifies their availabilities and capabilities. Further, it considers context information in order to design an optimized strategy. Context-aware reasoning is performed by IoT middleware on the cloud. However, the technical details related to this reasoning process are out of the scope of this chapter. At the end of this phase, a comprehensive plan (i.e. sensing schedule) for each individual sensor is designed.

8. Configure: Sensors as well as cloud-based IoT software systems are configured based on the strategy designed in the previous phase. Schedules, communication frequency, and sampling rates that are custom-designed for each sensor are pushed into the individual sensors. The connections between sensors and the cloud-based IoT software system are established through direct wireless communication or through intermediate devices such as MOSDEN [34] so the cloud can retrieve data from sensors. The configuration details (e.g. IP address, port, authentication) required to accomplish the above task are also provided to the sensor.

6 Design Decisions and Applications

We made a number of design decisions during the development of the CADDOT model. These decisions address the challenges we highlighted in earlier sections.

Security Concerns and Application Strategies: There are different ways to employ our proposed model CADDOT as well as the tool *SmartLink* in real world deployments. Figure 6 illustrates two different application strategies. It is important to note that neither our model nor the software tool is limited to a specific device or platform. In this paper, we conduct the experimentations on an Android-based mobile phone, as detailed in Sect. 7. In strategy (a), a Raspberry Pi (raspberrypi.org) is acting as the *SmartLink* tool. This strategy is mostly suitable for smart home and office environments where WiFi is available. Raspberry Pi continuously performs the discovery and configuration process, as explained in Sect. 5. Finally, Raspberry Pi provides the authentication details to the sensor which is connected to the secure home/office WiFi network. The sensor is expected to send data to the processing server (local or on cloud) directly over the secured WiFi network. In this strategy,

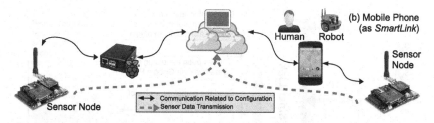

Fig. 6 Application strategies of CADDOT model and *SmartLink* tool. **a** usage of static *SmartLink* **b** usage of mobile *SmartLink*

SmartLink is in static mode. Therefore, several *SmartLink* installed Raspberry Pi devices may be required to cover a building. However, this strategy can handle a high level of dynamicity.

The *strategy (b)* is more suitable for situations where WiFi is not available or less dynamic. Smart agriculture can be considered as an example. In this scenario, sensors are deployed over a large geographical area (e.g. Phenonet [11]). Mobile robots[6] (tractors or similar vehicles) with a *SmartLink* tool attached to them can be used to discover and configure sensors. *SmartLink* can then help to establish the communication between sensors and sinks. The permanent sinks used in the agricultural fields are usually low-level sinks (such as Messhablium [27]). Such sinks cannot perform sensor discovery or configuration in comparison to SmartLink. Such sinks are designed to collect data from sensors and upload to the cloud via 3G.

Many more different strategies can be built by incorporating the different characteristics pointed out in the above two strategies. This shows the extensibility of our solution. For example, Raspberry Pi, which we suggested for use as a *SmartLink* in strategy (a), can be replaced by corporate mobile phones. So, without bothering the owner, corporate mobile phones can silently perform the work of a *SmartLink*.

System Architecture: The CADDOT model consists of three main components: sensors, a mobile device (i.e. *SmartLink*), and the cloud middleware. All three components need to work collectively in order to perform sensor discovery and configuration successfully. Figure 7 illustrates the interactions between the three components. The phases we explained earlier relating to the CADDOT model in Fig. 5 can be seen in Fig. 7 as well. As we mentioned before, *SmartLink* is based on a plugin architecture. The core *SmartLink* application cannot directly communicate with a given sensor. A plugin needs to act as a mediator between the sensor and the *SmartLink* core application, as illustrated in Fig. 7. The task of the mediator is to translate the commands back and forth. This means that in order to configure a specific sensor, the *SmartLink* core application needs to employ a plugin that is compatible with both the *SmartLink* application itself and the given sensor. We discuss this matter in the programming perspective later in this section.

Sensor-level Program Design: One of the most important components in the CADDOT model is the sensor. Sensors can be programmed in different ways. In this

---

[6] In small agricultural fields, farmers themselves can carry the *SmartLink* over the field.

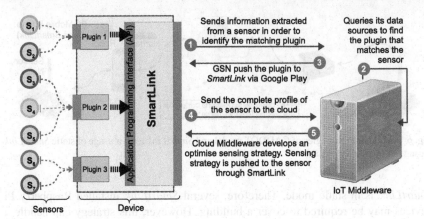

Fig. 7 System architecture of the CADDOT model which consists of three main components: sensors, *SmartLink* tool, and the cloud middleware. Interactions are numbered in order

```
1. Include Libraries    2. Definitions    3. Global variables declaration
void setup()
{
    4. Modules initialisation [Communication]    }  // This code only runs once
}
void loop()
{
    5. Connect to an access point
       [IP address, Port number]
    6. Sense the phenomenon                       }  // This code runs continuously,
    7. Send information to the cloud                   forming an infinite loop.
    8. Sleep [Communication Frequency]
}
```

Fig. 8 A simple sensor-level program design (*SPD*) that sends and transmits data to the cloud. It does not support dynamic discovery and configuration

chapter, we propose a program design that supports all the functional requirements identified in Sect. 3. The program we propose may not be the only way to support these requirements. Further, we do not intend to restrict developers to one single sensor-level program design. Instead, our objective is to demonstrate one successful way to program a sensor in such a way that it allows sensors to be re-configured at runtime (i.e. after deployment) depending on the requirements that arise later. Developers are encouraged to explore more efficient program designs. However, in order to allow *SmartLink* to communicate with a sensor which runs different program designs, developers need to develop a plugin that performs the command translations. We explain the translation process using both sensor-level program code as well as plugin code later in this section. First, we illustrate the simplest sensor-level program that can be designed to perform the task of sensing and transmitting data to the cloud in Fig. 8. We refer to this program design as *SPD* (Simple Program Design) hereafter. The basic structure of a sensor-level program is explained in [27].

The main problem in this program design is that there is no way to configure (i.e. sampling rate, communication frequency, data acquisition method) the sensor after deployment other than by re-programming (e.g. Over the Air Programming). However, such re-programming approaches are complex, labour-intensive and time consuming. In Fig. 9, we designed a sensor-level program that supports a comprehensive set of configuration functionalities. We refer to this design as *CPD* (Configurable Program Design) hereafter. In order to standardize the communication, we also defined a number of command formats. However, these messaging formats do not need to be followed by the developers as long as they share common standardised command formats between their own sensor-level program and the corresponding plugin. Different command formats used to accomplish different tasks in our approach are illustrated in Fig. 10. In comparison to *SPD*, *CPD* provides more configuration functionalities. With the help of the command formats illustrated in Fig. 10, *SmartLink* can configure a given sensor at any time.

Each command comprises several different segments, as depicted in Fig. 10. The first segment denotes whether the command is related to configuration or a data request. In our approach, [CON] denotes configuration and [DAR] denotes a data request. The CPD is designed to forward the command appropriately through IF-ELSE branches. The CPD accepts five different types of commands under the [CON] branch. Commands are classified based on the second segment. The following list summarises these commands. The first segment of every command contains only three letters which makes it easy to process. The commands can be sent using frames[7] or plain strings.

- **C1**: This command has only one segment. This segment always contains three letters [WHO]. This command is sent by *SmartLink* to a sensor. To support CAD-DOT, every sensor should be able to handle command C1. Then the sensor needs to respond with message **M1**. This is the only constraint that the sensor-level program developers are required to adhere to.
- **M1**: This message is sent by the sensor to *SmartLink* in response to C1. M1 contains information that helps to identify the sensor in *key-value pair* format. The information contained in this message is sent to the cloud IoT platform, as explained in phase (4) in the CADDOT model illustrated in Fig. 5. Detailed explanation of this message is out of the scope of this chapter.
- **C2**: This command consists of two segments. The first segment [DAR] denotes that this is a data request. The second segment [PL] denotes that the command is a pull request which the sensor is expected to respond to with sensors data once.
- **C3**: This command consists of five segments. The first segment [DAR] denotes that this is a data request. The second segment [PS] denotes that the sensor is expected to push data according to the information provided in the rest of the segments. The third segment specifies the sample rate and the fourth segment specifies the data communication frequency rate. The final segment specifies the duration for which the sensor needs to push data to the cloud.

---

[7] http://www.libelium.com/uploads/2013/02/data_frame_guide.pdf

```
    Include Libraries      Definitions      Global variables declaration
void setup()
{
        Read the parameters[SR, CF] from a file In SD card
        Modules initialisation [Communication]
}
void loop()
{
    while(isConnected()){
    // This method tries to discover an open access point. Next, it also
       attempts to establish a connection between the sensor and SmartLink.

       RequestType = ReadSegmentOne()
    // Listen to the communication channel (e.g. WiFi)

    IF (RequestType[segment 1] == DAR){

        IF (RequestType[segment 2] == PL){
            // Sense data and send back to the requester
        }ELSE IF (RequestType[segment 2] == PT){
            // Sense and send data back continuously but temporarily
               until the request expires. Request contains the information
               such as sampling rate, communication frequency, duration

        }ELSE IF (RequestType[segment 2] == PS){
            // Sense and send data back continuously according to the
               schedule specified. contains the information such as
               sampling rate, communication frequency, start time, end time
        }
    }ELSE IF(RequestType[segment 1] == CON){

        IF(RequestType[segment 2] == SMP){
        // Change the sampling rate as specified
        }ELSE IF(RequestType[segment 2] == DCF){
        // Change the data communication frequency as specified
        }ELSE IF(RequestType[segment 2] == SCH){
        %% Download the schedule file from the given location
        }ELSE IF(RequestType[segment 2] == CPR){
        // Send the complete sensor profile information
        }ELSE IF(RequestType[segment 2] == NET){
        // Store network setting such as access point, authentication
           key, IP address and port numbers
        }ELSE{
            // Send back an error message
        }
    }ELSE IF (RequestType[segment 1] == WHO){
        // Send back the identification details (e.g. SensorID)
    }ELSE{
        // Send back an error message
    }
  }
}
```

Fig. 9 A configurable sensor-level program design (*CPD*) that supports dynamic discovery and configuration after deployment at runtime

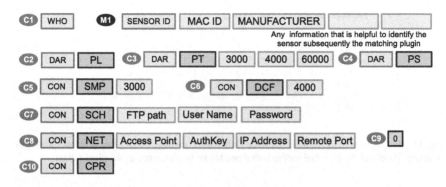

Fig. 10 Command formats used to perform sensor configuration

- **C4**: This command consists of two segments. The first segment [DAR] denotes that this is a data request. The second segment [PS] denotes that the sensor is expected to perform sensing and data transmitting tasks according to a sensing schedule specified in the sensing schedule file. It is expected to push data to the cloud.
- **C5**: This command consists of three segments. The first segment [CON] denotes that this is a configuration command. The second segment [SMP] denotes that this command configures the sampling rate. The third segment holds the actual sampling rate value that the sensor needs to sense in the future.
- **C6**: This command consists of three segments. The first segment [CON] denotes that this is a configuration command. The second segment [DCF] denotes that this command configures the data communication frequency. The third segment holds the actual data communication frequency rate value that the sensor needs to transmit data to the cloud in the future.
- **C7**: This command consists of five segments. The first segment [CON] denotes that this is a configuration command. The second segment [SCH] denotes that this command configures the sensing schedule. The rest of the segments contain information that is essential (i.e. FTP server path, user name, password) to download a sensing schedule file from an FTP server, as depicted in Fig. 10.
- **C8**: This command consists of seven segments. The first segment [CON] denotes that this is a configuration command. The second segment [NET] denotes that this command configures the network settings. The rest of the segments contain the information that is essential to connect to a secure network (i.e. access point name, authentication key, IP address, remote port) so the sensor can directly communicate with the cloud IoT platform.
- **C9**: This command stops the sensor completely and pushes it back to a state where the sensor listens for the next command.
- **C10**: This command consists of two segments. The first segment [CON] denotes that this is a configuration command. The second segment [CPR] denotes that the sensor is expected to reply with the complete sensor profile.

```
package au.csiro.smartlink;

import au.csiro.smartlink.beans.SensorProfile;

interface IPlugin {
    boolean setSamplingRate(int rate);
    boolean setCommunicationFrequency(int frequency);
    boolean setSchedule(in Map ftpSettings);
    boolean setNetworkSettings(in Map netSettings);
    SensorProfile getSensorProfile();
}
```

Fig. 11 IPlugin written in Android Interface Definition Language (*AIDL*) that governs the plugin structure. It defines the essential methods that need to be implemented in the plugin class

Scalable and Extensible Architecture: As we mentioned earlier, the reason for employing a plugin architecture is to support scalability and extensibility. Plugins that are compatible with *SmartLink* can be developed by anyone as long as they follow the basic design principles and techniques explained below. Such a plugin architecture allows us to engage with developer communities and support a variety of different sensors through community-based development. We expect to release our software as free and open source software in the future. We provide the main SmartLink application as well as the standard interfaces which developers can use to start to develop their own plugins to support different sensors. We provide sample plugin source code where developers only need to add their code according to the guidelines provided. The plugin architecture will enable more number of sensors to be supported by *SmartLink* over time. Applications stores (e.g. *Google Play*) built around the Android ecosystem provide an easy way to share and distribute plugins for *SmartLink*. The pluggable architecture dramatically reduces the sensor configuration time.

Let us explain how third party developers can develop plugins in such a way that their plugins are compatible with *SmartLink* so that *SmartLink* can use the plugins to configure sensors at runtime when necessary. In plugin development, there are three main components that need to be considered: (1) the plugin interface written in the Android Interface Definition Language (AIDL), (2) the plugin class written in Java, and (3) the plugin definition in the AndroidManifest file. Figure 11 shows the plugin interface written in AIDL. *IPlugin* is an interface defined in AIDL. Plugin developers should not make any changes in this file. Instead they can use this file to understand how the *SmartLink* plugin architecture works. *IPlugin* is similar to a Java interface. It defines all the methods that need to be implemented by all the plugin classes.

Figure 12 presents the basic structure of a *SmartLink* plugin. Each plugin is defined as an Android service. *SmartLink* plugin developers need to implement five methods: *setSamplingRate(int rate)*, *setCommunicationFrequency(int frequency)*, *setSchedule(in Map ftpSettings)*, *setNetworkSettings(in Map netSettings)* and *getSensorProfile()*. The methods are briefly explained below.

- *setSamplingRate(int rate)*: This method needs to send a command specifying the required sampling rate. For example, in our approach, we defined such a command, *C5*, in Fig. 10.

```
public class [Class] extends Service implements [Any Interface]{
    public int onStartCommand(Intent intent, int flags, int
    startId) {...}
    public void onDestroy() {...}

    public IBinder onBind(Intent intent) {...}

    private final IFunction.Stub mulBinder = new IPlugin.Stub(){
        public boolean setSamplingRate(int rate) throws
        RemoteException {...}

        public boolean setCommunicationFrequency(int frequency)
        throws RemoteException {..}

        public boolean setSchedule(Map ftpSettings) throws
        RemoteException {}

        public setNetworkSettings(Map netSettings) throws
        RemoteException {..}

        public SensorProfile getSensorProfile() throws
        RemoteException {}
    }
}
```

Fig. 12 *SmartLink* plugin is an Android service. This is the basic structure of a *SmartLink* plugin. The body of each method needs to be added by the developer based on the sensor-level program design

- *setCommunicationFrequency(int frequency)*: This method needs to send a command specifying the required communication frequency. For example, in our approach, we defined such a command as *C6* in Fig. 10.
- *setSchedule(in Map ftpSettings)*: This method needs to send a command specifying details (e.g. user-name, password, FTP path) that are required to connect to an FTP server and download the schedule. For example, in our approach, we defined such a command as, *C7*, in Fig. 10.
- *setNetworkSettings(in Map netSettings)*: This method sends a command specifying the details that are required to connect to a secure network so that direct communication between the sensor and the cloud IoT platform can be established. For example, in our approach, we defined such a command, *C8*, in Fig. 10.
- *getSensorProfile()*: This method sends a command to the sensor by asking for profile information. The sensor is expected to reply by providing information such as the data structure it produces, measurement units, and so on. Details of the sensor profiling are out of the scope of this chapter.

Figure 13 shows how the plugins need to be defined in the AndroidManifest so that the *SmartLink* application can automatically query and identify them. The Android plugin must have an intent filter which has action name *au.csiro.smartlink.intent. action.PICKPLUGIN*. Developers can provide any category name.

Support and Utilize Existing Solutions: Our model utilizes a few existing solutions. We employed Global Sensor Network [1] as the cloud IoT middleware. In CADDOT, GSN performs phases 3, 4, and 7. GSN is a widely used platform in the sensor data processing domain and is used in several European projects, including

```
<service
  android:name=[Plugin name]
  android:exported="true" >
  <intent-filter>
    <action android:name="au.csiro.smartlink.intent.action.PICK_PLUGIN"/>
    <category android:name="au.csiro.smartlink.intent.category.[PLUGIN_NAME]"/>
  </intent-filter>
</service>
```

Fig. 13 Code snippet of the plugin's *AndroidManifest* file

OpenIoT [31]. MOSDEN [34] is middleware that collects sensor data. MOSDEN is ideal for the application strategies we discussed in Sect. 6 (Fig. 6) for use in conjunction with *SmartLink*. *SmartLink* only performs the configuration. Sensor data collection needs to be performed by either cloud IoT middleware or solutions like MOSDEN. The proposed CADDOT model as well as the *SmartLink* tool complement the other solutions proposed by us as well as other researchers. Together, these solutions enable smooth data flow from sensors to the cloud autonomously.

7 Implementation and Experiment Testbed

We deployed the *SmartLink* application in a Google Nexus 4 mobile phone (Qualcomm Snapdragon S4 Pro CPU and 2 GB RAM), which runs the Android platform 4.2.2 (Jelly Bean). We deployed 52 sensors on the third floor of the CSIT building (#108) at the Australian National University. All sensors we employed in our experiment are manufactured by Libelium [27]. The sensors we used sense a wide variety of environmental phenomena, such as temperature, proximity and presence, stretch, humidity and so on [27]. *SmartLink* supports sensor discovery and configuration using both WiFi and Bluetooth. Other communication technologies such as ZigBee and RFID are supported through Libelium *Expansion Radio Boards* [27]. In order to simulate the heterogeneity of the sensors (in terms of communication sequence), we programmed each sensor to behave and respond differently. As a result, each sensor can only communicate with a plugin that supports the same communication sequence.

8 Evaluation of the Prototype

In this section, we explain how we evaluate the proposed CADDOT model and *SmartLink* tool using prototype implementations. We identified ten steps performed in the dynamic discovery and sensor configuration process. We measured the average amount of time taken by each of these steps (average of 30 sensor configurations). Figure 14 illustrates the results and the following steps are considered: Time taken

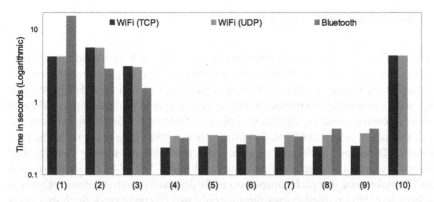

Fig. 14 Time taken (y-axis) to discover and configure a sensor step-by-step (x-axis). The experiments were conducted using three protocols: TCP, UDP, and Bluetooth

to (1) set up the sensor, (2) initiate connection between the sensor and *SmartLink*, (3) initiate communication between sensor and *SmartLink*, (4) extract sensor identification information, (5) retrieve the complete profile of the sensor, (6) configure the sampling rate, (7) configure the communication frequency, (8) configure the sensing schedule, (9) configure the network and authentication details (so the sensor can directly connect to the cloud), and (10) connect to the secure network using the provided authentication details.

Results: According to the results, the actual configuration tasks take less than 1 s. There is a slight variation in completion time in configuration step (4)–(9). This is due to storage access and differences in processing of configuration commands. Sensors takes comparatively longer time to connect to a network as well as to discover and connect to *SmartLink*. Especially, Bluetooth takes much longer to scan for devices in a given environment before it discovers and connects to *SmartLink*. Configuration is slightly faster when using TCP in comparison to UDP and Bluetooth. This is mainly due to reliability. However, the time differences are negligible. FTP is used to retrieve a scheduling file from a file server. This can take 15–25 s depending on the network availability, traffic, and file size. If a sensor cannot access a server via the Internet, a file can be transferred from *SmartLink* to the sensor as typical commands. Sensors generate the scheduling file using the data it receives from *SmartLink*. When using WiFi, a sensor may takes up to 4.5 s to connect to a secure network (e.g. WPA2). In contrast, sensors can connect to *SmartLink*'s open access point in less than 4 s. Despite the protocol we use, sensors take 5–15 s to boot and setup themselves. The setup stage consists of activities such as reading default configuration from files, and switching necessary modules and components (communication modules, real-time clock, SD card, sensor broads and so on).

9 Discussion and Lessons Learned

In what follows, we discuss major lessons we learned along with limitations. According to our results, it is evident that a single sensor can be configured in less than 12 s (i.e. assuming sensors are already booted, which takes an additional 5–15 s depending on the communication protocol). This is a significant improvement over a manual labour intensive sensor configuration approach. Additionally, *SmartLink* can engage with number of sensor configuration processes at a given time in parallel. The proposed CPD has not made any negative impact towards the sensing functionality though it supports advance configuration capabilities. The IF-ELSE structure used in CPD makes sure that each request gets to the destination with minimum execution of lines (e.g. 'PL' request passes through only two IF conditions). Such execution reduced the impact on sensing tasks while configuration tasks are also supported efficiently. Even though a detailed discussion on data acquisition methods is out of scope, it is important to note that pull, temporary push, and schedule based push add a significant amount of flexibility where each of the techniques is suitable to be used in different circumstances [36]. The cloud server has the authority to decide which method to be used based on the context information. This increases the efficiency and application scenario where the sensors can be used in sustainable (i.e. in term of energy) manner. Once the initial discovery and configuration of smart things are done, further configuration can be done in more user friendly manner by using techniques such as augmented reality [19].

10 Open Challenges

In this section, we briefly introduce some of the major open research challenges in the domain that are closely related to this work. We identify four main challenges that provide different research directions.

Sensing strategy optimization: We briefly highlighted the importance of optimizing sensing schedules based on context information in Sect. 3. Sensing strategy development encapsulates a broad set of actions such as deciding the sensing schedule, sampling rate, and network communication frequency for each sensor. Such a development process needs to consider two main factors: user requirements and availability of sensors. In IoT, there is no single point of control or authority. As a result, different parties are involved in sensor deployments. Such disorganized and uncoordinated deployments can lead to redundant sensor deployment. In order to use the sensor hardware in an optimized manner, sensing strategies need to be developed by considering factors such as sensor capabilities, sensor redundancies (e.g. availability of multiple sensors that are capable of providing similar data), and energy availability. Energy conservation is a key in sustainable IoT infrastructure because the resources constrained nature of the sensors. We provided such an example in Sect. 3 related to the agricultural domain. We believe that sensing as a service is a major business model that could drive IoT in the future. In such circumstances, collecting data from all the available sensors has no value. Instead, sensor data should be collected and processed only in response to consumer demand [37].

Context discovery: This is an important task where discovered information will be used during a reasoning process (e.g.sensing strategy development). *"Context is any information that can be used to characterise the situation of an entity. An entity is a person, place, or object that is considered relevant to the interaction between a user and an application, including the user and applications themselves"* [2]. Further discussion on context information and its importance for the IoT is surveyed in [36]. Context-based reasoning can be used to improve the efficiency of the CADDOT model where a matching plugin can be discovered faster, especially in situations where a perfect match cannot be found. For example, the location of a given sensor,[8] sensors nearby, details of the sensors configured recently, historic data related to sensor availability in a given location, etc. can be fused and reasoned using probabilistic techniques in order to find a matching plugin in an efficient manner. After integrating sensors into cloud-based IoT, the next phase is collecting data from the sensors. Annotating context information to retrieve sensor data plays a significant role in querying and reasoning them in later stages. Especially, in the sensing as a service model, sensor data consumers may demand such annotation so that they can feed data easily into their own data processing applications for further reasoning and visualization tasks. Some context information can be easily discovered at sensor-level (e.g. battery level, location) and others can be discovered at the cloud-level by fusing multiple raw data items (e.g. activity detection). Such context annotated data help to perform more accurate fusing and reasoning at the cloud level [30].

Utilization of heterogeneous computational devices: Even though the IoT envisions billions of *'things'* to be connected to the Internet, it is not possible and practical to connect all of them to the Internet directly. This is mainly due to resource constraints (e.g. network communication capabilities and energy limitations). Connecting directly to the Internet is expensive in terms of computation, bandwidth use, and hardware costs. Enabling persistent Internet access is challenging and also has a negative impact on miniaturization and energy consumption of the sensors. Due to such difficulties, IoT solutions need to utilize different types of devices with different resource limitations and capabilities. In Fig. 15, we broadly categorise these devices into six categories (also called levels or layers). Devices on the right side may use low-energy short distance wireless communication protocols to transmit the collected sensor data to the devices on the left. Devices on the left can use long distance communication protocols to transmit the data to the cloud for further processing. However, the more devices we use in smart environments, the more difficult it becomes to detect faults where an entire system could fail [44]. Providing a unified middleware support across heterogeneity of devices with wider rage of capabilities is an open challenge [10, 20].

---

[8] Location can be represented in many ways: GPS coordinate (e.g. $-35.280325, 149.113166$), name of a building (e.g. CSIT building at ANU), name of a city (e.g. Canberra), part of a building (e.g. living room), floor of a building (e.g. 2nd floor), specific part of a room (e.g. kitchen-top).

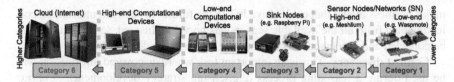

Fig. 15 Categorization of IoT devices based on their computational capabilities. The devices belonging to each category have different capabilities in terms of processing, memory, and communication. They are also different in price, with devices becoming more expensive towards the *left*. The computational capabilities also increase towards the *left*

Security and privacy: In this work, we considered some degree of security as briefly discussed in Sect. 6. However, research on security in the IoT is largely unexplored. Security and privacy need to be provided at both sensor-level and cloud-level. It is critical to develop a security model to protect the sensor configuration process, considering questions such as (1) *when to allow reconfiguration of a sensor*, (2) *who has the authority to configure a sensor at a given time*, (3) *how to change ownership of a sensor*, (4) *how to detect sensors with harmful programs installed on them that may cause security threats to a network*. Security and privacy concerns related to the IoT are presented in [41]. Additionally, security challenges unique to the sensing as a service model are discussed in [37].

11 Conclusions and Outlook

In this chapter, we addressed the challenge of integrating sensors into cloud-based IoT platforms through context-aware dynamic discovery and configuration. Traditionally, integration of '*things*' to software solutions is considered a labour-intensive, expensive and time-consuming task that needs to be carried out by technical experts. Such challenges hinders the non-technical users from adopting IoT to build smart environments. To address this problem, we presented the CADDOT model, an approach that automates the sensor discovery and configuration process in smart environments efficiently and effortlessly by handling key challenges such as a higher number of sensors available, heterogeneity, on-demand sensing schedules, sampling rate, data acquisition methods, and dynamicity. It also encourages non-technical users to adopt IoT solutions with ease by promoting automatic discovery and configuration IoT devices.

In this work, we supported and evaluated different types of communication technologies (i.e. WiFi and Bluetooth), application strategies, and sensor-level program designs, each of which has their own strengths and weaknesses. We validate the CADDOT model by deploying it in an office environment. As CADDOT required minimum user involvement and technical expertise, it significantly reduces the time and cost involved in sensor discovery and configuration. In the future, we expect to address the open challenges discussed in Sect. 10. In addition, we expect to integrate our solution with other existing solutions such as MOSDEN [34] and OpenIoT [31]. The functionality provided by CADDOT can improve these solutions in a major way.

Acknowledgments Authors acknowledge support from SSN TCP, CSIRO, Australia and ICT Project, which is co-funded by the European Commission under seventh framework program, contract number FP7-ICT-2011-7-287305-OpenIoT. The Author(s) also acknowledge help and contributions from The Australian National University.

References

1. Aberer, K., Hauswirth, M., Salehi, A.: Infrastructure for data processing in large-scale interconnected sensor networks. In: International Conference on Mobile Data Management, pp. 198–205, May 2007
2. Abowd, G.D., Dey, A.K., Brown, P.J., Davies, N., Smith, M., Steggles, P.: Towards a better understanding of context and context-awareness. In: Proceedings of the 1st International Symposium on Handheld and Ubiquitous Computing, HUC '99, pp. 304–307. Springer-Verlag, London (1999)
3. Ashton, K.: That 'internet of things' thing in the real world, things matter more than ideas. RFID J. http://www.rfidjournal.com/article/print/4986 (2009). Accessed 30 Jul 2012
4. Atzori, L., Iera, A., Morabito, G.: The internet of things: a survey. Comput. Netw. **54**(15), 2787–2805 (2010)
5. BCC Research. Sensors: technologies and global markets. Market forecasting, BCC Research. http://www.bccresearch.com/report/sensors-technologies-markets-ias006d.html (2011). Accessed 05 Jan 2012
6. Bizer, C., Boncz, P., Brodie, M.L., Erling, O.: The meaningful use of big data: four perspectives—four challenges. SIGMOD Rec. **40**(4), 56–60 (2012)
7. Botts, M., Robin, A.: Opengis sensor model language (sensorml) implementation specification. Technical report. Open Geospatial Consortium Inc. https://portal.opengeospatial.org/modules/admin/license_agreement.php?suppressHeaders=0&access_license_id=3&target=http://portal.opengeospatial.org/files/%3fartifact_id=12606 (2007). Accessed 15 Dec 2011
8. Brush, A.B., Filippov, E., Huang, D., Jung, J., Mahajan, R., Martinez, F., Mazhar, K., Phanishayee, A., Samuel, A., Scott, J., Singh, R.P.: Lab of things: a platform for conducting studies with connected devices in multiple homes. In: Proceedings of the 2013 ACM Conference on Pervasive and Ubiquitous Computing Adjunct Publication, UbiComp'13 Adjunct, pp. 35–38. ACM, New York (2013)
9. Carlson, D., Schrader, A.: Dynamix: an open plug-and-play context framework for android. In: Internet of Things (IOT), 2012 3rd International Conference on the, pp. 151–158. (2012)
10. Chaqfeh, M., Mohamed, N.: Challenges in middleware solutions for the internet of things. In: Collaboration Technologies and Systems (CTS), 2012 International Conference on, pp. 21–26. (2012)
11. Commonwealth Scientific and Industrial Research Organisation (CSIRO), Australia. Phenonet: Distributed sensor network for phenomics supported by high resolution plant phenomics centre, csiro ict centre, and csiro sensor and sensor networks tcp. http://phenonet.com (2011). Accessed 20 Apr 2012
12. Compton, M., Barnaghi, P., Bermudez, L., Garcfa-Castro, R., Corcho, O., Cox, S., Graybeal, J., Hauswirth, M., Henson, C., Herzog, A., Huang, V., Janowicz, K., Kelsey, W.D., Phuoc, D.L., Lefort, L., Leggieri, M., Neuhaus, H., Nikolov, A., Page, K., Passant, A., Sheth, A., Taylor, K.: The SSN ontology of the w3c semantic sensor network incubator group. Web Seman. Sci., Serv. Agents World Wide Web **17**, 25–32 (2012)
13. Cook, D., Das, S.: Smart Environments: Technology, Protocols and Applications (Wiley Series on Parallel and Distributed Computing). Wiley-Interscience, London (2004)

14. Cosm.: Cosm platform. https://cosm.com/ (2007) Accessed 05 Aug 2012
15. Dixon, C., Mahajan, R., Agarwal, S., Brush, A., Lee, B., Saroiu, S., Bahl, V.: An operating system for the home, In: Symposium on Networked Systems Design and Implementation (NSDI), USENIX, Apr 2012
16. EPCglobal.: Epc tag data standard version 1.5. Standard specification, EPCglobal. http://www.gs1.org/gsmp/kc/epcglobal/tds/tds_1_5-standard-20100818.pdf (2010). Accessed 16 Aug 2011
17. GSN team: global sensor networks project. http://sourceforge.net/apps/trac/gsn/ (2011). Accessed 16 Dec 2011
18. Guillemin, P., Friess, P.: Internet of things strategic research roadmap. Technical report. The Cluster of European Research Projects. http://www.internet-of-things-research.eu/pdf/IoT_Cluster_Strategic_Research_Agenda_2009.pdf (2009)
19. Heun, V., Kasahara, S., Maes, P.: Smarter objects: using ar technology to program physical objects and their interactions. In: CHI' 13 Extended Abstracts on Human Factors in Computing Systems, CHI EA'13, pp. 961–966. ACM, New York (2013)
20. Hong, Y.: A resource-oriented middleware framework for heterogeneous internet of things. In: Cloud and Service Computing (CSC), 2012 International Conference on, pp. 12–16. (2012)
21. Hu, P., Indulska, J., Robinson, R.: An autonomic context management system for pervasive computing. In: Pervasive Computing and Communications, 2008. PerCom 2008. Sixth Annual IEEE International Conference on, pp. 213–223, Mar 2008
22. IEEE Instrumentation and Measurement Society. IEEE standard for a smart transducer interface for sensors and actuators wireless communication protocols and transducer electronic data sheet (teds) formats. IEEE Std 1451.5-2007, pp. C1-236. 5-2007
23. International Data Corporation (IDC) Corporate USA. Worldwide smart connected device shipments. http://www.idc.com/getdoc.jsp?containerId=prUS23398412 (2012). Accessed 01 Aug 2012
24. Jung, M., Reinisch, C., Kastner, W.: Integrating building automation systems and ipv6 in the internet of things. In: Innovative Mobile and Internet Services in Ubiquitous Computing (IMIS), 2012 Sixth International Conference on, pp. 683–688. (2012)
25. Kiljander, J., Takalo-Mattila, J., Etelapera, M., Soininen, J.-P., Keinanen, K.: Enabling end-users to configure smart environments. In: Applications and the Internet (SAINT), 2011 IEEE/IPSJ 11th International Symposium on, pp. 303–308. (2011)
26. Kovatsch, M., Lanter, M., Duquennoy, S.: Actinium: a restful runtime container for scriptable internet of things applications. In: Internet of Things (IOT), 2012 3rd International Conference on the, pp. 135–142. (2012)
27. Libelium Comunicaciones Distribuidas. libelium. http://www.libelium.com/ (2006). Accessed 28 Nov 2012
28. Manyika, J., Chui, M., Brown, B., Bughin, J., Dobbs, R., Roxburgh, C., Byers, A.H.: Big data: the next frontier for innovation, competition, and productivity. Technical report, McKinsey Global Institute, 2011. http://www.mckinsey.com/Insights/MGI/Research/Technology_and_Innovation/Big_data_The_next_frontier_for_innovation [Accessed on: 2012–06-08]
29. Naphade, M., Banavar, G., Harrison, C., Paraszczak, J., Morris, R.: Smarter cities and their innovation challenges. Computer 44(6), 32–39 (2011)
30. Oh, Y., Han, J., Woo, W.: A context management architecture for large-scale smart environments. Commun. Mag. IEEE 48(3), 118–126 (2010)
31. OpenIoT Consortium. Open source solution for the internet of things into the cloud. http://www.openiot.eu (2012). Accessed 08 Apr 2012
32. Pereira, P., Eliasson, J., Kyusakov, R., Delsing, J., Raayatinezhad, A., Johansson, M.: Enabling cloud connectivity for mobile internet of things applications. In: Service Oriented System Engineering (SOSE), 2013 IEEE 7th International Symposium on, pp. 518–526. (2013)
33. Perera, C., Jayaraman, P., Zaslavsky, A., Christen, P., Georgakopoulos, D.: Dynamic configuration of sensors using mobile sensor hub in internet of things paradigm. IEEE 8th International Conference on Intelligent Sensors. Sensor Networks, and Information Processing (ISSNIP), pp. 473–478. Melbourne, Australia, Apr 2013

34. Perera, C., Jayaraman, P.P., Zaslavsky, A., Christen, P., Georgakopoulos, D.: Mosden: an internet of things middleware for resource constrained mobile devices. In: Proceedings of the 47th Hawaii International Conference on System Sciences (HICSS). Hawaii, USA, Jan 2014
35. Perera, C., Zaslavsky, A., Christen, P., Compton, M., Georgakopoulos, D.: Context-aware sensor search, selection and ranking model for internet of things middleware. In: IEEE 14th International Conference on Mobile Data Management (MDM), Milan, Italy, June 2013
36. Perera, C., Zaslavsky, A., Christen, P., Georgakopoulos, D.: Context aware computing for the internet of things: a survey. IEEE Commun. Surv. Tut. 16(1), 414–454 (2014). doi:10.1109/SURV.2013.042313.00197
37. Perera, C., Zaslavsky, A., Christen, P., Georgakopoulos, D.: Sensing as a service model for smart cities supported by internet of things. Trans. Emerg. Telecommun. Technol. **25**(0):81–93 (2014)
38. Perera, C., Zaslavsky, A., Christen, P., Salehi, A., Georgakopoulos, D.: Capturing sensor data from mobile phones using global sensor network middleware. In: IEEE 23rd International Symposium on Personal Indoor and Mobile Radio Communications (PIMRC), pp. 24–29. Sydney, Australia, Sept 2012
39. Perera, C., Zaslavsky, A., Liu, C.H., Compton, M., Christen, P., Georgakopoulos, D.: Sensor search techniques for sensing as a service architecture for the internet of things. IEEE Sens. J. **14**(2):406–420 (2014)
40. Postscapes.com. A showcase of the year's best Internet of Things projects. http://postscapes.com/awards/winners (2012). Accessed 10 Jan 2013
41. Roman, R., Najera, P., Lopez, J.: Securing the internet of things. Computer **44**(9), 51–58 (2011)
42. Sheng, X., Tang, J., Xiao, X., Xue, G.: Sensing as a service: challenges, solutions and future directions. Sens. J. IEEE **13**(10), 3733–3741 (2013)
43. Shon, T., Park, Y.: Implementation of rf4ce-based wireless auto configuration architecture for ubiquitous smart home. In: Complex, Intelligent and Software Intensive Systems (CISIS), 2010 International Conference on, pp. 779–783. (2010)
44. Son, J.-Y., Lee, J.-H., Kim, J.-Y., Park, J.-H., Lee, Y.-H.: Rafd: resource-aware fault diagnosis system for home environment with smart devices. Consum. Electron. IEEE Trans. **58**(4), 1185–1193 (2012)
45. Sundmaeker, H., Guillemin, P., Friess, P., Woelffle, S.: Vision and challenges for realising the internet of things. Technical report, European Commission Information Society and Media. http://www.internet-of-things-research.eu/pdf/IoT_Clusterbook_March_2010.pdf (Mar 2010). Accessed 10 Oct 2011
46. Weiser, M., Gold, R., Brown, J.S.: The origins of ubiquitous computing research at parc in the late 1980s. IBM Syst. J. **38**(4), 693–696 (1999)

Simultaneous Analysis of Multiple Big Data Networks: Mapping Graphs into a Data Model

Ahmad Karawash, Hamid Mcheick and Mohamed Dbouk

Abstract Network analysis is of great interest to web and cloud companies, largely because of the huge number of web-networks users and services. Analyzing web networks is helpful for organizations that profit from how network nodes (e.g. web users) interact and communicate with each other. Currently, network analysis methods and tools support single network analysis. One of the Web 3.0 trends, however, namely personalization, is the merging of several user accounts (social, business, and others) in one place. Therefore, the new web requires simultaneous multiple network analysis. Many attempts have been made to devise an analytical approach that works on multiple big data networks simultaneously. This chapter proposes a new model to map web multi-network graphs in a data model. The result is a multidimensional database that offers numerous analytical measures of several networks concurrently. The proposed model also supports real-time analysis and online analytical processing (OLAP) operations, including data mining and business intelligence analysis.

1 Introduction

By its very nature, cloud network connection shares big data. The amount of data crossing networks will continue to explode. By 2020, 50 billion devices will be connected to networks and the internet [9] and the absolute volume of digital information

A. Karawash (✉) · H. Mcheick
Department of Computer Science, University of Quebec at Chicoutimi (UQAC),
555 Boulevard de l'Université Chicoutimi, Chicoutimi G7H2B1, Canada
e-mail: ahmad.karawash1@uqac.ca

H. Mcheick
e-mail: hamid_mcheick@uqac.ca

A. Karawash · M. Dbouk
Department of Computer Science, Ecole Doctorale des Sciences et de Technologie (EDST),
Université Libanaise, Hadath-Beirut, Lebanon
e-mail: mdbouk@ul.edu.lb

N. Bessis and C. Dobre (eds.), *Big Data and Internet of Things:*   243
A Roadmap for Smart Environments, Studies in Computational Intelligence 546,
DOI: 10.1007/978-3-319-05029-4_10, © Springer International Publishing Switzerland 2014

is predicted to increase to 35 trillion gigabytes, much of it comes from new sources including blog networks, social networks, internet search, and sensor networks. The network can play a valuable role in increasing big data's potential for enterprises. It can assist in collecting data and providing context at high velocity and it can impact the customer's experience.

As the number of online-network communications is increasing sharply, it is difficult to access or analyze relevant information from the web. One possible solution to this problem offered by Web 3.0 is web personalization [7]. Personalization aims at alleviating the burden of information overload by tailoring the information presented to individual and immediate user needs [16]. One of the personalization requirements, which can affect a large part of the network data, is the combination of user web accounts to constitute a personal profile for each user.

In response to emerging trends, this capter studies how to deal with multiple networks using the data model view. We treat the multiple network idea from the graph model perspective. Indeed, network graphs have been growing rapidly and showing their critical importance in many applications such as the analysis of XML, social networks, the web, biological data, multimedia data and spatial-temporal data.

This chapter proposes a model which merges multiple network graphs and then maps the obtained graph in the data model, thereby achieving a multidimensional database which enables better network analysis. As a result, an OLAP (online analytical processing) approach (data-mining and business-intelligence analysis) can be applied in numerous networks at the same time. The network's data are collected in such a way that analysis measures are requested by a database query for several networks at the same time. There are many network analysis measures, but this chapter studies only centrality measures. To explain the proposed idea, this chapter also consists of a case study simulation of three social network groups. Also, to illustrate the importance of our study, we show two real examples from different domains in which the model is applicable.

The chapter is organized as follows. Section 2 explains the problem. Section 3 describes the background to the chapter and the state of the art. Section 4 discusses the proposed model for mapping multi-network graphs in a multidimensional database. Section 5 explains the simulation steps and describes some results. Section 6 highlights the benefits of the proposed model. Section 7 concludes and outlines our future work.

2 Problem Statement

The huge number of random web and cloud connections and the unorganized storage of big data in Web 2.0 motivated computer scientists to develop Web 3.0. The new web is based on a wide arrangement of data. One of the problems with Web 2.0 is the random distribution of multi-accounts of users (social, business or other). Web 3.0 proposed the idea of personalization that meant web concepts shifted from working with words to dealing with personal profiles. To achieve a personal profile, all the user's accounts are treated as one block (account aggregation). Although personalization concept can solves many problems, including random accounts and search engine difficulties, it could affect negatively in the analysis phase. Before personalization, analytical methods were easier to apply because the target was one network. In the new web, however, the goal is multi-network analysis (or multidimensional network graph analysis). For example, in the social network case it is easy to apply analysis to one network as a calculation of centrality measures, but how can we analyze several graphs with a different purpose for one person at the same time (e.g. calculating the degree of centrality of a person in both Facebook and Twitter networks at the same time and with one request)?

Currently the available methods and tools deal with one-dimensional graphs. Thus, the challenge to the new web is to analyze the multi-network (multidimensional) graphs simultaneously. What is the degree of online network analysis that can be achieved with Web 3.0?

3 Background and Related Works

"Network" is a heavily overloaded term, and "network analysis" means different things to different people. Specific forms of network analysis are used in the study of diverse structures such as the internet, transportation systems, web graphs, electrical circuits, project plans, and so on [9]. Numerous network analysis measures have been developed since the mid-twentieth century: for example, Katz [14], Hubbell [13], Hoede's [12], Taylor's [19] and Freeman's closeness and betweenness [12], flow betweenness [11], and Bonacich's eigenvector [1, 2], etc.

Although studies on network analysis have been around for decades, and a surfeit of algorithms and systems have been developed for multidimensional analysis in relational databases, none has taken both aspects into account in the multidimensional network scenario.

Ulrik Brandes proposed algorithms to compute centrality indices on large network graphs (2001). Costenbader et al. discussed, in [5], how to analyze a research network and they used bootstrap sampling procedures research network to determine how sampling affects the stability of several different network centrality measures. In [4], Chen et al. developed a graph OLAP framework, which presents a multi-dimensional and multi-level view over graphs. In [20], Tore et al. proposed generalizations that combined centrality measures. Also in Manuel et al. [15] devised ManyNets to analyze several networks at the same time with visualization. Xi-Nian et al. investigated a broad array of network centrality measures to provide novel insights into connectivity within the whole-brain functional network [23]. Also in 2012, the HMGraph OLAP was developed by Mu et al., that provide more operations on a multi-dimensional heterogeneous information network. In Daihee et al. [6] devised the NetCube network traffic analysis model using online analytical processing (OLAP) on a multidimensional data cube, which provides a fast and easy way to construct a multi-dimensional analysis of long-term network traffic data. Wararat et al. proposed a framework to

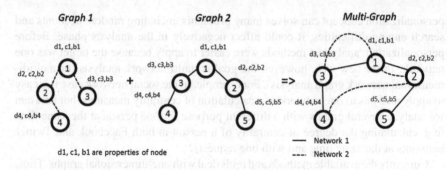

Fig. 1 Merging multiple network graphs in one multi-network graph

materialize this combination of information networks and discussed the main challenges [22].

4 Multi-Network Graph and Data Model (Proposed Model)

This section highlights the relationship between the graph model and the data model. The new web trend is to use a multi-network model instead of a graph model to deal with the explosive growth of online networks. A graph is a representation of a set of objects wherein some pairs of objects are connected by links. The interconnected objects are represented by mathematical abstractions called vertices, and the links that connect some pairs of vertices are called edges. Typically, a graph is depicted in diagrammatic form as a set of dots for the vertices, joined by lines or curves for the edges [21]. The edges may be directed or undirected. A multi-network graph is generally understood to mean a graph in which multiple edges are allowed.

Figure 1 shows an example of how a multi-graph is obtained from several graphs. Graph1 and Graph2 represent the node connections in two different networks.

A multi-graph is based on *vertices*, *edges*, *belonging network* and *vertex properties*. A multi-graph is an ordered set $M = (V, L, N, P)$ such that:

– V is a set of *vertices*,
– $L = \{\{p, q\} : p, q \varepsilon V\}$ is a set of links (*edges*) between two *vertices* which are subsets of,
– $N = \{n1, n2, \ldots, nk\}$ is the set of *belonging networks* that node belongs to and
– $P = \{degree\ centrality,\ closeness,\ betweenness, \ldots, etc\}$ is the set of *properties* of a node.

In order to talk about the relationship between the multi-graph model and the data model, it is necessary first to introduce the entity relationship (*ER*) model. *ER* is the most widespread semantic data model. It was first proposed by Chen [3] and has become a standard, extensively used in the design phase of commercial applications.

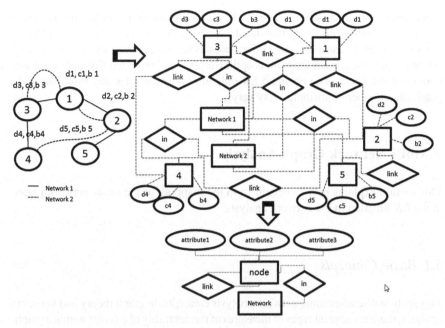

Fig. 2 Mapping a multi-network graph into ER diagram

The entity relationship set $ER = (E, R, A)$ is composed of three basic types of sets: *entities*, *relationships*, and *attributes*. An entity set E denotes a set of objects, called *instances*, which have common properties. Element properties are modeled through a set of *attributes* A, whose values belong to one of several predefined domains, such as integer, string, or boolean. Properties that are caused by relations to other entities are modeled through the participation of the *entity* in *relationships*. A *relationship* set R denotes a set of tuples, each of which represents an association among a different combination of instances of the *entities* that participate in the *relationship*.

Let $g:V->E$ and $h : N->E$ be two functions mapping the values in set *V and N* to set E, in which if $x\varepsilon V$, then $g(x)\varepsilon E$. Facts $g(V)$ and $h(N)$, derived from the multi-graph M are defined as follows: every *vertex* (node) x in the set of vertices V and every *belonging network* y in the set N is mapped by g and h respectively into *entities* in the set E.

Let $k:L->R$ be a function such that $k(i)\varepsilon R$, where $i\varepsilon L$. This means that every *edge* belonging to set L is mapped to *relationship* by k.

Let $w:P->A$ be a function such that $w(c)\varepsilon A$, where $c\varepsilon P$. This means that every *property* in the multi-graph is mapped in *attribute* in the *ER* diagram.

Figure 2 shows how a multi-network graph is mapped in the *ER* diagram. The multi-network graph consists of five nodes each with specific properties. Also, as in the graph in Fig. 2, some of the nodes belong to network 1 (lined link) whereas others belong to network 2 (dotted line), and some may belong to both networks at the same time. As shown in Fig. 2, the top **ER** diagram forms the result of the translation, in which nodes are translated to entities, properties to attributes and links to relationships. Because the same information is repeated (node name, network type and attributes) the top **ER** diagram is optimized into an optimized **ER** diagram at the bottom of the figure. The obtained **ER** diagram is the same for any multi-network graph (the number of attributes may vary).

5 Multi-Network Graph Analysis

This section explains how to benefit from the mapping of the multi-network graph in the **ER** diagram in the network analysis.

5.1 Basic Concepts

This section discusses some network analysis concepts. In graph theory and network analysis, there are several types of measures of the centrality of a vertex within a graph that determine the qualified status of a vertex within the graph (e.g. how important a person is within a social network, how important a room is within a building or how well-used a road is within an urban network). Many of the centrality concepts were first developed in social network analysis, such as degree centrality, betweenness, and closeness.

Degree Centrality: The first and conceptually simplest concept, which is defined as the number of links incident upon a node. It is the number of nodes adjacent to a given node (sent = out a degree or received = in degree). The measure is entirely local, saying nothing about how one is positioned in the wider network. Degree centrality is defined by a degree of unit $x : c_D(x) = degree\ of\ unit\ x$. Relative degree centrality is:

$C_D(x) = c_D(x)/highest\ degree - 1 = c_D(x)/n - 1$, if n is the number of units in a network, the highest possible degree (network without loops) is $n - 1$.

Closeness Centrality: Measures how many steps away from others one is in the network. Those with high closeness can reach many people in a few steps. Technically it is the sum of network distance to all others. This is not just a local measure, but uses information from the wider network. Sabidussi [18] suggested a measure of centrality according to the closeness of unit x: $C_c(x) = 1/\sum_{y \varepsilon U} d(x, y)$, where $d(x; y)$ is the length of the shortest path between units x and y, and U is the set of all units. Relative closeness centrality is defined by: $C_c(x) = (n - 1) * C_c(x)$, where n is the number of units in the network.

Betweenness Centrality: Betweenness centrality measures how often a given actor sits "between" others, "between" referring to the shortest geodesic. It detects the actor that has a higher likelihood of being able to control the flow of information in the network. Freeman [10] defined the centrality measure of unit x according to betweenness in the following way:

$$c_B(x) = \sum_{y<z} \frac{\text{\# of shortest paths between } y \text{ and } z \text{ through units } x}{\text{\# of shortest paths between } y \text{ and } z}$$

Suppose that communication in a network always passes through the shortest available paths: the betweenness centrality of unit x is the sum of probabilities across all possible pairs of units that the shortest path between y and z will pass through unit x.

In network analysis, relative betweenness centrality is used; it has two formulas according to the type of network. For undirected graphs of relative betweenness, we have $C_B(x) = c_B(x)/((n-1)*(n-2)/2)$. For direct graphs of relative betweenness, we have $C_B(x) = c_B(x)/(n-1)*(n-2)$.

5.2 Analyzing Multi-Network Graphs Using OLAP

This part maps the obtained *ER* diagram in Fig. 2 to a multi-dimensional database (cube). In this mapping, we study the three centrality measures (degree centrality, closeness and betweenness) explained in Sect. 5.1.

Every data analysis is based on a dataset, which is stored in a database. But in our case, we have a multi-dimensional graph. Therefore, we propose to map this type of graph in a multidimensional database. The functions and notations in this part depend on the previous definitions in Sect. 4 above. Let $L_M(s, x)$ denote a link between s and x where $s, x \varepsilon V$ and $\varphi_s = |\sum_{i \varepsilon IN} L_M(s, x_i)/n - 1|$, where n is the number of nodes. Let function $d_M(s, t)$ calculate the shortest path distance between $s, t \varepsilon V (G_M)$.

Let $S_{st} = |n - 1/d_M(s, t)|$, where n is the number of nodes. Let P_{st} denote a set of different shortest paths between s *and* t (such that $s, t \varepsilon V$) and $\beta_{st} := |P_{st}|$. For every $v \varepsilon V$ let $P_{st}(v)$ denote the set of different shortest paths containing v with $s \neq v \neq t$, *and* $\beta_{st} := |P_{st}(v)|$.

Let $D_{i*j*k}(R_{i*k}(D), C_{j*k}(D))$ be a multidimensional database (cube) of order 3, which represents a node in a multi-network graph, as shown in Fig. 3. $R_{i*k}(D)$ denotes the row i at the k level of the cube and $C_{j*k}(D)$ denotes the column j at the level k of the cube, and $i, j, k \varepsilon IN^+$.

Let $M_k(a_{i,j})$ be a matrix of i, j dimensions, where $a_{i,j}$ is a value of the matrix entity at row i and column j with $i, j, k \varepsilon IN^*$. $M_k := U_{0 \leq i \leq n} Ri * k_D := k, n \epsilon N$

$U_{0 \leq i \leq n} Cj * k(D)$, which means matrix M_k is formed by the union of cube $k, n \epsilon N$
rows or column at a specific level k. Let denote the set of networks to be studied such that $R_D = \{R_0(D), R_1(D), \ldots, R_N(R_D)\} = \{Networkname_1, Networkname_2, \ldots, Networkname_N\}$. Let set C_D denote the set of node names such that $C_D = \{C_{0*k}(D),$

Fig. 3 A structure of a cube
with three faces and "k" levels
of analysis measures

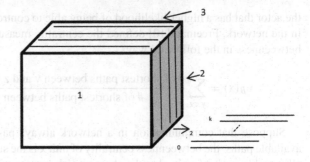

Table 1 Matrix example represents level 0 of the cube

	Nodename1 (n_1)	Nodename2 (n_2)	Nodename3 (n_3)
Network1 (r_1)	$\phi_{sn1}{}^{r1}$	$\phi_{sn2}{}^{r1}$	$\phi_{sn3}{}^{r1}$
Network2 (r_2)	$\phi_{sn1}{}^{r2}$	$\phi_{sn2}{}^{r2}$	$\phi_{sn3}{}^{r2}$
Network3 (r_3)	$\phi_{sn1}{}^{r3}$	$\phi_{sn2}{}^{r3}$	$\phi_{sn3}{}^{r3}$

Table 2 Matrix example represents level 1 of the cube

	Nodename1 (n_1)	Nodename2 (n_2)	Nodename3 (n_3)
Network1 (r_1)	$S_{sn1}{}^{r1}$	$S_{sn2}{}^{r1}$	$S_{sn3}{}^{r1}$
Network2 (r_2)	$S_{sn1}{}^{r2}$	$S_{sn2}{}^{r2}$	$S_{sn3}{}^{r2}$
Network3 (r_3)	$S_{sn1}{}^{r3}$	$S_{sn2}{}^{r3}$	$S_{sn3}{}^{r3}$

$C_{1*k}(D), \ldots, C_{N*k}(D)\} = \{C_0(D), C_1(D), \ldots, C_N(D)\} = \{nodename_1, nodename_2, \ldots, nodename_n\}$ (or $= \{A, B, \ldots, Z\}$ sorted by first letter). Let set C_{D0} denote the set of number of links divided by $n - 1$ ($\varphi_{sx} = |\sum_{i \varepsilon IN} L_M(s, x_i)/n - 1|$) between a studied node and the other nodes named in C_D, such that $C_{D0} = \{C_{0*0}(D), C_{1*0}(D), \ldots, C_{N*0}(D)\}$ or in other words C_{D0} represents the face of the cube at level zero. Let set C_{D1} denote the set of the distances ($S_{xt^i} = |n - 1/d_{GM}(x, t^i)|$) from a studied node "$x$" to all the other nodes "t^{i}", such that $C_{D1} = \{C_{0*1}(D)C_{1*1}(D), \ldots, C_{n*1}(D)\}$. For all the other columns C_{Di}, where $i >= 2$, let set C_{Di} denote the set of different paths between any two nodes passing through a specific node v which is studied by the cube ($\beta_{st} := |P_{st}(v)|$) divided by the sum of different paths between any two nodes ($\beta_{st} := |P_{st}|$), such that $C_{Di} = \{C_{0*i}(D) C_{1*i}(D), \ldots, C_{n*i}(D)\}$.

Table 1 explains how the node's cube is structured as a three-dimensional cube of three faces that are divided into "K" number of levels (0,1,..., k).

Table 2 represents the level 0 of the node's cube "s" as a matrix, in which the columns show the other node's name on the graph and the rows show the networks that a node appears in. The values in the matrix entries contain the *degree of centrality* φ that node "s" has with the other nodes.

Table 3 Matrix example represents level 2 of the cube

	Nodename1 (n_1)	Nodename2 (n_2)	Nodename3 (n_3)
$Net_1(r_1)$	$\sum\limits_{\forall xi,yi \in V} \dfrac{\beta xiyi[r1]}{\beta xiyi\,(n1)[r1]}$	$\sum\limits_{\forall xi,yi \in V} \dfrac{\beta xiyi[r1]}{\beta xiyi\,(n2)[r1]}$	$\sum\limits_{\forall xi,yi \in V} \dfrac{\beta xiyi[r1]}{\beta xiyi\,(n3)[r1]}$
$Net_2(r_2)$	$\sum\limits_{\forall xi,yi \in V} \dfrac{\beta xiyi[r2]}{\beta xiyi\,(n1)[r2]}$	$\sum\limits_{\forall xi,yi \in V} \dfrac{\beta xiyi[r2]}{\beta xiyi\,(n2)[r2]}$	$\sum\limits_{\forall xiyi \in V} \dfrac{\beta xiyi[r2]}{\beta xiyi\,(n3)[r2]}$
$Net_3(r_3)$	$\sum\limits_{\forall xi,yi \in V} \dfrac{\beta xiyi[r3]}{\beta xi\,yi\,(n1)[r3]}$	$\sum\limits_{\forall xi\,yi \in V} \dfrac{\beta xiyi[r3]}{\beta xiyi(n2)[r3]}$	$\sum\limits_{\forall xi\,yi \in V} \dfrac{\beta xi\,yi[r3]}{\beta xi\,yi(n3)[r3]}$

Table 3 represents level 2 of the node's cube "s" as a matrix. The values in the matrix entries, however, contain the result of calculating the number of different paths between any two nodes passing through a node "s" ($\beta_{st}(S) := |P_{st}(S)|$) divided by the sum of different paths between any two nodes ($\beta_{st}(S) := |P_{st}(S)|$).

A database cube is obtained that represents a multi-network graph at the same time. As a result, it is easy to calculate centrality measures for each node depending on its cube (D_{i*j*k}) and by directly applying queries on cube values. In order to calculate the degree centrality and the closeness centrality, the contents of cube levels $k = 0$ and $k = 1$ are invoked, respectively. For betweenness centrality, the cube level $k = 2$ is invoked. If the studied graph is undirected, then we divide the result by $((n - 1) * (n - 2)/2)$; otherwise the result is divided by $(n - 1) * (n - 2)$.

6 Simulation and Results

In this section, a simulation of a real multi-network graph example is applied to show the analytical benefits of the proposed mapping of models. In the network analysis domain, social networks retain the first level of importance. There's absolutely no doubt that social networks continue to play an increasingly important part in many people's lives. By 2017, the worldwide social network users will total 2.55 billion [8]. Figure 4 shows the distribution of internet users through the social network services. As shown, more than 50 % of internet users are Facebook users.

With regard to the importance of social networks in global network data analysis, we have collected and studied three small sets of social data (Facebook, Twitter and Google+). The first is an undirected graph of a Facebook group of 104 members, the second a directed Twitter graph of 76 members and the third an undirected graph of a Google+ group network of 61 members. We tried to gather the same people together in a large part of the data set that shared accounts in different networks. We initially applied one of the traditional network analysis tools, Gephi 0.8.2, to analyze each of the networks alone to get the centrality results.

Figures 5, 6 and 7 present the results of analyzing the studied social network groups (Google+, Facebook and Twitter groups respectively). The top left diagram shows the connections between nodes through networks. The top right scatter diagram shows the degree distribution over the group. The bottom left scatter diagram shows the closeness centrality distribution over the group. The bottom right scatter diagram shows the betweenness centrality distribution over the group.

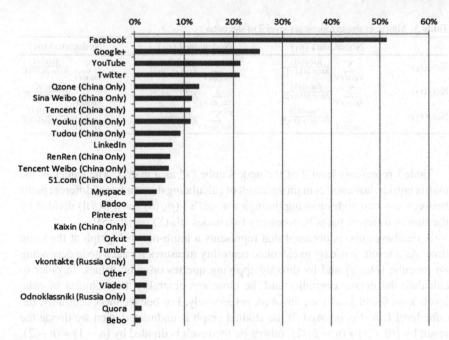

Fig. 4 Percentage (%) of global internet users (*Source* GlobalWebIndex.com, 2013)

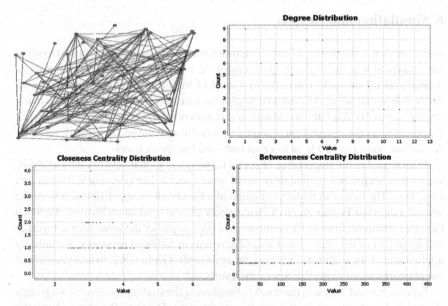

Fig. 5 Centrality measure and graph of the Google+ group

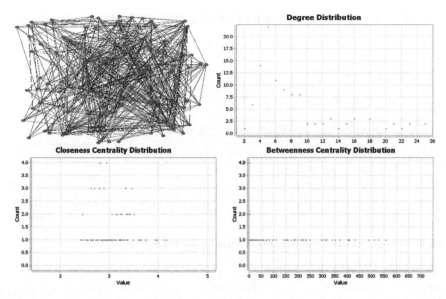

Fig. 6 Centrality measure and graph of Facebook group

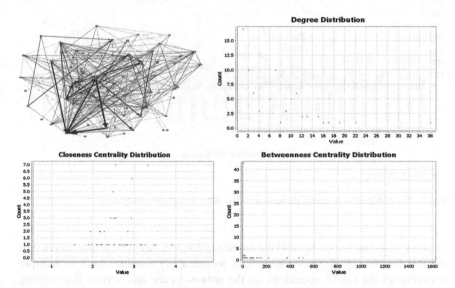

Fig. 7 Centrality measure and graph of a Twitter group

Figure 8 shows how to apply the traditional multidimensional database in multi-networks. First, we extract the centrality measures from the given graphs using specific Java codes; this step is similar to the extraction stage in the *OLTP* (online transaction processing). The second step is summarized by building the multi-database schema for the networks using the *SQL server 2008* operations. Then, we customize this

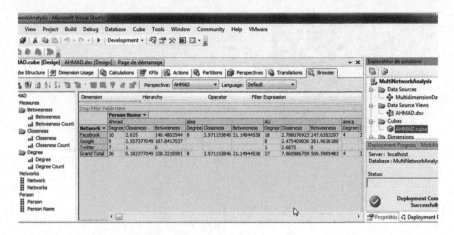

Fig. 8 Building data warehouse using business intelligence and SQL server 2008

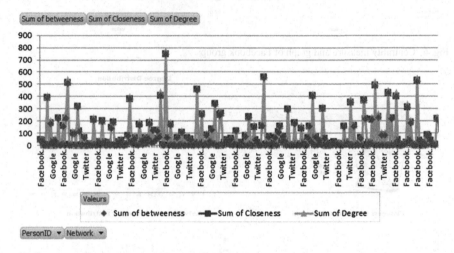

Fig. 9 The distribution of centrality measures over three social network groups

database as a data source to build the required cube measures and dimensions using the *SQL business intelligence studio*. Now we can apply simultaneously in multiple networks all the *OLAP* operations on the obtained cube and several data mining algorithms such as: *decision tree, clustering, association rules* and *neural network*.

In order to obtain a visual analytical report about the obtained multidimensional database, we have imported the obtained cube in the *SQL business intelligence studio* to *Microsoft Excel 2012*. Figure 9 shows a line chart of the distribution of the centrality measures as a function of the network name.

Figure 10 shows the distribution of centrality measures as a function of both networks and node id. The above results give some idea how analysis can benefit from simultaneous multi-network analysis. This is not everything, however; in fact

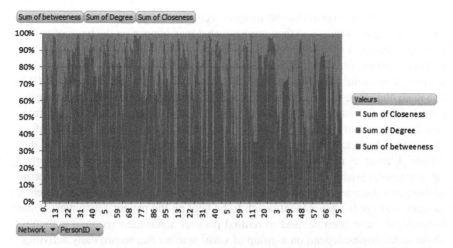

Fig. 10 Area chart reflecting the distribution of analysis measures for the three social network groups concurrently

the analyzer can now apply several queries (see below), which cannot be achieved when each network is analyzed separately:

- Give the name of the person who has the highest degree of centrality in all his/her social network services.
- Give the name of the most important person (according to the centrality measure) in all the social network services.
- Give the name of the person who has a centrality degree greater than 5 in all his/her social network services.
- Give the number of friends that the most important person has through all his/her social network services.

More advanced queries can be applied by means of the proposed model to understand node behaviors in several networks.

7 Benefits and Facilities

The proposed model has attained a multidimensional database for three measures of network analysis and it can be extended to all network analysis measures. Given the large amount of cloud network data, it is hard to analyze the database directly, however. Therefore, a data warehouse has been built. The data warehouse (and its steps extract, transform and load) facilitates reporting and analysis and provides access to structured and unstructured information and operational and transactional data in real time. The obtained data warehouse allows the analyzer to study relations among different networks. It also makes it easier to access simultaneously the node behaviors in multiple networks and answers such plain-language questions as "What happened?" and "Why?" and then predicts what may happen on the basis of strong analytical results. To show the importance of the proposed model in the big network analysis domain, we offer two real examples where the multi-network graph data warehouse is useful. The first example is the analysis of the US election in 2012. The main challenge was to circumscribe web content (web sites, RSS feeds, tweets) coming from the States under scrutiny [17]. The analysis team collected information from the social networks (multiple cloud networks) over three months before the election. In fact, building a complete multi-network graph data warehouse, that studies all analysis measures, could result a better analyze of election than applying a temporal study. The second example is from Bioinformatics, and concerns Alzheimer's disease. Alzheimer's is a widespread disease that affects the patient's memory and needs a permanent watch to be kept on the patient's activities. Some technologies have been devised to remind patients about their required activities. These technologies depend on a group of smart sensors that record daily activities. Every sensor recognizes the patient as an object and provides an *activity network* that reflects how s/he deals with the other objects (people, machines,... etc.). Each environment, however, has its own *activity network* and to offer a unique solution suitable for all environments, scientists analyze multiple networks, but not simultaneously. Thus, the solution is to map all the *activity networks* (as a multi-network graph) from sensors to a data warehouse.

8 Conclusion

Web researchers make strenuous efforts to convert the information retrieval web (Web 2.0) into a semantic web (Web 3.0). In parallel to web upgrade, cloud computing is gradually adopted by web companies; however the consequence is a big network data that needs analysis. One of the new concepts of Web 3.0 is personalization that requires aggregation of web user accounts. Indeed, every web user has several web accounts (social, business, study … etc.). If all the networks of these accounts are treated as one, without effect on the individual characteristics of each network, then a multi-network graph of big data web accounts can be achieved for every user. Also, from the network analysis perspective, it is harder to analyze multi-networks concurrently. To solve this problem we have proposed a novel model that maps multi-networks graphs in a multidimensional database. To validate our idea, we applied our model to some network analysis measures of centrality: degree centrality, closeness and betweenness. By means of a simulation, we have discussed the simultaneous analysis of three social networks. In future work, we hope to expand this model to cover ontology and apply it in real complex networks such as Bioinformatics networks.

Acknowledgments This work has been supported by the University of Quebec at Chicoutimi and the Lebanese University (AZM Association).

References

1. Bonacich, P.: Power and centrality: a family of measures. Am. J. Sociol. **92**, 1170–1182 (1987)
2. Bonacich, P.: Simultaneous group and individual centralities. Soc. Netw. **13**, 155–168 (1991)
3. Chen, P.: The entity-relationship model-toward a unified view of data, ACM Trans. Database Syst. **1**(1), 9–36 (1976)
4. Chen, C., Yan, X., Feida, Z., Jiawei, H.: Graph OLAP: towards online analytical processing on graphs, data mining (ICDM '08). Eighth IEEE International Conference, Pisa, (2008). doi:10. 1109/ICDM.2008.30, pp. 103 – 112
5. Costenbader, E., Valente, T.W.: The stability of centrality measures when networks are sampled. Soc. Netw. **25**(4), 283–307 (2004.). Elsevier
6. Daihee, P., Jaehak, Y., Jun-Sang, P.: NetCube: a comprehensive network traffic analysis model based on multidimensional OLAP data cube. Int. J. Netw. Manage. **23**(2), 101–118 (2013)
7. Eirinaki, M., Vazirgiannis, M.: Web mining for web personalization. ACM Trans. Internet Technol. **3**(1), 1–27 (2003)
8. eMarketer report.: Worldwide Social Network Users: 2013 Forecast and Comparative Estimates. Freeman, L.C., Borgatti, S.P., White, D.R., 1991. Centrality in valued graphs: a measure of Betweenness based on network flow. Social Networks **13**, 141–154 (2013)
9. Evans, D.: The internet of things: how the next evolution of the internet is changing everything. Cisco IBSG. 1–11 (2011)
10. Freeman, L.C.: A set of measures of centrality based on betweenness. Sociometry **40**, 35–41 (1977)
11. Freeman, L.C., Borgatti, S.P., White, D.R.: Centrality in valued graphs: a measure of betweenness based on network flow. Soc. Netw. **13**, 141–154 (1991)
12. Hoede, C.: A new status score for actors in a social network. Department of Mathematics, Twente University, unpublished manuscript, (1978)
13. Hubbell, C.H.: An input-output approach to clique identification. Sociometry **28**, 377–399 (1965)
14. Katz, L.: A new index derived from sociometric data analysis. Psychometrika **18**, 39–43 (1953)
15. Manuel, F., Catherine, P., Ben, S., Jen, G.: ManyNets: an interface for multiple network analysis and visualization, ACM CHI (2010)
16. Mobasher, B., Cooley, R., Srivastava, J.: Automatic personalization based on web usage mining. Commun. ACM **43**(8), 142–151 (2000)
17. Papadopoullos, A.: CASE STUDY Social Network Analysis of the 2012 US Elections. Chief Technology Officer, Semeon (2013)
18. Sabidussi, G.: The centrality index of a graph. Psychomatrika **31**, S81–603 (1966)
19. Taylor, M.: Influence structures. Sociometry **32**, 490–502 (1969)
20. Tore, O., Filip, A., John, S.: Node centrality in weighted networks: generalizing degree and shortest paths. Soc. Netw. **32**(3), 245–251 (2010). Elsevier
21. Trudeau, R.J.: Introduction to Graph Theory. New York: Dover Publishers. pp. 19. ISBN 978-0-486-67870-2. http://store.doverpublications.com/0486678709.html
22. Wararat, J., Cécile, F., Sabine, L.: OLAP on Information Networks: a new Framework for Dealing with Bibliographic Data, 1st International Workshop on Social Business Intelligence (SoBI 2013). Genoa, Italy (2013)
23. Xi-Nian, Z., Ross, E., Maarten, M., Davide, I., Xavier, F.C., Olaf, S., Michael, P.M.: Network centrality in the human functional connectome. J Life Sci. Med. Cereb. Cortex **22**(8), 1862–1875 (2012). Oxford

Toward Web Enhanced Building Automation Systems

Gérôme Bovet, Antonio Ridi and Jean Hennebert

Abstract The emerging concept of Smart Building relies on an intensive use of sensors and actuators and therefore appears, at first glance, to be a domain of predilection for the IoT. However, technology providers of building automation systems have been functioning, for a long time, with dedicated networks, communication protocols and APIs. Eventually, a mix of different technologies can even be present in a given building. IoT principles are now appearing in buildings as a way to simplify and standardise application development. Nevertheless, many issues remain due to this heterogeneity between existing installations and native IP devices that induces complexity and maintenance efforts of building management systems. A key success factor for the IoT adoption in Smart Buildings is to provide a loosely-coupled Web protocol stack allowing interoperation between all devices present in a building. We review in this chapter different strategies that are going in this direction. More specifically, we emphasise on several aspects issued from pervasive and ubiquitous computing like service discovery. Finally, making the assumption of seamless access to sensor data through IoT paradigms, we provide an overview of some of the most exciting enabling applications that rely on intelligent data analysis and machine learning for energy saving in buildings.

G. Bovet
Telecom Paris Tech, 46 rue Barrault, 75013 Paris, France
e-mail: gerome.bovet@hefr.ch

G. Bovet · A. Ridi · J. Hennebert (✉)
University of Applied Sciences Western Switzerland, Bd de Pérolles 80,
1700 Fribourg, Switzerland
e-mail: jean.hennebert@hefr.ch

A. Ridi · J. Hennebert
University of Fribourg, Bd de Pérolles 90, 1700 Fribourg, Switzerland
e-mail: antonio.ridi@hefr.ch

N. Bessis and C. Dobre (eds.), *Big Data and Internet of Things:*   259
A Roadmap for Smart Environments, Studies in Computational Intelligence 546,
DOI: 10.1007/978-3-319-05029-4_11, © Springer International Publishing Switzerland 2014

1 Introduction

In the last decade, we have become more and more concerned by the environmental dimension resulting from our behaviours. Further than the ecological trend, the interests are also economics-centred due to the raising cost of the energy. Representing 20–40 % of the global energy bill in Europe and USA, buildings are a major source of energy consumption, actually more important than industry and transportation [1]. In a building, half of the energy consumption comes from the Heating, Ventilation and Air Conditioning systems (HVAC), followed by lighting and other electrically operated devices. In offices, HVAC, lighting and electrical appliances together reach about 85 % of the total energy consumption.

For these reasons, the reduction of building energy consumption has become an important objective. Many works are currently undertaken towards renovation and improvement of building insulation. The other facet is to leverage on energy efficiency that involves better usage of HVAC equipment taking into account local production and storage capacity as well as temporal occupation of the rooms. In simplified terms, the aim is to optimize the ratio between the energy savings and user comfort.

This objective requires a clear move from state-of-the-art conventional building automation systems to advanced information systems leveraging on (1) a variety of interconnected sensors and actuators, (2) a unified management of heating, lighting, local energy production or storage and (3) data modelling capacities to model room usage and predict user comfort perception.

Most automated buildings are currently working with dedicated building networks like KNX [2] or EnOcean [3]. These networks are specifically conceived for the purpose of building automation, including all layers of the OSI model starting from the physical to the application one. In such settings, a central Building Management System (BMS) is typically connected to the network and manages the equipments by implementing the operation rules of the building.

In many buildings, we observe the coexistence of different network technologies, often caused by the installation of new equipments answering specific physical constraints, for example wiring or power supply. The protocols are often relying on proprietary layers and this heterogeneity actually leads to two situations. In the first one, several BMS are coexisting and share the management of independent equipments, making difficult any global optimisation. In the second one, a unique but more complex and costly BMS is used where bridges to the different protocols are integrated. Without prejudging on the strategies of technology providers, we can reasonably converge to the fact that there is a lack of standardisation in building automation systems, at the network level and, probably more importantly at the application level. BMS could largely benefit of a common protocol stack compatible with any device.

Thanks to the miniaturization of electronics and increasing computing power, devices offering native IP connectivity are appearing. Sensor networks based on 6LoWPAN or Wi-Fi are nowadays in competition with classical building networks. Everyday objects are now able to connect to IPv4 and IPv6 networks, offering new functionalities like machine-to-machine communications. This has led to the emergence of *Internet-of-Things* paradigms (IoT), a research field trying to find answers in how to connect objects to the Internet from the network point of view, i.e. covering the first four layers of the OSI model.

Going up to the application layer, the heterogeneity problem is even worse as there are currently no strong standards defining the *semantic* of the building resources, i.e. an expression of device and service capabilities. The paradigms of the Web-of-Things (WoT), which can be viewed as the natural extension on top of the Internet-of-Things, are here proposing to rely on web application standards. Arguably, these standards are more like a set of best practices than real standards. In the vision of IoT and WoT, any device embeds a Web server offering lightweight Web services for interaction. The Application Programming Interfaces (APIs) of the services often rely on RESTful principles, providing natural ways to embed the semantics in the communication protocol.

In this chapter we will present the actual state-of-the-art in the field of IoT in smart buildings, putting into evidence remaining ongoing challenges. Some propositions overcoming open problematic will be discussed, especially regarding the integration of existing building automation systems in the core IoT and their respective discovery. Finally, making the assumption of a building where IoT paradigms are fully integrated, we will provide an overview of some enabling applications that rely on data analysis and self-learning algorithms for energy saving in buildings.

2 Integrating Building Automation Systems in the IoT

Many new or renovated buildings are nowadays equipped with automation networks. We can here mention office buildings, factories and even private households. The relative high investment costs have an impact on the payback period which is rather high, often around ten years. A sudden change of technology is therefore not conceivable. We envision here the IoT as adapting itself to existing installations and thus encouraging a smooth transition until building automation systems natively support it. Meanwhile, a mix of different technologies will probably coexist in buildings.

In this section, after reviewing existing building automation systems and technologies, we propose a Web-oriented protocol stack able to solve the heterogeneity problem between sub-systems. Concrete application scenarios will serve as basis of discussion.

2.1 Related Work

Historically, buildings are equipped with networks especially designed for automation purposes and offering services tailored to buildings. We can here mention several technologies like BACnet, LonWorks, KNX and EnOcean. The physical mediums

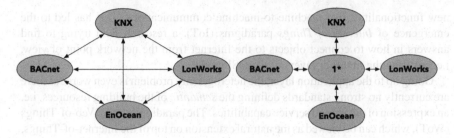

Fig. 1 N-to-N (*left*) and N-to-1* (*right*) approaches for protocol mapping in buildings

are typically not restricted to certain types, like KNX can support twisted pair, RF, power line and even Ethernet. Because of the custom protocols used for the transport and networks layers, it is not conceivable to shift the application layer to a more common network like IPv6. Some works have proposed to rely on multi-protocol devices [4]. The drawback of this approach resides in the integration cost of such devices that are, anyway, quite non-existent on the market. Another solution consists of providing gateways. As illustrated in Fig. 1, gateways can operate according to two modes, a *N-to-N* protocol mapping or a *N-to-1** approach mapping all sub-systems to only one central protocol. The *1** refers to a new protocol stack suited for IoT interactions. In the N-to-N case, gateways between BAS translate telegrams from the originating network to its destination. Those gateways have knowledge about each protocols composing the stacks of each network. Although this approach solves the heterogeneity across networks, it induces some limitations. First, it is possible that not all capabilities of a BAS can be mapped to another one, thus restricting functionalities. Secondly, this approach requires $\frac{n*(n-1)}{2}$ mappings between BAS, representing a considerable effort.

In contrast to the N-to-N approach, the N-to-1* one considerably simplifies the number of gateways needed to n by introducing a common technology. The key challenge resides in the 1* technology where no standard is currently defined. Its components have still to be identified, according to Internet-of-Things constraints. A remaining decision has to be taken when integrating BAS into the IoT regarding the gateway position in the network. There are two extremes, either centralizing the services on a single node, or migrating them as close as possible to field devices. Centralizing the access at a backbone server brings some advantages in terms of maintenance, even if scalability problems may arise. Putting the services at the field level requires devices with more computational power but allows a direct native interaction between sensors and actuators with Web services over IP. More specifically, we can describe four different integrations styles, as illustrated in Fig. 2.

1. **On a centralized server**: this approach is currently the most used one where a server at the IP backbone handles all the interaction with the BAS. It allows an integration into enterprise systems.
2. **On a bridge**: by encapsulating field telegrams into IP packets one can interact with the BAS. Although being a more decentralized approach, the same problem regarding the application level remains open.

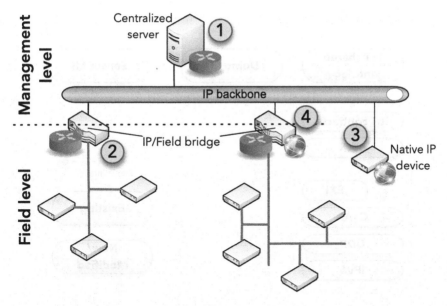

Fig. 2 Building automation systems integration styles

3. **On field devices**: IP enabled devices offering Web services for intercommunication with other participants of the network.
4. **Multi-stack bridges**: allow a more decentralized approach. By implementing the field stack and the IoT stack, mappings between Web services and endpoints are possible. By proceeding this way remote devices acting as clients are not aware that the final device resides on another type of network even having no IP connectivity.

A few research and proofs of concepts have already been proposed for IoT gateways. We can here cite the Universal Device Gateway developed in 2008 for enabling interoperability among heterogeneous communication protocols like ZigBee, X10 and KNX with an IPv6 compliant architecture [5]. More recently, a mapping from KNX endpoints, also known as datapoints, to oBIX objects has been proposed [6, 7]. The oBIX framework was developed for facilitating data interchange with BAS, relying on a XML object model representing devices. Those endpoints are accessible over Web services. The oBIX approach is quite similar to the WS-* stack of classical SOAP Web services. The oBIX HTTP server waits for requests and then translates them to KNX telegrams. Here, depending on the type of request (GET or POST), the action performed will be either a query for an endpoint status, or an update of a state. Although providing a common mapping, the proposed approach leaves certain issues open as for example the integration in existing installations.

In another work, the full paradigm of the Web-of-Things for smart buildings has been investigated in the context of KNX and EnOcean gateways in [8, 9]. This work shows that Web resources can be mapped to Building Automation System

IoTSyS candidate

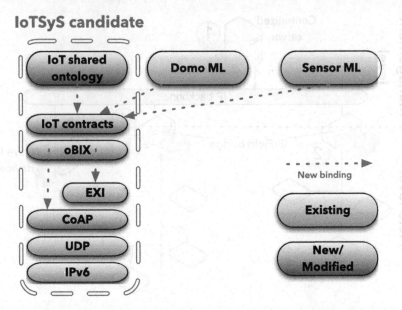

Fig. 3 Protocol stack for IoT building automation systems as proposed by [10]. Blocks that are introduced in this proposition or requiring some adaptations relative to their traditional use are presented in *pink*. Existing blocks requiring no adjustment are presented in *blue*. New bindings between blocks are shown with *pink arrows*

(BAS) endpoints using RESTful APIs, which are lightweight Web services based on loose-coupling. In addition to this, a semantic for URLs pointing to resources is proposed by including the location of the device inside the DNS name of the resource, thus providing a clear way for identifying them.

Devices becoming more and more IoT aware, the question regarding the standardization of the application layer and semantics remains open. A concrete proposition of an application layer compatible with IoT's paradigms was proposed in [10, 11]. The main contribution resides on the proposition of IoTSyS, an entire protocol stack for integrating BAS in the Internet-of-Things. IoTSyS relies on technologies such as Constrained Application Protocol (CoAP), oBIX and EXI, as illustrated in Fig. 3. CoAP, which is a lightweight version of HTTP pursuing the RESTful concept, is used as application transport protocol. The advantage of CoAP over HTTP is to be optimized for constrained devices by having much less overhead. Then, at the application layer, the oBIX XML schema is used to describe endpoints and to define contracts according to device functionalities, such as, for example, push button, relay or temperature sensor. The contracts can be annotated by using some data models as sensorML or domoML allowing a better processing by machines. As XML payload is not suited for small devices or constrained devices, the EXI compressor is used, which allows to reduce the size of the exchanged data. In order to perform the best possible compression, a schema describing the contracts is used on both the server and the client.

Although the IoTSyS solution fulfils some requirements of the IoT, it induces one major drawback, which is being not loosely-coupled. Indeed, every client must have knowledge about the IoT contracts schema in order to be able to decompress the XML. This approach restricts the possible evolution of the protocol stack because of the difficulty to update the schema on existing devices. This situation would lead in multiple versions of contracts distributed across networks, resulting in incompatibilities. In addition, using a lightweight protocol (CoAP) while adding some complexity by relying on oBIX could be seen as a contradiction. Additionally, many applications are nowadays built upon Web technologies like HTML and AJAX following the Web 2.0 concept. Integrating specific technologies like oBIX and EXI is actually adding more complexity to those applications and require special knowledge from developers. Despite the fact that some semantics are needed at least for describing resources, we believe that existing Web technologies have to play a more important role.

2.2 A Web-Oriented Protocol Stack

The device accessibility can be decomposed into four sub-layers, which are resource identification, linking, resource description and operation on resources [12].

Resource identification: The location of a device, which is often an important information in BAS can be included in the DNS name describing the resource. For example *coap://temp.office05.00.c.company.org* would represent the temperature sensor in room 05, on the ground floor of building C of a company. The final endpoint can then be further accessed by specifying the resource or even a sub-resource. We can here illustrate this with a temperature sensor offering two contracts, one in Celsius *../temp/celsius* and the second one in Fahrenheit *../temp/fahrenheit*. We here reuse the concept of Web resource and URI for identifying unique device endpoints. In this approach, a traditional DNS architecture can be used. However we recommend to rely on the distributed approach of DNS, namely mDNS that is also compatible with classical DNS clients. The difference lies in the queries that are sent over multicast instead of unicast. This approach is more scalable and fault-tolerant as the knowledge about the DNS is distributed across nodes.

Linking: In the concept of WoT and Resource Oriented Architecture (ROA), an important aspect is the ability to discover resources by following links. A resource should provide links back to its parent and forward to its children. This allows to crawl knowledge about resource hierarchy. When applying this concept to the field of building automation, it comes out that devices should be linked to their location and endpoints. To achieve this, we propose to use an implicit and an explicit way. By decomposing the host part of the URI and discarding the first part, one can go back in the building location hierarchy (e.g. *coap://kitchen.ground.home.com -> coap://ground.home.com*). This way is especially thought for humans. Machine-to-machine interaction could prefer the explicit way. There, each resource description, i.e. location, device or endpoint, must provide absolute URLs to its parent or children.

Nonetheless, linking is closely related to resource description as the format and semantic used will influence how linking is provided.

Resource description: Discovering the capabilities of a resource is also a very important aspect in machine-to-machine communications. Not only the semantic but also the format is decisive to increase the chance of self-understanding. Regarding the format, a frequent suggestion is to use JSON instead of XML. First, JSON is more lightweight than XML, and that in terms of message size and parsing time [13], therefore better suited for devices with limited capabilities. Moreover, it can directly be parsed in JavaScript, allowing a faster integration into Web application and supporting the concept of mashups. Concerning the semantics, we here propose to rely on the concept of RDF and ontologies forming together the semantic Web. Further details about semantics and ontologies are given in Sect. 2.3.

Operation on resources: The last step is the execution of an operation on the resource. For devices with limited resources, the CoAP protocol is probably one of the best candidate. CoAP is a simplified version of HTTP, aiming at reducing the overhead size. Gateways translating HTTP to CoAP, and vice-versa, are available, exposing CoAP devices over the HTTP protocol. In a similar way as for HTTP, CoAP offers GET, POST, PUT and DELETE operations that have a standard semantic according to REST services, as for example:

- **GET** is used for retrieving the current state of a resource, e.g read the actual temperature.
- **PUT** is used to update the state of a resource, e.g. switch a power outlet.
- **POST** is used to create a new resource, e.g. register for event notification.
- **DELETE** is used to delete a resource, e.g. delete a threshold on a device.

Figure 4 summarizes the protocol stack for building automation systems relying on the Web-of-Things paradigm. The strengths of our WoTBAS (Web-of-Things Building Automation Stack) proposition reside in taking part of the best practices of the Web, being lightweight and loosely-coupled.

2.3 Semantics and Ontologies

Still a topic of research, semantics and ontologies are potential technologies to define common languages and representations facilitating machine exchange. The Web Service Description Language (WSDL) is currently the most used language for describing Web Services. WSDL is often combined with the Simple Object Access Protocol (SOAP) allowing the exchange of structured information about Web services. Although WSDL brings some standardization in terms of method prototyping by describing the attributes and return value, it is not intended for describing the *semantic* of Web resources. For this reason, many XML schemas and other data models were developed trying to describe processes and physical environments [14]. Nowadays the RDF language is used together with ontologies [15]. Ontologies represent a vocabulary and association of definitions, while RDF is the

WoTBAS candidate

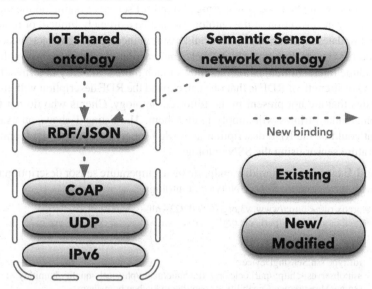

Fig. 4 Protocol stack for WoT building automation. Blocks that are introduced in this proposition or requiring some adaptations relative to their traditional use are presented in *pink*. Existing blocks requiring no adjustment are presented in *blue*. New bindings between blocks are shown with *pink arrows*

commonlanguage to express the meta-data. Statements are expressed with *triples* composed of subject-predicate-object. For example the notion "The sensor is of type temperature" in RDF is represented with the following triple: a subject denoting "The sensor", a predicate "is of type" and an object denoting "temperature". In order to have standardized triples, the use of ontologies is strongly recommended. One can find various ontologies developed for specific contexts. The Semantic Sensor Network Ontology (SSN) is based around concepts of systems, processes and observations [16]. It supports the description of the physical and processing structure of sensors. However notions of units and locations are not part of it. One can include other ontologies like DOLCE Ultra Lite (DUL) [17]. Before defining the SSN ontology, the working group of the W3C performed an analysis of all existing data models representing sensors. The best practices of each data model were included in the specification of the SSN ontology. Even if the ontology is quite complete and allows to express a lot of properties, there is always the possibility to expand it with new classes and definitions. For example one could extend the sensor class in order to define a concrete sensor type with its own specific properties as for example temperature or presence. When further pushing this concept, one could imagine to create sub-classes of concrete sensor type representing a sensor model. This leads to an important question: should the ontology implement a low or a high level of abstraction? Having a low level of abstraction by defining sensor models in the ontology seems here not a good choice as it would result in a heavy-coupling between devices. Also, the ontology would be huge considering all types of sensors existing on the market. Additionally the evolution of the ontology would be extremely limited, or requiring frequent updates. Such an approach would result in incompatibilities across devices of different versions. Keeping a high level of abstraction certainly barely complicates the machine intercommunication but allows much more scalability in terms of evolution. One strength of RDF is that one can expand the RDF description with its own properties that are not present in the followed ontology. Clients who do not know the non-standard triples will simply ignore them. The listing 1 shows an example of what could be the RDF description in Turtle format of a humidity endpoint on a temperature sensor using the SSN ontology.

Listing 1 Example of a humidity endpoint on a temperature sensor description using RDF with the Semantic Sensor Network Ontology

```
@prefix rdf: <http://www.w3.org/1999/02/22−rdf−syntax−ns#>.
@prefix ssn: <http://purl.oclc.org/NET/ssnx/ssn>.

<coap://temp.kitchen.home>
  rdf:type ssn:SensingDevice ;
  ssn:observes <http://purl.oclc.org/NET/muo/ucum/physical−quality/humidity>;
  ssn:hasMeasurementCapability <coap://temp.kitchen.home/hum>.

<coap://temp.kitchen.home/hum>
  rdf:type ssn:MeasurementCapability ;
  ssn:hasMeasurementProperty <coap://temp.kitchen.home/hum/accuracy>;
  ssn:hasMeasurementProperty <coap://temp.kitchen.home/hum/sensivity>.

<coap://temp.kitchen.home/hum/accuracy>
  rdf:type ssn:Accuracy ;
  rdf:value 1 .

<coap://temp.kitchen.home/hum/sensivity>
  rdf:type ssn:sensivity ;
  rdf:value 0.1 .
```

Another important concept of RDF is its associated query language SPARQL recognized as a key technology of the Semantic Web [18]. With SPARQL one can crawl over resources for retrieving only the interesting ones meeting some criteria. Queries are expressed with a dedicated language also expressing triple composed of conjunctions-disjunctions-patterns. The query can filter results according to literal, numeric data type properties, which makes it very similar to SQL. SPARQL offers four types of queries: LIKE for extracting raw values, CONSTRUCT for obtaining valid RDF, ASK for obtaining a simple true/false response on an endpoint, and finally the DESCRIBE for obtaining an RDF graph. In our context we will only use the SELECT type of query to return the URL of the resource. This information alone is sufficient as the client can then ask for the whole description by accessing the resource directly. However the client can also specify other attributes than the URL in the SELECT section of the query if needed. To illustrate what a SPARQL query for discovering devices could look like, we rely on a practical example as follows:

what are the resources of type temperature with a value higher than 25 °C located in the kitchen? Listing 2 shows the resulting SPARQL query.

Listing 2 Example of a SPARQL query to look after temperature sensors with a value higher than 25 °C located in the kitchen

```
@prefix ssn: <http://purl.oclc.org/NET/ssnx/ssn>.
SELECT ?node WHERE {
    ?node a ssn:Sensor ;
    ssn:observes <http://purl.oclc.org/NET/muo/ucum/physical−quality/temperature>.
    dul:hasLocation "kitchen" ;
    rdf:value ?lv ;
    FILTER(?lv > 25).
}
```

3 Discovery

Discovery of device and service is crucial in the IoT approach that rely on participating objects distributed all over the network. This case is also true for building automation. For example, we want the overall system able to discover if a temperature sensor located in a certain room is available. We can here distinguish two discovery strategies. Either by following links across resources, or by asking clients to provide their capabilities through requests. As the first one has already been discussed in the previous sections, we will here focus on the latter.

3.1 Related Work

A key concept for Web-of-Things discovery is defined with CoAP through the use of the *./well-known/core* resource which will respond with all accessible services on the device [10, 11]. One major drawback of this concept is in the fact that previous knowledge about the device's DNS or IP address is required. Furthermore the response is limited to few predefined attributes giving little information about the services themselves. Regarding service discovery, well-spread protocols like UPnP's SSDP [19], SLP [20] or DNS-SD [21] have emerged in the last years. They are mostly implemented in operating systems and computer peripherals like printers or PDAs. They are based on a multicast approach allowing to discover the existence of devices according to user role and network scopes. DNS-SD, for example, extends the classical DNS or mDNS naming by recording services as DNS entries which can then be discovered by DNS query messages. Service discovery protocols that are specifically conceived for very constrained devices have also been proposed [22, 23]. Many of these protocols only consider static service attributes that are provided during service registration. In an ideal case, services should be discovered according to dynamic and contextual properties [24].

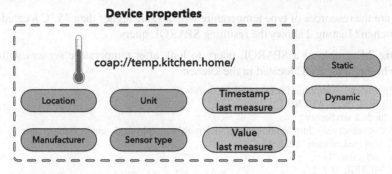

Fig. 5 Static and dynamic properties of a resource

3.2 Building Automation System Requirements

From the perspective of discovery, building automation systems are significantly different to classical pervasive devices for which location services are most of the time sufficient. Actually, an extension from simple discovery to query is here required. Indeed the context of a device or, in our case, a resource is decisive during discovery. Many building management systems have to look for devices being in a certain state or context for regulation or alarming purposes. For example one BMS could search for temperature sensors on the ground floor having measured in the last five minutes a value higher than 25 °C. This request must also consider devices that have not yet reported and are not known by the system. Relying on a plug-and-play approach allows to have highly dynamic systems requiring no previous knowledge of the available resources. We introduce here the concept of static and dynamic resources properties as illustrated in Fig. 5. If one wants to discover all the available devices in a given installation, a simple query specifying no properties would be sufficient. The next step would consist of retrieving the description for resources of interest. Working with constrained devices and low-power networks, the energy efficiency of the new layer is also key factor that has to be optimized in order to not affect life span of battery-operated devices.

Discovery in building automations systems should fulfil the following minimal requirements:

1. Be optimized for constrained devices
2. Allow a plug-and-play installation of new devices
3. Allow a discovery of the entire network
4. Allow a precise discovery and selection of devices according to some contextual parameters
5. Be scalable and fault tolerant.

3.3 Architecture

In order to fulfil the above mentioned requirements, we have to make several choices regarding architecture and technologies. A potential answer can be proposed by dividing the problem into four sub-problems: infrastructure, application layer, data format and query engine.

Infrastructure: A first question is about the use of central or distributed directories of resources descriptions. Using repositories is more suitable for environments with thousands of services, where queries can be processed more efficiently. However a centralized approach has some drawbacks. The first one is about fault tolerance (see requirement 5 above). A central directory failure would result in the inability for clients to discover services. Further to this, the repository has to be frequently updated with context and device availability that would generate a lot of traffic. For these reasons, a de-centralized approach is probably recommendable, where every device will be a discovery endpoint, and thus guarantee a reliable availability. In addition, the traffic of messages is kept very low by using multicast techniques. Only one message is sent by the client, receiving unicast response from matching endpoints.

Application layer: For the application layer, we suggest to rely on CoAP for transporting the requests and responses. CoAP is indeed already present on devices implementing Web-of-Things stacks and more specifically building automation stacks. Additionally, CoAP provides a multicast communication in the *Groupcomm* draft, which can be used to address the entire network with only one packet [25]. With this method, devices offering discovery capability expand their interface with a new service responding to multicast requests. Leveraging on CoAP allows already to satisfy requirements 1 and 3 listed above. The use of multicast requests allows limiting the number of packets transiting over the network. The number of packets sent for discovering a resource with multicast can actually be computed with $N_p = n + 1$, with n representing the number of devices matching the query.

Query engine: The query engine is at the heart of the discovery process. It finds matches according to the properties specified by the client. Many systems for querying structured data exist. As explained in Sect. 2.3, SPARQL and RDF comes out to be a promising alternative. In this context, when receiving a discovery request, the SPARQL engine retrieves the current up-to-date description of each resource containing the static and dynamic property values. The next step consists of applying the query to the RDF document. In the case of matching results, they will be piggybacked within the response packet. If no resource meets the criteria, no response will be sent to the client. From a conceptual point of view, the coupling of the resource context with RDF and SPARQL makes the discovery process closer to a distributed query engine, which meets requirement number 4.

Data format: Following the same principle as for the building automation stack presented earlier, a reasonable proposition is to rely on JSON as exchange data format to minimise the network traffic. However, translating the SPARQL query expressions to JSON would add complexity and overhead as SPARQL isalready free of tags and

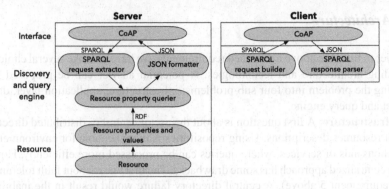

Fig. 6 Client and server discovery architecture

rather compact. Therefore, the original SPARQL format could be perfectly used for expressing the queries. On the other side, query responses are usually provided in XML format for which a translation to JSON could be recommended. Using SPARQL for queries and JSON for responses would allow minimizing the exchanged data and therefore comply with the requirement 1 stated earlier.

An implementation based on the proposed architecture for discovery is depicted in Fig. 6. The different modules can be grouped in interface, discovery and query engine, and finally the resource.

In more details, the flow of operations is the following. The CoAP server listens on a multicast socket with preconfigured port and IP address. Its role is to dispatch incoming discovery requests to the discovery and query engine. If matches are found, it will respond to the source of the request with the according payload data. Going one level down in the server, the discovery and query engine is responsible for evaluating requests and finding corresponding matches. It starts with a SPARQL request extractor that will take out the query from the CoAP messages and perform some validations regarding the format. Once the request is considered as well-formed, it is passed to the querier. This module will gather the RDF representations of all the resources present and available on the device. Once the collection complete, the SPARQL query is applied to each RDF representation of the collection. Each match is then stored in a collection of results. If the collection of results is empty after querying, the process can stop at this step. Otherwise, the collection is forwarded up to the JSON formatter. This module iterates through the results collection and formats SPARQL responses to JSON. The last step consists of responding to the client over the CoAP interface with the results.

We believe that having a modular approach as described above reduces the complexity of the discovery architecture and allows for future evolutions. Additionally, the different modules can rely on existing libraries offering the desired functionalities therefore reducing the implementation time.

4 Bridging Automation and Adaptation

So called *building automation systems* will perform more and more complex tasks, going way further than simple HVAC automation and light actuation. Future applications will rely on holistic optimisation leveraging on the access to data from the entire set of sensors. Applications will also use external data coming, for example, from weather web services, or even from neighbouring smart buildings, leading to the concept of smart neighbourhood or smart city. Regarding algorithmic, systems will have to incorporate more advanced data analytics technologies. Threshold based rules, often a priori set, will be replaced by complex parametric models implementing self-learning capacities and allowing for dynamic rules as a function of the context of use of the building. In this section, we take a side step to analyse the system architectures that will, at the same time, leverage on the strengths of Internet and Web technologies, and on new emerging intelligent services, enabling to move from *building automation systems* to *building intelligence systems*.

4.1 Related Work

Classical building automation systems are composed of three tiers, as depicted on the left part of Fig. 7 [26]. At the bottom we find the *field level* consisting of the network topology and the attached sensors and actuators providing for measurements and actuations. The *automation layer* provides control logic for driving actuators, providing some kind of intelligence to the building. The purpose of the automation layer is to optimize the comfort inside the building by using rules of actuation typically based on predefined threshold values. At the top, the *management level* offers applications for configuration and data visualization. With this architecture, regulation algorithms are mostly centralized on a single node and rely on few historical data. In the case of large buildings, multiple computers are used with a repartition of the logic corresponding to parts of the building to avoid dependencies between the sub-systems.

Nowadays, an evolution of the automation can be observed such as the inclusion of genetic algorithms [27], artificial neural networks [28] and empirical models [29], among many others. Such models have the ability to capture through historical sensor data information about the physics of the building and therefore to elaborate automation rules that are more precise. New approaches are also attempting to optimize the energy consumption according to a modelling of the user behaviour (see Sect. 5).

Despite a lack of literature related to system architecture in buildings, we can reasonably argue that the classical three-tiers architecture currently proposed by industries is not well suited for future developments. Indeed, the arrival of IoT, inherently relying on distributed nodes, as well as the emergence of these new modeling strategies are advocating for new architectures that we present in the next sections.

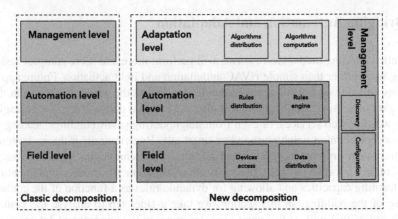

Fig. 7 Level based architecture of building automation systems. The *left* part shows the classic state-of-the-art decomposition. The *right* part shows the potential evolution of the architecture needed for future building applications

4.2 Evolution of Building System Architecture

We illustrate on the right part of Fig. 7 the perceived evolution of building system architectures. We first introduce an *adaptation level* that will dynamically feed the *automation level* with control logic, i.e. rules. In the IoT approach, the *management level* has also to be made available transversally as configuration, discovery and monitoring services must be made accessible to all levels.

4.3 Adaptation Layer

The current trend is to use intelligent data analysis technologies such as machine learning. Such algorithms need historical data from sensors to generate dynamic rules. Following IoT and WoT concepts, the data storage, as well as the automation and adaptation layers should be decentralized across the network and become ubiquitous. Consequently, historical data will be spread over the network and stored close to the devices. Algorithms and rules have also to be considered as Web resources in a similar way as for sensors and actuators. Figure 8 compares the repartition of roles for a classical building automation system (left) to the new WoT-enabled architecture (right). In this context, future works will have to be carried on to find solutions to minimize the transfer of data and the distribution of algorithms.

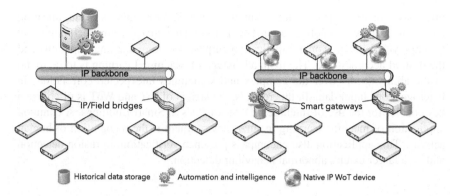

Fig. 8 Role distribution for a classical building automation system (*left*) and for a web-of-things architecture (*right*)

4.4 Automation Layer

The automation layer sends messages to the actuators to control the building equipments. The control is done according to rules which, in the classical approach, are most of the time static and a priori defined. In the new approach, rules will be generated by the adaptation layer, potentially evolving with time. A WoT compatible approach will be to distribute the rules over the network on devices with computational capabilities. We can already distinguish two types of automation rules that will probably be handled differently: event based and data oriented. In the first category, the rules will only be triggered as answers of building events. The second category of rules will require access to historical data such as, for example control loops. In the similar way as for the adaptation layer, future work will have to define strategies in order to reduce the traffic of data and to optimise the overall efficiency of the distributed system.

Although the adaptation and automation layers are quite similar from a WoT perspective, their purpose is not the same. For instance, buildings may not implement an adaptation layer and take part only of the automation one. This is actually a reasonable argument in favour of a separation of both layers.

5 Intelligent Data Analysis for Smart Environments

In this section, we focus on providing further discussion on the *Adaptation Level* as described in Fig. 7. One could qualify an environment as being smart whenever it is able to make the right decision at the right moment and with a satisfying rate of success regarding the outcome of the decision. A recent trend to build smart systems is to rely on so-called *Intelligent Data Analysis* approaches. Such approaches are often based on machine learning techniques able to infer prediction or classification models from large set of observation data. Different machine learning techniques can be employed depending on the application purpose, the computational capability and the desired accuracy rate. The main advantage of Machine Learning is to be found in the ability to discover the complex and sometimes unexpected correlations in heterogeneous input data. In this regards, the arrival of IoT and WoT is of course a key enabler for the use of Intelligent Data Analysis in Smart Buildings. A common application of machine learning in smart environment is the recognition of human activity. This application also encompass presence verification, intrusion detection and, to a larger extent, abnormal behaviour detection.

5.1 Evolution of Control Algorithms

Many algorithm strategies have been applied to smart buildings. A classification from the simplest algorithm with no or few adaptation, to the most complex approaches mostly relying on machine learning can be given as follows.

- **Fixed threshold based algorithms**. Such algorithms implement simple rules typically using fixed thresholds to control the equipments. These rules are often set without considering the real needs and dynamics of the users. HVAC systems are typically controlled depending on target temperature values conditioning the air by injecting heat or cold. The rules are sometimes using a schedule of predefined values to comply with known cyclic energy needs, typically day/night or seasonal cycles.
- **Physics based algorithms**. Such control algorithms are using mathematical models of the physics of the building. For example, the thermal inertia is computed and used to avoid undershoot or overshoot of temperatures. More sophisticated models will take into consideration the building orientation, its geometry, size of windows, among others for computing optimal blind control. Such models, while improving significantly the energy usage, are nevertheless targeting a comfort level for an "average user", disregarding the dynamics of the real use of the building. Furthermore, such systems are typically costly in terms of setup and need careful tunings at the commissioning phase.
- **Self adapting algorithms**. Machine learning technologies are typically used here to compute data-driven models for prediction of variables and classification of events. Such algorithms have the possibility to adapt continuously to building characteristics, building use, and environmental conditions.

We are here specifically interested into the third category of self adapting algorithms that have the most potential regarding smart applications. The arrival of IoT/-WoT architectures are also enabling the use of more complex algorithms having to deal with heterogeneous and asynchronous data. In the scientific literature, many models can be found such as predictive models, artificial neural networks, fuzzy logic and many others. Recently, stochastic state based data-driven models have also been proposed to capture the spatial and temporal statistics of building usages.

Self adapting algorithms present several advantages. First, they are independent of the physic of the building or, more precisely, the building characteristics are learned intrinsically from the observation data. Second, the optimization has the potential to become global instead of local, especially in the case of unified IoT/WoT architectures where all sensors and actuators are exposed from the sub-systems. Third, time-varying usage of buildings will be tracked as self-learning algorithms have the ability to incrementally adapt their parameters with new observation data. Such approaches allow an adaptation to real user, and not just an average user, thus minimizing the risk of rejection. Finally, new building configurations can be automatically learned, hence reducing setup costs. This last point has actually to be weighted by the fact that self adapting algorithms need some time to converge, creating potential loss of comfort during learning.

We have to point out that these advantages are still theoretical for smart buildings. Indeed, at the time of writing this text, we did not yet observe a smart buildings implementation relying fully on IoT/WoT and self-adapting algorithms. The vision is actually to go for a system where, typically, the installation of a new sensor would be automatically known from the rest of the system and incorporated into the algorithms as new piece of information for a better global settings of the equipments.

5.2 Adaptation Level Architecture

In this section, we propose a global and unified architecture for data modeling in the context of smart building. In many configurations, physics based algorithms or self adapting algorithms can usually be considered as extensions of threshold based systems, where the threshold values are continuously set from the models. Actually, the fixed threshold based algorithms could be seen as rules being processed in the *automation level* as described in Fig. 7. Physic or self adapting algorithms would be dynamically creating the rules in the *adaptation level* and feeding them into the *automation level*. The algorithms of the adaptation level can operate either continuously or at predifined periods of time, for example everyday at midnight. The frequency depends on the needs of the application and on system capabilities in terms of memory and computing power. In this sense, the adaptation level can actually be seen as decoupled from the automation level implementing the on-line working system.

Figure 9 illustrates the generic life-cycle of intelligent data analysis. At the field level, the sensor nodes provide information captured from the environment. The raw observations are communicated to the adaptation level. The adaptation level then performs a feature extraction. The purpose of this operation is to compute useful characteristics from the raw observations. Typically, features consist of normalized or filtered raw values. More complex feature extraction can also be applied such as frequency analysis with Fourier transform or computation of relevant features

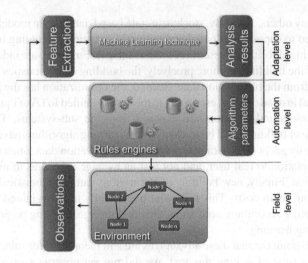

Fig. 9 Life-cycle of data processing with intelligent data analysis

with Principal Component Analysis. The next step is to model the features for a specific objective such as prediction of future values or classification of events. At this point a specific machine learning technique is applied to compute the models. From a general perspective, the parameters of a mathematical model are computed or updated from exemplary feature data, i.e. data for which the expected output of the model is known. These data are refered to as *training data* or *ground truth*. Once the models are ready, they are used to compute on-line rules that are fed into the automation level. The execution of a rule in the automation level will finally send commands to the actuators of the field level.

5.3 Adaptation Level and State Modelling

In the case of smart buildings, the signals to be modelled are often time series representing the evolution of some observations measured at the level of a sensor, or at the level of a combination of sensors located in a room, or even, at the most general level, the whole set of sensors in a given building. These multi-level and time dimensions are suggesting that the signals can be modelled using state-based models such as Hidden Markov Models. Figure 10 illustrates this approach. Such modelling usually implies that a state is tied to a stability of the statistical properties of the signal. Typically, the task will be to discover the most probable state in which the sytem currently is. The state label will be used to generate rules then fed into the automation level as explained before. As depicted in this Figure, the discovered state at one given layer can be fed to the above layer as an extra input signal. For sake of simplicity, only three layers are shown, but more or less layers could be present

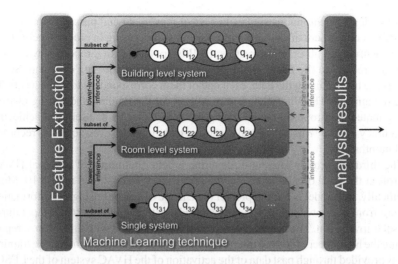

Fig. 10 Example of adaptation layer based on state modelling

depending on the context of the application. The generic principle is to include the information coming from the more granular to the less granular levels, from sensor information, to room information and up to floor and building knowledge.

5.4 Examples of Application

We describe here three applications that were implemented according to the architecture proposed in the previous section. Hidden Markov Models (HMMs) were used as modelling tool.

The first application is linked to single system modelling, i.e. the lowest level as illustrated in Fig. 10. A single system consisting of a smart plug is used to measure the electricity consumption of appliances at the frequency of 10^{-1} Hz. The target application is the automatic identification of electric appliances so that building residents can have detailed information on their electricity bill. The HMMs are trained using a publicly available database of electric signatures called *ACS-F1* for Appliance Consumption Signature Fribourg 1 [30]. The database consists of two acquisition sessions of 1 h for 100 appliances spread uniformly into 10 categories. Two evaluation protocols are proposed with the database distribution. The first *intersession* protocol consists in attempting to recognize signatures from appliances already seen in the training phase [31]. The second protocol, called *unseen instance*, aims at recognizing signatures coming from appliances never seen in the training phase. In this more difficult case, the system has to be able to generalize to new brands or models [32].

The second application aims at automatically recognizing activities at the level of a full floor in a residence. A set of presence and door sensors is spread in the residence and provides asynchronuous signals in case of movement and passage. The feature extraction consist here of a simple shifted window processing in order to reconstruct time synchronized observation vectors. HMMs are built with a topology where the states are related to the zones of the floor where the activities are expected. Seven HMMs are trained using the data, each one of them corresponding to a given activity to be recognized. The Viterbi algorithm is used for the training and testing phases. We evaluated the proposed models using the WSU CASAS dataset [33], achieving an accuracy of 98.9 % using the leave-one-out technique on seven activities recorded for 8 months [34].

The third application aims at discovering a better definition of seasonal HVAC controls at the level of a building, i.e. the upper level as illustrated in Fig. 10. More specifically, the model is built to predict the change of seasons using the information coming from window openings, window blinds, external and internal temperatures and solar irradiation. The model is based on a HMM defined with three states, representing the heating, cooling and intermediate seasons. In this experiment, the training data is provided through past data of the activation of the HVAC system of the LESO-PB building in EPFL, Lausanne. For example, when the system is in heating mode, the associated state is known and the parameters of the models can be trained with the input data. In the testing phase, the Viterbi algorithm is used to determined the state label at a given time and compared to the ground truth. The results are showing a correct season identification with an accuracy up to 91 % for the heating and cooling seasons, while the most difficult intermediate season shows a rate of 69 % correct detection [34]. While this experiment is still a bit theoretical, it shows that a machine learning based modelling can capture the actual controls of a well-tuned HVAC system. It can be reasonably expected that the learned model can be re-used in a similar building configuration.

The three preceding applications show the feasibility of the proposed state-based modelling as algorithmic approach to implement the adaptation level of Fig. 10. A missing step which is not explored in this work, is the inference of relevant rules to be injected in the automation level.

6 Conclusion

Buildings automation systems have in the last years not followed the trend of modernization and are always relying on isolated networks. The emerging technologies of sensors and especially the Internet-of-Things can provide many advantages to those buildings. In this chapter we investigated the main issues that have to be solved for augmenting traditional building automation systems with IoT capabilities. The main challenge lies in the natural heterogeneity of building networks working on different protocols. A successful homogenization of BAS can only be achieved by implementing a standard and open protocol stack. Multi-protocol gateways hiding the complexity of BAS by mapping devices capabilities to Web services highly simply the integration of existing networks. The emergence of IPv6 solves the address limitation of IPv4 and allows nowadays any device to be directly connected to IP networks. Trendy Web technologies like CoAP following the REST architectural style already contribute to the standardization process with their lightweight Web services. A key role in homogenisation is played by the acceptance of the application layer. Some propositions have already been done in this area converging to the use of data models for sensor properties. Until today none of them managed imposing itself as the de-facto standard. We can explain this by the heavy-coupling they impose using XML schemas that have to be known by every device. In our point of view only a loosely-coupled solution offering enough flexibility has a chance of being accepted by the scientific community and especially the industry. Here the contributions of the World Wide Web Consortium with RDF and ontologies have all their importance. Working in pervasive and ubiquitous environments requires approaching discovery of devices with another point of view. The need of a solution allowing to crawl devices with no prior knowledge appears inevitable. Distributing SPARQL queries over the network via multicast appears to be a clear response to this constraint. Building automation systems can benefit from advances made in the field of machine learning techniques. A new dimension of energy saving opens to IoT-enabled building management systems. Anticipating the users behaviour and actuating HVAC systems according to the buildings use has a real potential reducing the overall energy consumption. This target is only achievable by introducing a new level to the classical three-tiers decomposition of BAS.

IoT and WoT are expected to be key enablers for advanced building controls based on intelligent data analysis. In this direction, different machine learning techniques can potentially be used for managing a smart environment. The life-cycle of machine learning applications starts with an acquisition of raw data from the sensors distributed in the building, continues with an extraction of relevant features, follows with the identification or prediction of some events that are then, in turns used to modify the rules of the automation level of the building management system. The modification of the rules can take place continuously or off-line, at regular interval. A potential powerful modelling scheme can be found in state-based modelling such as Hidden Markov Models where the statistic of the features are captured according to states representing the piece-wise evolution of the building context. In the description of the adaptation level provided in this chapter, we have left apart important practical considerations, for example the ones regarding the computing power, data storage and the energy consumption of the adaptation layer itself. The limitation of resources and lack of energetic availability in IoT installations limits de facto the complexity of the machine learning techniques that may be used.

Although building automation being quite a recent research field, certain directions still emerge. The near future will give us more insights about the practical feasibility and the acceptance of Web technologies in buildings. Moreover we can already distinguish two imminent key points that will limit the spreading of IoT building automation systems. The first concern is about the energetic impact of such installations. The overall IoT approach should not consume more energy than traditional BAS and therefore counter the efforts done optimizing the building. Secondly, the very restricting open issue that has not been tackled is about security. One can easily imagine the consequences resulting in a misuse of the automation system. Cybercrime has become a major problem of today's information systems. Solutions limiting mashups between devices explicitly authorized and ensuring the privacy of historical data are essential before deploying IoT systems in buildings.

References

1. Perez-Lombard, L., Ortiz, J., Pout, C.: A review on buildings energy consumption information. Energy Build. **40**, 394–398 (2008)
2. Knx association. http://www.knx.org/
3. Enocean. http://www.enocean.com/en/home/
4. Granzer, W., Kastner, W., Reinisch, C.: Gateway-free integration of bacnet and knx using multi-protocol devices. In: Proceedings of the 6th IEEE International Conference on Industrial Informatics (INDIN '08) (2008)
5. The universal device gateway. http://www.devicegateway.com/
6. oBIX 1.1 draft committe specification. The universal device gateway. https://www.oasis-open.org/committees/download.php/38212/oBIX-1-1-spec-wd06.pdf
7. Neugschwandtner, M., Neugschwandtner, G., Kastner, W.: Web services in building automation: Mapping knx to obix. In: Proceedings of the 5th IEEE International Conference on Industrial Informatics (INDIN '07) (2007)
8. Bovet, G., Hennebert, J.: Web-of-things gateway for knx networks. In: Proceedings of the IEEE European Conference on Smart Objects, Systems and Technologies (Smart SysTech) (2013)
9. Bovet, G., Hennebert, J.: Offering web-of-things connectivity to building networks. In: Proceedings of the 2013 ACM International Joint Conference on Pervasive and Ubiquitous Computing (2013)
10. Jung, M., Weidinger, J., Reinisch, C., Kastner, W., Crettaz, C., Olivieri, A., Bocchi, Y.: A transparent ipv6 multi-protocol gateway to integrate building automation systems in the internet of things. In: Proceedings of the IEEE International Conference on Green Computing and Communications (2012)
11. Jung, M., Reinisch, C., Kastner, W.: Integrating building automation systems and ipv6 in the internet of things. In: Proceedings of the 6th International Conference on Innovative Mobile and Internet Services in Ubiquitous, Computing (2012)
12. Guinard, D.: A web of things application architecture—integrating the real world into the web. PhD thesis, ETHZ (2011)
13. Nurseitov, N., Paulson, M., Reynolds, R., Izurieta, C.: Comparison of JSON and XML data interchange formats: A case study. In: CAINE (2009)
14. W3C Semantic Sensor Network Incubator Group: Review of sensor and observation ontologies. http://www.w3.org/2005/Incubator/ssn/wiki/Incubator_Report/Review_of_Sensor_and_Observation_ontologies
15. W3C. Resource description framework. http://www.w3.org/RDF/
16. W3C Semantic Sensor Network Incubator Group: Semantic sensor network ontology. http://www.w3.org/2005/Incubator/ssn/ssnx/ssn
17. Dolce ultralite ontology. http://ontologydesignpatterns.org/wiki/Ontology%3aDOLCE%2BDnS_Ultralite
18. W3C. Sparql query language for rdf. http://www.w3.org/TR/rdf-sparql-query/
19. Goland, Y., Cai, T., Leach, P., Gu, Y., Albright, S.: Simple service discovery protocol/1.0 operating without an arbiter. Technical report, IETF (2000)
20. Guttman, E., Perkins, C., Veizades, J., Day, M.: Service location protocol, version 2. Technical report, IETF (1999)
21. Cheshire, S., Krochmal, M.: Dns-based service discovery. Technical report (2013)

22. Kovacevic, A., Ansari, J., Mahonen, P.: Nanosd: A flexible service discovery protocol for dynamic and heterogeneous wireless sensor networks. In: Proceedings of the 6th International Conference on Mobile Ad-hoc and Sensor, Networks, pp. 14–19 (2010)
23. Butt, T., Phillips, I., Guan, L., Oikonomou G.: Trendy: An adaptive and context-aware service discovery protocol for 6 lowpans. In: Proceedings of the 3rd International Workshop on the Web of Things (2012)
24. Mayer, S., Guinard, D.: An extensible discovery service for smart things. In: Proceedings of the 2nd International Workshop on the Web of Things (WoT) (2011)
25. Rahman, A., Dijk, E.: Group communication for coap. Technical report, IETF (2013)
26. Iso 16484-2:2004 building automation and control systems (bacs). Technical report, International Organization for Standardization (2004)
27. Huang, W., Lam, H.N.: Using genetic algorithms to optimize controller using genetic algorithms to optimize controller parameters for hvac systems. In: Energy and Buildings, pp. 277–282 (1997)
28. Kanarachos, A., Geramanis, K.: Multivariable control of single zone hydronic heating systems with neural networks. In: Energy Conversion and Management, pp. 1317–1336 (1998)
29. Yao, Y., Lian, Z., Hou, Z., Zhou, X.: Optimal operation of a large cooling system based on an empirical model. Appl. Therm. Energy **24**, 2303–2321 (2004)
30. Gisler, C., Ridi, A., Zufferey, D., Khaled, O.A., Hennebert, J.: Appliance consumption signature database and recognition test protocols. In: Proceedings of the 8th International Workshop on Systems, Signal Processing and their Applications (Wosspa'13), pp. 336–341 (2013)
31. Ridi, A., Gisler, C., Hennebert, J.: Automatic identification of electrical appliances using smart plugs. In: Proceedings of the 8th Internation Workshop on Systems, Signal Processing and their Applications (Wosspa '13), pp. 301–305 (2013)
32. Ridi, A., Gisler, C., Hennebert, J.: Unseen appliances identification. In: Proceedings of the 18th Iberoamerican Congress on Pattern Recognition (Ciarp '13), to appear (2013)
33. Cook, D.J.: Learning setting-generalized activity models for smart spaces. IEEE Intell. Syst. **27**, 32–38 (2012)
34. Ridi, A., Zakaridis, N., Bovet, G., Morel, N., Hennebert, J.: Towards reliable stochastic data-driven models applied to the energy saving in buildings. In: Proceedings of the International Conference on Cleantech for Smart Cities and Buildings (Cisbat '13) (2013)

Intelligent Transportation Systems and Wireless Access in Vehicular Environment Technology for Developing Smart Cities

Jose Maria León-Coca, Daniel G. Reina, Sergio L. Toral,
Federico Barrero and Nik Bessis

Abstract This chapter is focused on Intelligent Transport Systems (ITS) and wireless communications as enabling technologies for future Smart Cities. The chapter first reviews the main advances and achievements in both fields, highlighting the major research projects developed in Europe and USA as well as the suitability of Wireless Access in Vehicular Environment (WAVE) technology. These advances are contextualized with the notion of SmatCity, as a new emergent paradigm within the information and communication technologies. The chapter highlights the main contributions that ITS can provide in the development of Smart Cities as well as the future challenges.

1 Introduction: The Smart City Paradigm and Enabling Technologies

The smart city paradigm envisions one way of making the most with the city resources at all levels and areas, such as management, organization, technology, governance, policy, context-awareness, people, communities, economy, exist-

J. M. León-Coca · D. G. Reina · S. L. Toral (✉) · F. Barrero
Electronic Engineering Department, University of Seville, Seville, Spain
e-mail: storal@etsi.us.es; toral@esi.us.es

J. M. León-Coca
e-mail: jleon10@us.es

D. G. Reina
e-mail: dgutierrezreina@us.es

F. Barrero
e-mail: fbarrero@esi.us.es

N. Bessis
School of Computing and Maths, University of Derby, Derby, UK
e-mail: n.bessis@derby.ac.uk

N. Bessis and C. Dobre (eds.), *Big Data and Internet of Things:*   285
A Roadmap for Smart Environments, Studies in Computational Intelligence 546,
DOI: 10.1007/978-3-319-05029-4_12, © Springer International Publishing Switzerland 2014

ing infrastructure, and natural environment, among others, using new wireless technologies and cooperative communications. There is not a fully-defined methodology that encompasses how a smart city must be created, but the primary goals of a smart city could be stated in two ideas: a sustainable and enhanced city. The continuous population growth in the urban area is forcing an optimal use of the existing infrastructures, where the management and monitoring of transportation systems are capital factors. The smart city concept is making in recent times an important contribution in this field, allowing cities to self-manage their own resources.

Much effort has been focused on developing ITS infrastructure in America, Europe and Asia. Many projects have been proposed and carried out to demonstrate the feasibility of the deployment of an infrastructure capable of establishing real time communications between vehicles, and vehicles and the fixed infrastructure. In this context, WAVE technology has also played an important role since it has been envisioned as the baseline technology for developing the ITS infrastructures. Due to the different architectures proposed in the different continents, it is important to understand the similarities and dissimilarities of these architectures.

These ITS infrastructure will be the core for the creation of new Smartcities. Many of the challenges proposed in the ITS context are also challenges in the creation of Smartcities. For these reasons, the successful creation of Smartcities is strongly connected to the successful implementation of ITS infrastructure. ITS applications such as warning of congested roads and traffic accident avoidance are clear example of how a Smartcity, that is a better city, can be achieved through ITS infrastructure. In addition, this infrastructure can be extended to support new added multimedia services such as video exchange.

In this chapter, we aim to offer a state-of-the-art review of intelligent transportation systems and its new available technologies. For this purpose, we start with a review of related deployed projects in USA and Europe. Section 2 describes the wireless technology that supports the new advances in vehicular communications. Section 3 is focused on the Smartcity concept and in the potential benefits of ITS. Section 4 presents the future challenges to get over in order to deploy the intelligent transportation system infrastructure, and finally the conclusion of this chapter are found in Sect. 5.

2 Intelligent Transportation Systems

Throughout history the increment of population has been driving the expansion and growth of urban areas, being necessary the construction of new infrastructures to offer new transport systems for citizens and to accomplish their mobility needs. The preferred mobility system is the road transport with motor vehicles. However, its massive usage produces traffic congestion, reducing the efficiency of mobility systems, increasing travel times, air pollution and fuel consumption [7]. These new challenges related to sustainable and efficient public infrastructure are leading the idea of ITS. The ITS purpose consists of integrating new technologies with the transport infrastructure in order to enhance safety, mobility, environmental efficiency, and sustainability and economic performance in the interest of citizens. Since then, information and communications technologies (ICTs) have been used to develop new intelligent transport applications. In the last two decades, the cooperative driving has captured a lot of attention from the research community. This concept establishes that all transport system elements are able to cooperate and communicate by exchanging real-time information, avoiding accidents and improving the safety behind the wheel. Moreover, it collects important information to manage infrastructures. This enabling technology, which also makes possible vehicle-to-vehicle (V2V) communications and vehicle-to-infrastructure (V2I) communications, has been the focus of researches and projects around the world.

2.1 Deployed ITS Projects

This section presents the major deployed ITS projects focused on vehicular networking. These projects have been classified chronologically distinguishing between their geographical region of development, USA and Europe. In general, these projects have agglutinated efforts from national governments, car manufacturers, consortiums and research centers to achieve the global standardization of architectures and protocols in these issues.

2.1.1 ITS Projects in USA

The strategic plan on this field can be found with foundation of the National ITS Programme of the US, which was a joint effort of ITS America and the USDOT. It was created by Congress in the Intermodal Surface Transportation Efficiency Act (ISTEA) of 1991. When this initiative began, it was called Intelligent Vehicle-Highway Systems Program (IVHS) and later on it was known by its current name. The National ITS Programme describes a long-term plan for using modern communications, information processing, control, and electronic technologies to improve the operation of surface transportation system [18]. First researches about wireless communications for vehicles generated the Dedicated Short-Rage Communications (DSRC). DSRC began as a vehicle to roadside communication technology using the 915 MHz band and designed for wireless toll applications. Later on, the potential benefits of V2V communications for information exchange suggested its use for that, but DSRC cannot get over the technological requirements for the communications between vehicles and it needed to be improved. Throughout the exposed projects of this section, the new DSRC was developed and tested. Figure 1 highlights the projects order in a timeline.

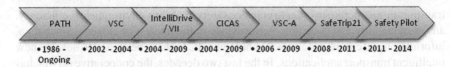

• 1986 -   • 2002 - 2004   • 2004 - 2009   • 2004 - 2009   • 2006 - 2009   • 2008 - 2011   • 2011 - 2014
Ongoing

Fig. 1 US project timeline

PATH (1986: Ongoing)

The California Partners for Advanced Transportation Technology (PATH) [43] program was born in 1986 and is composed of Institute of Transportation Studies of University of California, Berkeley, the California Department of Transportation (Caltrans) and in 2011, the California Center for Innovative Transportation (CCIT). PATH's purpose is to develop innovative ITS strategies and technologies to improve the safety, flexibility, mobility, stewardship and delivery of transportation systems in California, the United States and the world [60]. It has also played an important role in the creation of the ITSA. It was the first ITS research center in the US and has become in one of the most important in the world. Internally, it is divided into three research programs: traffic operations, transportation safety, and modal application. The PATH consortium has also collaborated in several projects under the National ITS programs such as SafeTrip21, CICAS, VII or DSRC communications.

VSC (2002–2004)

Vehicle Safety Communications (VSC) project was promoted by the Crash Avoidance Metrics Partnership (CAMP) Vehicle Safety Communications Consortium (VSCC) which was composed of BMW of North America, LLC, DaimlerChrysler Research and Technology North America, Inc., Ford Motor Company, General Motors Corporation, Nissan Technical Center North America, Inc., Toyota Technical Center USA, Inc., and Volkswagen of America, Inc., in partnership with USDOT. The USDOT research about countermeasures of rear-end, lane change and road departure crashes have demonstrated the benefits of using collision warning systems supported by V2V communications and vehicle positioning. These systems also were combined with traditional active safety equipment (autonomous sensing, lidar, radar, etc.). Following that idea, it was defined the VSC goals: to estimate the potential benefits of communication-based vehicle safety applications and to define their communications requirements; to ensure that proposed DSRC communications protocols meet the needs of vehicle safety applications; to investigate specific technical issues that may affect the ability of DSRC to support deployment of vehicle safety applications; and to estimate the deployment feasibility of communications-based vehicle safety applications [59].

One of the major conclusions of this project was that the potential ability of 5.9 GHz DSRC to enable a huge amount of applications and to offer significant vehicle safety benefits. These applications needed to be complied, evaluated and classified to identify the scenarios and communications requirements that cannot be archived by several evaluated wireless technologies. In contrast, DSRC gets over the requirements offering latencies of less than 100 ms, with a maximum range of 1,000 m, and also offering broadcast capability and high-availability communications. In addition, VSCC was actively participating in DSRC standardization activities [58]. As an initial work this project established the bases to be continued in the VSC-A project.

IntelliDrive/VII (2004–2009)

IntelliDrive or as it was also formally known Vehicle Infrastructure Integration (VII) is a joint effort of USDOT, associations and universities (ITS America, PATH, Virginia Tech, etc.), and industry (VSCC, communication service providers, etc.) whose research efforts are focused on V2V and V2I communications [39] in order to enable advanced crash avoidance applications and communications with the broader infrastructure.

The VII Working Group selected twenty vehicular safety use case applications (See Table 1) called "VII Day 1 Applications" which were considered the most important for stakeholders. They established a four-phase action plan that follows an application research, Proof of Concept (POC), prototype development and full-scale deployment.

It is important to highlight the realized POC testing which had the purpose of validating DSRC standards (802.11p, 1609, SAE J2735), providing core services as a full network architecture (VII Concept), supporting the vehicle safety applications and scenarios, and demonstrating security and privacy for each application. To sum up, the VII POC system successfully demonstrated the core DSRC-based V2V and V2I communications functions [39]. However, final results for future works recommended that, although DSRC radio range can be achieved in most conditions, the effects of multipath needed to be further studied to determine its full impact. Regarding security, they stated the necessity for two security-related IP protocols (V-DTLS and V-HIP) that needed also to be further developed by standards bodies. In general, some refinements are given for the tested protocols which will help the standardization bodies to improve the trial WAVE standard and to create the National ITS Architecture.

CICAS (2004–2009)

CICAS, which stands for Cooperative Intersection Collision Avoidance Systems, was carried out by USDOT with others stakeholders involved as CAMP, State Departments of Transportation and Automotive manufacturers. It was aimed to reduce cross-path crashes at intersection for both violations and gaps; to assess the value and acceptance of cooperative collision avoidance systems; and to develop and provide

Table 1 Day 1 uses cases applications

Number	Use case
1	Emergency brake warning
2	Traffic signal violation warning
3	Stop sign violation warning
4	Curve speed warning
5	Display local signage
6	Present OEM off-board navigation
7	Present OEM reroute information
8	Present traffic information
9	Electronic payments: parking/general
10	Electronic payments: gasoline
11	Electronic payment: toll roads
12	Traveler information
13	Ramp metering
14	Signal timing optimization
15	Pothole detection
16	Winter maintenance
17	Corridor management planning assistance
18	Corridor management load balancing
19	Weather information: traveler notification
20	Weather information: improved forecasting

tools to support industry deployments [6]. In addition and in collaboration of VII, it was developed and tested a viable system prototype of both Road Side Unit (RSU) and On Board Unit (OBU). The major CICAS contribution was a solid understanding of safety benefits and user acceptance after series of coordinated field operational tests (FOT).

VSC-A (2006–2009)

The Vehicle Safety Communications-Applications (VSC-A) Project has got as the starting point the VSC Project and continued their work with almost the same collaborative agents: The USDOT and the CAMP-VSC2 Consortium (Second edition of CAMP-VSC with less partners). The goal of the VSC-A project was to develop and test communication-based vehicle safety systems to determine if DSRC at 5.9 GHz, in combination with vehicle positioning, can improve upon autonomous vehicle-based safety systems and/or enable new communications-based safety applications. To achieve successful objectives, it was faced through the guidelines of: Crash Scenarios Identification and Selection, DSRC+Positioning and Autonomous Sensing Safety System Analysis, DSRC+Positioning-Only Test Bed System Development, Objective Test Procedures, Communications and Standards, Relative positioning Technology Development, Security and, for finalize, Multiple-OBE Scalability Testing. It is important to highlight their major accomplishments, such as the definition of a firm ground (defining crashes scenarios and V2V Safety applications for them, with an efficient system architecture), and the test and POC necessaries for standardization activities (SAE J2735, IEEE 1609). The VSC-A project finalized with next steps analysis distinguishing in the technical and non-technical issues [1].

SafeTrip21 (2008–2011)

The Safe and Efficient Travel through Innovation and Partnerships for the 21st Century (SafeTrip21) initiative was founded by the USDOT Research and Innovative Technology Administration (RITA) in 2008 after VII programme. This initiative was aimed at testing and evaluating integrated, intermodal ITS applications, focusing on providing immediate benefits such as improved safety, reduced congestion, and advancing the nation's transportation system [42]. This celerity was looking for expanding and accelerating DSRC communication deployments and testing the V2V and V2I communications in real environments in order to explore and validate the potential safety enhancements of this technology. The USDOT gave four awards to explode specific ITS applications [19]: The California Connected Traveler Test Bed headed by Caltrans for improving the travel experience on the San Francisco Bay Area's congested highways; The I-95 Corridor Coalition Test Bed headed by University of Maryland for informing the travelers about traffic congestion along Eastern Seaboard; and the last two awards were issued to iCone Products LLC and TrafInfo Communication, Inc. for an independent ITS evaluation in order to improve the deployed test-beds and equipment. The results and evaluation of SafeTrip21 were made by an independent company [Science Applications International Corporation (SAIC)] and can be consulted in [42], where each project and application is deeply described. Finally, this initiative can be considered as the first federally funded ITS field test focused on market-ready consumer products which put in the showcase the technology bondages to be used in the real world and to be sold in the global market-place.

Safety Pilot (2011–2014)

The Safety Pilot is a V2V and V2I DSRC technology test-bed project led by the University of Michigan Transportation Research Institute (UMTRI) in collaboration with CAMP and inside the USDOT ITS programme [53]. Its purpose is to demonstrate in an urban scenario with real drivers the benefits of V2V communications and its acceptance. The deployed model includes more than 73 lane-miles of equipped roads in the city of Ann Arbor (see Fig. 2). They offer the drivers a full equipment V2V platform (more than 2,800 have been given) to realize its measurements and collect one year of data [19]. Recently, six motorcycles and a bicycle have been joined to the project to study how these vehicles interact with deployed infrastructure [38].

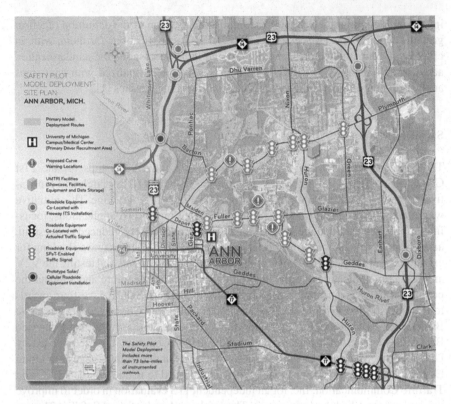

Fig. 2 Safety pilot model deployment site plan [47]

The project is planed following four phases: the vehicle builds and driver clinics, device development and certification, real-world testing, and, finally, independent evaluation. The obtained results from this project in addition to other ITS research projects will be used by National Highway Traffic Safety Administration (NHTSA) to determine the proceeding of V2V communication activities, including possible future rulemakings [54].

2.1.2 ITS Projects in Europe

European research and projects about ITS in Europe are mainly driven and funded by the European Commission (EC). A good starting point can be to consider the creation of eSafety [14] Working Group. It was created as a joint industry-public sector initiative in 2002 to explore ways on how to accelerate the ITS technologies in European roads. It is composed of representatives of the EC (Directorates-General Information Society, Transport and Energy, Enterprise), the automotive industry (European Automobile Manufacturers Association (ACEA) and European Council for Automotive

R&D (EUCAR), component suppliers of telecommunications and service industries, infrastructure operators and other stakeholders like ERTICO, public authorities and user associations [29]. Its main purpose is to coordinate, advice, and be the intermediary between the standardization bodies, establishing an ITS action plan. After eight fruitful years, eSafety decided to change its name to iMobility keeping its own mindset and way of proceeding, becoming part of the Commission's new Intelligent Car Initiative [28]. The eSafety forum also created a parallel project that collected all the results, presentations and information from every ITS European project. It was called COMeSafety, and the collected results are also known through this project name (e.g. COMeSafety architecture). Throughout its history, they have been established the road map for ITS in several projects and initiatives within the 6th and 7th Framework Programme for Research and Technological Development (FP6 and FP7) collected in the Directive 2010/40/EU [28] in order to face the transportation problems around all state members. This directive outlines a method to coordinate and accelerate deployments of ITS focusing on priority areas for the development and use of specifications and standards [10]. It has been categorized as priority areas the optimal use of road, traffic and travel data using a proper management thanks to ITS services to improve its safety and security allowing the linkage of vehicles and the transport infrastructure.

In this section, the studied projects are classified by the entity funding the project. First, the three major integrated projects within FP6 such as CVIS, SAFESPOT and COOPERS, are introduced. They are focused on the research and deployment of applicable technologies in cooperative systems and vehicular communications from three different and complementary points of view [40] (Fig. 3):

- CVIS works on the core architecture that enables vehicular communications V2V and V2I, developing prototypes and realizing tests.
- SAFESPOT analyses and defines critical tasks in the main scenarios for vehicular safety.
- COOPERS is focused on the road operators' viewpoint, offering the needed infrastructure to support vehicular applications.

Second, three FP7 projects are also described. The two first projects, PRECIOSA and PREDRIVE C2X, prepared the ground to DRIVE C2X and those that came later on. These projects continue building on the results and recommendations of previous projects.

- PRECIOSA is focused on security and privacy issues.
- PREDRIVE C2X has the purpose of enhancing materials and method to develop better tests.
- DRIVE C2X is aimed to implement the major tests ever done to validate new ITS.

There are several European standardization bodies aimed to create the foundation and management of the harmonization activities over new ITS technologies such as the European Committee for Standardization (CEN), the European Committee for Electrotechnical Standardization (CENELEC), and the European Telecommunications Standards Institute (ETSI). It is also worth highlighting the CAR 2 CAR

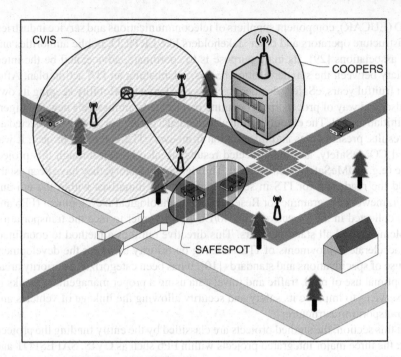

Fig. 3 CVIS, SAFESPOT and COOPERS vision

Communication Consortium (C2C-CC) initiated by European vehicle manufacturers, which represents the automotive industry activity, but it is also open for suppliers, research organizations and other partners [5].

CVIS (2006–2010)

Cooperative Vehicle and Infrastructure Systems (CVIS) project was aimed to create the core network of vehicular environment, developing a full architecture which serves as infrastructure for vehicular communications. CVIS also defined an open framework architecture to facilitate the creation of cooperative vehicular applications, supporting real-life applications and services for drivers, operators, industry and other key stakeholders [11]. This project was pioneer in the use of ISO "CALM" (Communications Access for Land Mobiles) standards, which was still under development, becoming within this project its own "proof of concept". To achieve its objectives the project was divided into three core sub-projects: IP management, core architecture, and deployment enablers. These subprojects were developed in parallel, coordinating their individual efforts in the production and demonstration phases. The results of the CVIS project can be evaluated by its yielded benefits collected

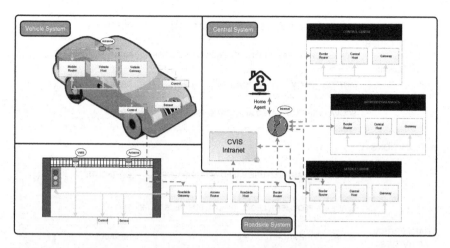

Fig. 4 The three CVIS technical sub-systems, based in [45]

between technologies and applications produced over its cooperative applications infrastructure. The output technologies are given below:

- COMM (Components for Communications and networking): the purpose was to fulfill the requirements of availability, connectivity, flexibility, and transparency proposed in the CALM draft standards. As a result, the definition of a fully architecture was designed with three different entities, where several communications technologies were considered (CALM infrared, CALM M5, CALM MM, CALM 2G/3G, CEN DSRC). It is depicted in Fig. 4.
- FOAM (Framework for Open Application Management): It is an open execution platform to support every phase of in-vehicle and road side applications during its lifecycle. These applications will improve both drivers and road side authority experience, offering new transport safety information and management control strategies. The base idea is that everyone can be connected through a cooperative transport architecture that is implementation-independent using a unified and adaptable middleware.
- POMA (Positioning Map and Local Referencing): a novel positioning and mapping solution was proposed in order to obtain the needs of accurate positioning requirements of vehicular applications, highlighting those related to active security and avoidance. On one hand, several positioning modules of different technologies were integrated, taking into account the signal robustness to describe the committed error rate. On the other hand, new map modules were developed with enhanced map content, map access APIs and map update mechanisms.

Using the described technologies, vehicular safety applications were implemented to demonstrate the system functionality in the test-beds.

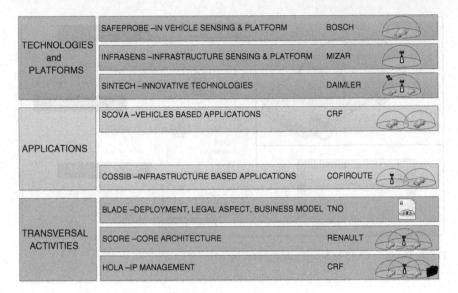

TECHNOLOGIES and PLATFORMS	SAFEPROBE –IN VEHICLE SENSING & PLATFORM	BOSCH	
	INFRASENS –INFRASTRUCTURE SENSING & PLATFORM	MIZAR	
	SINTECH –INNOVATIVE TECHNOLOGIES	DAIMLER	
APPLICATIONS	SCOVA –VEHICLES BASED APPLICATIONS	CRF	
	COSSIB –INFRASTRUCTURE BASED APPLICATIONS	COFIROUTE	
TRANSVERSAL ACTIVITIES	BLADE –DEPLOYMENT, LEGAL ASPECT, BUSINESS MODEL	TNO	
	SCORE –CORE ARCHITECTURE	RENAULT	
	HOLA –IP MANAGEMENT	CRF	

Fig. 5 SAFESPOT sub-projects, based on [46]

SAFESPOT (2006–2010)

The Cooperative vehicles and road infrastructure for road safety (SAFESPOT) was aimed to design systems for cooperative driving which makes use of V2V and V2I communications [52]. These systems will enable technologies that improve the safety in roads, such as a "Safety Margin Assistant" to detect potentially dangerous situations, advising drivers with a big amount of information. SAFESPOT was formed as a consortium of 51 partners led by FIAT RESEARCH CENTER—Italy with a budget of 38 million € [20]. Key challenges of SAFESPOT are based into three statements:

- Test and working with 802.11p, candidate radio technology for V2V and V2I communications.
- Developing and improving a real time positioning and mapping system.
- Aligning with C2C-C and CALM standardization group.

Different tasks and sub-projects were established to deal with these challenges, as shown in Fig. 5.

The results from the project are the establishment of the "Common European ITS Communication Architecture" for cooperative systems, which was created in cooperation with several European projects (CVIS, COOPERS, PRE-DRIVE C2X, etc.), the implementation of a high speed ad hoc network with a complete set of messages based in the IEEE 802.11p, and a report about the future perspectives and new services and business models supported.

COOPERS (2006–2010)

COOPERS stands for CO-OPerative SystEms for Intelligent Road Safety. This project was aimed to create a Cooperative Traffic Management system, applying telematics innovation to the road infrastructure [8]. This project uses the cooperative systems, supported by wireless connections (V2V and V2I) to collect useful information from vehicles (e.g. speed, location, journey time, weather conditions, vehicle status), creating a huge data base shared with traffic operators and ITS service providers. A proper analysis of this collected data base will provide new services that improve the safety on roads and the infrastructure optimization (e.g. incident warning, temporary traffic regulations, dynamic route guidance, road charging, etc.). It will yield an easier applications development, joining together telematics initiatives between car industry and infrastructure operators. Therefore, COOPERS project was established focusing on the statements of data collection, data analysis and in-vehicle information to develop their systems. The COOPERS concept was validated through four demonstration sites, where the challenges were:

- The rapidly adaptation of traffic demands to offer alternative routes.
- International handover of COOPERS-services.
- To validate the vehicle and infrastructure systems effectiveness in security issues from a traffic management.
- To enable real time information services and eCall integration.

Finally, the Coopers project was considered a success by the road operators involved in the field test [9]. First results showed the capability of their systems to provide warning of incidents more timely than roadside information systems [44].

Several different communication technologies and protocols were evaluated such as CALM-IR, DVB-H, DVB, WiMAX, GPRS, etc. and not 802.11p. The main reason is that they were focused on the core network using an established technology.

PRECIOSA (2008–2010)

PRivacy Enabled Capability In co-Operative Systems and Safety Applications (PRECIOSA) project was focused on security and privacy issues collected in the recommendations given by previous projects. The guidelines and objectives of PRECIOSA are [50]:

- To define an approach for the private evaluation of co-operative systems in terms of communication privacy and data storage privacy.
- To define a private aware architecture for co-operative systems which involves suitable trust models and ontologies, a V2V privacy verifiable architecture, and a V2I privacy verifiable architecture, including the architecture components for protection, infringement detection, and auditing.
- To define and validate guidelines for privacy aware co-operative systems.
- To investigate specific challenges for privacy.

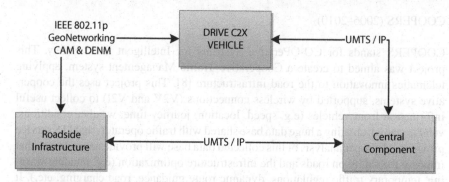

Fig. 6 DRIVE C2X Architecture

PREDRIVE C2X (2008–2010)

The PREparation for DRIVing implementation and Evaluation of C-2-X Communication technology (PREDRIVE C2X) project was aimed to prepare a large scale field trial for vehicular communication technology, following the COMe-Safety architecture and developing a robust enough prototype of V2X system and its verification method for its use in future field operational tests. Also, this project has done a simulation tool for modeling the car to car (C2C) applications. PREDRIVE can be considered a bridge project from FP6 vehicular projects and the oncoming projects of FP7 being the predecessor of DRIVE C2X, the last exposed project.

DRIVE C2X (2011–2014)

DRIVing implementation and Evaluation of C2X communication technology in Europe (DRIVE C2X) is the major pan-European field operational test on cooperative systems based on CAR-2-X communication using WLAN and 3G communication technologies [57]. It was built on foundations of PREDRIVE C2X project updating technology issues, such as the application of new standard series, the integration of a data backend (enabling the first commercial ITS services). The system architecture identifies the same three major aspects given by CVIS, as depicted in Fig. 6.

Figure 6 shows how to equip the vehicle with a platform that implements IEEE 802.11p and UMTS to respectively talk to the infrastructure and to the ITS central. Its protocol stack supports ad hoc communications based on GeoNetworking on one hand, and IPv6 on the other hand. The projects ends by the end of 2013, and the final report will be published by half of 2014 [13].

2.2 Reference ITS Architectures

The creation of ITS reference architectures was developed by standardization and harmonization bodies in parallel, or even aligned, with deployed ITS projects. These architectures pursuit to minimize the developed efforts made on middleware and adaptation layers, creating a common structure for all projects. As follows, the American and European ITS architectures are presented. Both architectures are not exactly the same, but they are very similar in the sense of looking for a global market (e.g. 3GPP).

2.2.1 National ITS Architecture

The National ITS Architecture (NITSA) is the North America ITS reference architecture from the National ITS Programme of the US. The NITSA was created due to the necessity of having a reference framework that conforms step by step how to plan, define and integrate ITS [49]. Currently, the NITSA is in its 7.0 revision which can be considered as a mature product. This architecture emerged from the joint actions of several stakeholders (transportation practitioners, systems engineers, system developers, technology specialists, consultants, etc.). The architecture of NITSA is divided into three layers: Institutional, Transportation and Communications layers.

- The Institutional Layer includes the institutions, rules, methods, process, and policies that are required for proper definition of any phases in an ITS. This layer is placed at the base because solid institutional support and effective decisions are prerequisite for an effective ITS program. This is where ITS challenges and requirements are established.
- The Transportation Layer is where the transportation solutions are defined in terms of the subsystems and interfaces as well as in terms of the underlying functionality and data definitions that are required for each transportation service. This is the core layer of the NITSA.
- The Communications Layer provides the accurate and timely exchange of information between systems to support the transportation solutions.

Figure 7 illustrates the NITSA from a high-level physical point of view. First, the four big boxes represent the categories in which the different subsystems are grouped. It distinguishes among Travelers, Centers, Vehicles and Field. Next, the subsystems are illustrated as white boxes and represent the physical elements reaching a total of 22 transportation subsystems. Finally, the oval shape elements correspond to the four general communication links used to exchange information between subsystems which are supported by the following technologies:

- Vehicle—Vehicle Communications: WAVE/DSRC 5.9 GHz.
- Field—Vehicle Communications: WAVE/DSRC 5.9 GHz, WiFi, WiMAX, and wireless mesh networks.

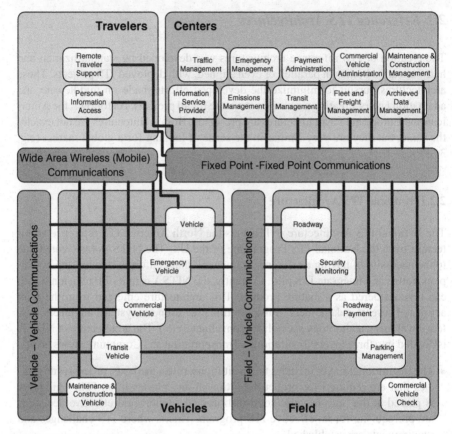

Fig. 7 The National ITS Architecture, based in [48]

- Wide Area Wireless (Mobile) Communications: Cellular networks, WiMAX, wireless mesh networking.
- Fixed Point—Fixed Point Communications: Public or private communications networks that assure a reasonable quality levels.

The NITSA is supported by RITA in collaboration with USDOT. An important milestone is a software application based on NITSA 7.0 as starting point. This software is called Turbo Architecture available for its free download [12].

2.2.2 ISO CALM Architecture

The European ITS Architecture is gathered under different names and regulations from each standardization body, but they make reference to the same contents. In the literature could be known such as ISO CALM Architecture and COMeSafety Architecture, and it refers to the open architecture created by the following groups:

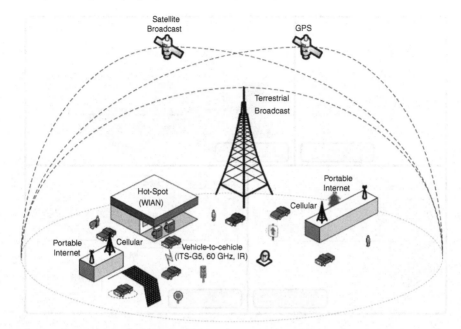

Fig. 8 ISO CALM scenario, based on [15]

- European Projects (COMeSafety, COOPERS, CVIS, SAFESPOT, PREDRIVE C2X).
- Industrial activity (C2C-CC).
- Standardization Work Group (ISO TC204, ETSI TC ITS, CEN TC278, IEEE 802.11p, IEEE 1609).

ISO CALM Architecture has been considered more wireless technologies than it was in NITSA, as it is shown in the introduction scenario of Fig. 8.

Every used wireless technology is collected into the ISO CALM protocol suite and adapted to be used by the uppers layers through:

- CALM 2G Cellular Systems [31].
- CALM 3G Cellular Systems [32].
- CALM Infra Red Systems [33].
- CALM M5 [34].
- CALM MM [35].
- CALM Mobile wireless broadband using IEEE 802.16 (WiMAX) [36].
- CALM Using broadcast communications [30].
- CALM Satellite networks [37].

ISO CALM Architecture also introduces four different subsystems the same than NITSA, but they have been called using different names (Fig. 9).

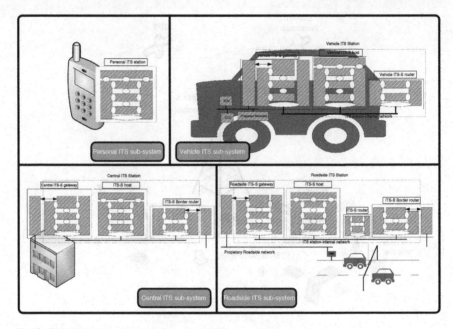

Fig. 9 ISO CALM Architecture, based on [15]

It can be noticed that the highest level view of ISO CALM architecture is very detailed because it also includes the stacks layers implemented by each entity from every subsystems:

- Central ITS sub-system contains a central ITS station that supports ITS services. It can be considered the backbone of the ITS network.
- Roadside ITS sub-system which offers primary ITS services and access to the ITS network.
- Vehicle ITS sub-system is a mobile on board unit that shows drivers ITS services and gathering relevant information to feedback the safety applications.
- Personal ITS sub-system provides the access to the ITS services in a hand-held devices allowing, for example, pedestrians to use offered ITS applications and communications.

3 WAVE

WAVE is a wireless technology specially designed to work in difficult and hash environments which allows a quick way of establishing communications between vehicles with very high mobility characteristics, enclosed threshold delays for safety messages with a severe QoS, optimal power consumption, and maintenance of privacy and anonymity of roaming users, among many other environmental challenges.

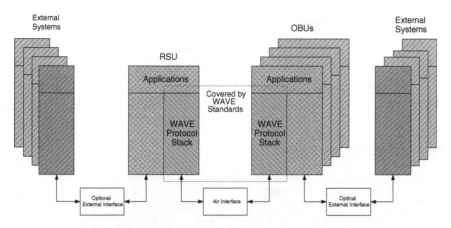

Fig. 10 Example of WAVE system component, based on [27]

Currently, as a global term, WAVE is used to define all around wireless vehicular communications, becoming to be arbitrarily used as a more general term such as DSRC. From the view point of standardization, WAVE is a term used to design the suite of IEEE 1609.x standards over the IEEE 802.11p standard.

3.1 The WAVE Architecture

There are two terms which are very frequently used in the specific literature. On one hand, the Road Side Units (RSU) is typically a fixed unit that connects moving vehicles as an Access Point (AP) to the network. They can be considered as the Field subsystems (see above in Fig. 7) of the National ITS Architecture. On the other hand, the On Board Unit (OBU) is a network device attached to a vehicle and/or also carried by pedestrians. In both cases, OBUs are always considered in motion. Figure 10 delimits the application scope of WAVE in a clearer way. It illustrates the fields covered by WAVE Standards in a possible case of use, where a communication is established between one RSU and many OBUs. Moreover, other communication technologies are also represented with external systems through different links.

WAVE standard is specifically designed to work in the range of 5.9 GHz frequency. More specifically, as it can be seen in Fig. 11, frequencies from 5.850 to 5.925 GHz in the USA [2], and from 5.855 to 5.925 GHz in Europe [16]. The bands are divided into seven channels, each channel with a bandwidth of 10 MHz. Generally, in the two considered areas, these channels have been dedicated for a specific purpose depending on the target application: general ITS applications, ITS road safety applications and ITS non-safety applications. Note that Fig. 11 also includes the Industrial, Scientific and Medical (ISM) band.

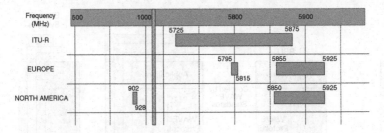

Fig. 11 ITS spectrum band frequency

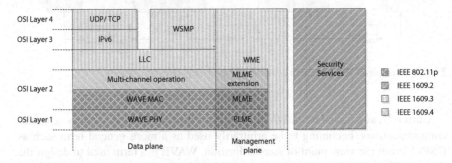

Fig. 12 The WAVE communication stack

Finally, WAVE has also been designed for safety applications such as hazardous location notification, stationary vehicle warning, roadwork warning, emergency vehicle warning, intersection collision avoidance, etc. All safety applications were collected in a basic set of applications for vehicular communications in [6] for USA and in [17] for Europe. Many of them were restricted by critical time latency of 100 ms and a minimum frequency of broadcast of 10 MHz.

3.2 The Suite of Standards WAVE

The historical evolution of WAVE has been decided by projects (POC and FOT). Therefore, between 2006 and 2007 a suite of standards under the name of IEEE 1609 was accepted for its trial use. Initially, the standards were developed for their use over IEEE 802.11p. The protocol layers are from bottom to top: IEEE 802.11p-2007, IEEE 1609.4-2006, IEEE 1609.3-2007, IEEE 1609.2-2006, and IEEE 1609.1.

In 2010, the full use standard was published. Some substantial changes from trial standard were made. The major appreciable change was the elimination of IEEE 1609.1 because it was found unnecessary (Fig. 12).

This stack offers two possibilities of climbing the stack, the first one is over IPv6, and the second one using short messages specifically designed for this environment WAVE Short Message Protocol (WSMP). As follows all full suite of WAVE will be exposed.

3.2.1 The IEEE 802.11p

This new standard was born with the purpose of providing a wireless communications for devices which are possible in a motion and a very short-duration communications exchange are required. Transactions must be completed in time frames shorter than the standard authentication and association to join a BSS on this environment. Currently, this standard can be consulted as an amendment of IEEE 802.11-2012 in his 11th part as an extension for devices which operate outside the context of a basic service set [26]. IEEE 802.11p has not added many changes since the trial version of WAVE. This standard collects the standardization methods applied in the 5.9 GHz band of OFDM for the PHY layer specifying channel bandwidths, transmit power classifications, transmission masks, etc. Inside of, the MAC sub-layer uses carrier sense multiple access with collision avoidance (CSMA-CD) without multi-channel operations functions. There are seven communications channels. In one side, the control channel (CCH) will be used for safety applications. In the other side, whereas the other six channels, called service channels (SCHs), will be used for other safety services or, also, infotainment or commercial applications. This wireless technology allows both ad-hoc and WLAN roles in the communications adjusting perfectly themselves to this sort of environment.

3.2.2 The IEEE 1609.4

The IEEE 1609.4 has been approved in 2010 and its major difference from the trial is the MAC address change mechanism [27]. This standard is aimed to complements the MAC sub-layer, allowing the support of multi-channel wireless connectivity without being necessary the knowledge of PHY layer from upper layers [25]. The multi-channel operations are realized with the management of both entities defined by IEEE 802.11p: the CCH and the SCH. The enabled transmit operations are data queuing and prioritization, channel routing, and channel coordination. It has also the possibility to distinguish between WSMP and IPv6 data flows, giving the proper priority to each flow. The WAVE Short Messages (WSM) associated with security applications has the higher priority and travel through the CCH. The other services and applications negotiate through the CCH and the SCH.

3.2.3 The IEEE 1609.3

The IEEE 1609.3 is the WAVE network layer and it can be distinguished between data plain and management plane services. On one hand, it offers the data services of Logical Link Control (LLC) sub-layer, IPv6 and higher layers as User Datagram Protocol (UDP) and Transmission Control Protocol (TCP) and WSMP. And on the other hand, it offers the management services of IPv6 configuration, WAVE Service Advertisement monitoring, management data delivery, service requests and channel access assignment [24]. Many things have been changed from the trial version, most important are: in the WSMP definition were removed the forwarding and added an extensible header; in the service channel quality evaluation were added the link quality indicator and deleted IDLE channel timeout [27].

3.2.4 The IEEE 1609.2

This standard is oriented to keep the privacy from users and create a security mechanism that guarantees communications. The necessity for a compromised rule between the safety-critical performance of WAVE applications which required low levels of latency and a robust secure method is necessary. Compared to the trial version, the differences are: the introduction of ToBeEncrypted as data type, the suppression of secured WSM for assuring the WSMP communications in the application layer, and the improvement of certificate management processes [27]. IEEE 1609.2 provides mechanisms to authenticate WAVE management messages, to authenticate messages from non-anonymous users, and to encrypt messages to a known recipient [23]. The major challenge of this standard was the needed for a compromise rule between the safety-critical nature of WAVE applications that low latency and the addition of headers and mechanism that engrosses the transmission load to create secure mechanism is necessary.

3.2.5 Uppers Layers

Several uppers layers has been defining to offers different standardization applications as the IEEE 1609.11 [21] for specifying a payment protocol referencing ISO standards and the IEEE 1609.12 [22] for recording the WAVE Provider Service Identifier (PSID) allocation. Other WAVE standards could be developed to specify new higher layer features.

4 Development of Smart and Safe Cities Using WAVE and ITS

The SmartCity has been a concept that has been theorized in the last decade. There are several models which try to explain a complete vision about the necessity of cities to be "smart". Many of these statements are related to the future vision encouraged by the new ITS advances. Next, it will be presented three vision models in the existent literature which explain the essence of a SmartCity:

- The Smart City Wheel: This model was presented by Boyd Cohen and it is defined as a holistic framework for considering every key component that makes a city to be a smart [4]. These indicators are: smart economy, smart government, smart people, smart living, smart mobility and smart environment. He also defined ruling actions for each indicator to promote the progress. This scheme follows the lean start up principles [51].

- The IBM model: This model was proposed by IBM which, among others things, is a leading provider of smart solutions. They establish three pillars: the people, infrastructure and the operations. They also define three basic services: first, human services as education, healthcare, and social issues; second, infrastructure services as water, energy and transportation, and third, the overall management of all services such as public safety, city governance, etc. [56]. In this model, citizens are considered in the center of ecosystem, and the city boundaries are directed to satisfy their needs and growth introducing the collaboration among citizen groups, universities and city leaders. This view classifies Smartcities as a complex infrastructure of systems which should be connected between them.

- The Hitachi's model: The major objective in this model is to create a city which has the ability of satisfying the desires and values of its residents using advanced information technologies. They catalogue several activities related to the people daily lifestyles as living, work, study, and travel. Thus, a SmartCity is defined as "an environmentally conscious city that uses information technology (IT) to utilize energy and other resources efficiently" [55]. In this way and following a green movement, the relationship between people and the Earth must be well-balanced, that is, a city in harmony with the natural environment. This city must be supported by convenient and safety infrastructures highlighting the field of energy, mobility, water, communications and IT platforms.

Taking into account these definitions, the future vision given by ITS researchers is in the same line. The new ITS services and wireless technologies like WAVE can solve or help to get over the challenges proposed by each model in order to create a SmartCity. WAVE communications represent a new reference for future ITS proposals. However, it is still pending the necessary infrastructure that must be deployed to support the models of a SmartCity. Currently, the ITS boundaries around cooperative systems can only be qualified as potential benefits, because they still need FOT data that assure its assessment. In this line, FOT based projects as DRIVE C2X are working on getting the data to validate the expected impacts of this technology [3]. There are several fields in which ITS and Wave technologies can support SmatCity models:

- Safety: Through V2X communications drivers can be advised in advance about potentially dangerous situations, so they can act accordingly. Simulation works show the difference in the speed adaptation of warned drivers with respect to those that were not warned. It has a direct effect in the avoidance of accidents or at least reducing its severity, and indirect effects in the cost related with vehicles accidents. The main safety applications that can be easily incorporated with current ITS are:

  - Traffic jam ahead warning: When a vehicle is moving closer to a traffic jam an advice message can be received from others vehicles or from the infrastructure preparing drivers to go slower and to avoid a possible rear end collision.

- Road works warning: the RSU send broadcast messages to arriving vehicles advising about a potentially dangerous situation.
- Car breakdown warning: In a similar manner than the RSU in the previous function, the stopped vehicle sends a warning to incoming vehicles, enhancing the security in low visibility roads.
- Emergency vehicle approaching: The emergency vehicles sends "I am coming" messages requesting the way priority to neighbor vehicles.
- Weather warning: Several weathers condition cannot be perceived by drivers and/or their driving can be inappropriate for these conditions.
- Emergency electronic brake lights: When a vehicle suddenly brakes hard, a message is sent and retransmitted by neighbor vehicles. This application is primarily thought for dense traffic situations where the braking vehicle can be directly seen in order to avoid rear ends and chain collisions.
- Slow vehicle warning: The purpose of this warning functionality is to advice drivers that a slow vehicle is in front of them. This system tries to mitigate the rear end collision and get over potentially dangerous situations caused by an unusual speed.
- Post crash warning: It is aimed to give relevant information about crashes vehicles in driver's route. The system tries to provide the information as soon as possible considering crash auto detection.
- Obstacle warning: These messages will be manually or automatically generated whenever an obstacle is in the vehicle's route. They can also be generated by road operators through a RSU. The main benefit is providing an early detection and preparing drivers.
- Motorcycle warning: The motorcycle is always sending its movement and position to neighbor vehicles allowing both vehicle and motorcycle to detect possible crossing.
- In-vehicle signage: This functionality is provided by RSUs on traffic signs and key points along roads, with the idea of informing drivers about signaling and dangerous situations.

• Efficiency: Thanks to the collected data from vehicles with OBUs which are continuously transmitting their positions and speeds in an anonymous way, the enhanced traffic centers can evaluate and process this data to improve traffic management and traffic information as well as other economic issues.

- Traffic management improvement: The real time knowledge of traffic situation and potential hazards on the roads allows traffic control centers to offer optimized routes and to balance the traffic performance for especial situations. For instance, the Green light optimized speed advisory (GLOSA) is intended to provide speed advices to get the "green wave" or at least the time to green traffic light.
- Traffic information improvement: The traffic centers can also improve the information about routes such as travel time, distance to destination and fuel consumption, avoiding dense traffic areas or informing about points of interest (parking areas, restaurants, fuel stations, etc.).

- Economic issues: Other potential benefits are related with monetary aspects. Important cost savings can be achieved by reducing congestions, time of journey, CO2 emissions, and crashes.

- Sustainability: The sustainability refers to the optimal utilization of existing resources on roads. Through V2X communications, traffic control centers can manage better the traffic balance and offer intelligent route guidances. Moreover, traffic emissions and travel time will be improved by adapting the vehicles' speed to the green light cycles. Sustainability issues also require a change of mentality in drivers, companies and governments.
- Convenience: The V2X communications are not only empowering safety and efficiency applications, but they are also making use of mobile communications to support several services from commercial providers. Therefore, it is enabling a new business model which can attract the attention of considerable investments.

It can be clearly observed that the new ITS technologies enabled by the WAVE protocol can help in the creation of Smartcities. Each vision alludes to the optimization of transport infrastructures that solves the citizen needs. They also promote a green ecosystem with CO_2 reduced emissions and safe cities. All these statements are also part of the key challenges of ITS.

5 Future Challenges

An imaginative idea about zero accidents on roads and no more death behind fly-wheels is probably a utopian for now and evermore. But visionaries who work on transportation systems to create a better world try to get this idea as a final challenge. Their efforts have developed great inventions throughout history for improving drivers' conditions and, also, for saving their lives such as three-point safety belt, electronic stability control, Airbag, Anti-Lock Braking System (ABS) and, now, the cooperative systems. Its potential ability in the prevention of driver issues and crash avoidances have been probed through the upper commented projects and initiatives. However, these benefits still need to be reaffirmed through FOT and validated data with the aim of reaching an agreement in its implantation. The current status of ITS technology makes affordable the creation of ITS infrastructure and the implementation of new value added services. This infrastructure should not be considered as part of the department of transportation or to the traffic control centers, but as strategic infrastructure for the city and citizens in the way for creating a SmartCity. But to achieve this vision, several challenges and barriers must be overtaken:

- Economic challenges: The global crisis has caused a slowdown in the implantation of this kind of systems. First, the governments' subsidies and investments have decreased becoming unfeasible the full deployment of RSU along roads. Second and due to the same reasons, car manufactures cannot add fully equipped OBUs as there is not any available infrastructure to communicate with, and they are waiting governments and public authorities to make the first steps. The problem of enabling cooperative systems is that every considered sub-system needs to incorporate this new technology. Otherwise, it is useless to have a reduced number of fully equipped vehicles if they cannot communicate with the environment. That is the reason why new market strategies and business models consider deployment as the top priority. Several studies such as asking people about how much money they consider an OBU should cost have been done. They look for a first phase of deployment where vehicle holders buy their own OBU in order to generate a new market niche for hardware manufacturers. In the same way, this ITS infrastructure creation will also be money attractive for new service providers that could get benefits and contribute with new functionalities.

- Security challenges: The security issues and information privacy in the delivered messages enabled by V2X communications also constitutes a big challenge. These types of messages are required to be anonymous, but it is also imperative quick communication establishments in order to obtain low latency rates. The security in these applications can be supported by digital certificates forming a public key infrastructure (PKI). This certificate hierarchy has not been completely defined yet and an authentication established method like PKI from National eID Cards could be used for personal identification, following the scheme of [41].

6 Conclusions

This chapter analyzes the contributions of ITS and Wave to the aim of creating new Smart Cities. The benefits and current status of ITS have been reviewed through the major research projects developed during the last years in Europe and USA as well as the role of Wave as a facilitating technology. Several fields in which ITS and WAVE can support the models of Smart Cities have also been identified and highlighted through the chapter. Although the current status of technology is able to support intelligent environments, it is still necessary to advance in the deployment of the necessary infrastructure to create real smart cities.

References

1. Ahmed-Zaid, F., Bai, F., Bai, S., Basnayake, C., Bellur, B., Brovold, S., Brown, G., Caminiti, L., Cunningham, D., Elzein, H., Hong, K., Ivan, J., Jiang, D., Kenney, J., Krishnan, H., Lovell, J., Maile, M., Masselink, D., McGlohon, E., Mudalige, P., Popovic, Z., Rai, V., Stinnett, J., Tellis, L., Tirey, K., VanSickle, S.: Vehicle safety communications-Applications (VSC-A) final report, Sept 2011, NHTSA Headquarters and Research and Innovative Technology Administration (2011).
2. ASTM E2213–03: Standard specification for telecommunications and information exchange between roadside and vehicle systems 5 GHz band dedicated short range communications

(DSRC) medium access control (MAC) and physical layer (PHY) specifications, ASTM Standard (2010).
3. Benefits of ITS by Drive C2X, available at http://www.drive-c2x.eu/benefits
4. Boyd Cohen website available at http://www.boydcohen.com/smartcities.html
5. C2C-CC website available at http://www.car-to-car.org/
6. CICAS website available at http://www.its.dot.gov/cicas/
7. Commission, European: Intelligent Transport Systems-EU Funded Research for Efficient. Clean and Safe Road Transport. Publications Office of the European Union, Luxembourg (2010). ISBN 978-92-79-16401-9
8. COOPERS website available at: http://www.coopers-ip.eu
9. COOPERS: Final report on demonstration, available at: http://www.coopers-ip.eu/fileadmin/results/deliverables/D6100_-_Final_report_on_demonstrations_V1_0.pdf
10. Council, European Parliament: Directive 2010/40/EU of the European Parliament and of the Council of 7 July 2010 on the framework for the deployment of intelligent transport systems in the field of road transport and for interfaces with other modes of transport Text with EEA relevance, Aug 2010. Official J. Eur. Union (2010).
11. CVIS website available at http://www.cvisproject.org/
12. Download site of Turbo Architecture available at: http://local.iteris.com/itsarch/html/turbo/turboform.php
13. DRIVE C2X website available at: http://www.drive-c2x.eu
14. eSafety website available at: http://ec.europa.eu/information_society/activities/esafety/index_en.htm
15. ETSI EN 302 665:2010–09, Intelligent Transport Systems (ITS): Communications Architecture, ETSI Standard.
16. ETSI TR 102 492: Electromagnetic compatibility and radio spectrum matters (ERM); intelligent transport systems (ITS), ETSI Standard (2005).
17. Etsi, T.R.: 102 638: intelligent transport systems (ITS). Vehicular communications: basic set of applications: definitions, ETSI Standard (2009).
18. Euler, G.W., Robertson, H.D.: Intelligent transportation system-National ITS program plan-Synopsis, March 1995. USDOT and ITS America (1995).
19. Hector-Hsu, J., Ritter, G.T., Sloan, S., Waldon, L., Thornton, P., Blythe, K.: SafeTrip-21-Federal ITS field tests to transform the traveler experience, June 2011, John A. Volpe National Transportation Systems Center, ITS Joint Program Office (2011).
20. http://ec.europa.eu/information_society/activities/esafety/doc/esafety_2006/spectrum_28feb2006/6_safespot_applications.pdf
21. IEEE 1609.11: IEEE Standard for wireless access in vehicular environments (WAVE)-Over-the-Air electronic payment data exchange protocol for intelligent transportation systems (ITS), Jan 2011, IEEE Standard (2010).
22. IEEE 1609.12: IEEE Standard for wireless access in vehicular environments (WAVE)-Identifier allocations, Sept 2012, IEEE Standard (2012).
23. IEEE 1609.2: IEEE Standard for wireless access in vehicular environments (WAVE)-Security services for applications and management messages, April 2013, IEEE Standard (2013).
24. IEEE 1609.3: IEEE Standard for wireless access in vehicular environments (WAVE)-Networking services, Dec 2010, IEEE Standard (2010).
25. IEEE 1609.4: IEEE Standard for wireless access in vehicular environments (WAVE)-Multi-channel operation, Feb 2011, IEEE Standard (2010).
26. IEEE 802.11: Telecommunications and information exchange between systems local and metropolitan area networks-specific requirements Part 11: wireless LAN medium access control (MAC) and physical layer (PHY) specifications, March 2012, IEEE Standard (2012).
27. IEEE Draft Guide for Wireless Access in Vehicular Environments (WAVE)-Architecture: IEEE P1609.0/D6.0, June 2013, IEEE Draft Standard (2013).
28. iMobilitysupport website available at http://www.imobilitysupport.eu
29. Information Society Technologies: Research on integrated safety systems for improving road safety in Europe-The information society technologies (IST) programme 1998–2002, Sept 2002. Office for Official Publications of the European Communities, Luxembourg (2002).

30. ISO 13183:2012, Intelligent transport systems-Communications access for land mobiles (CALM)-Using broadcast communications, ISO Standard.
31. ISO 21212:2008, Intelligent transport systems-Communications access for land mobiles (CALM)-2G Cellular Systems, ISO Standard.
32. ISO 21213:2008, Intelligent transport systems-Communications access for land mobiles (CALM)-3G Cellular Systems, ISO Standard.
33. ISO 21214:2006, Intelligent transport systems-Communications access for land mobiles (CALM)-Infra Red Systems, ISO Standard.
34. ISO 21215:2010, Intelligent transport systems-Communications access for land mobiles (CALM)-M5, ISO Standard.
35. ISO 21216:2012, Intelligent transport systems-Communications access for land mobiles (CALM)-Millimetre wave air interface, ISO Standard.
36. ISO 25112:2010, Intelligent transport systems-Communications access for land mobiles (CALM)-Mobile wireless broadband using IEEE 802.16, ISO Standard.
37. ISO 29282:2011, Intelligent transport systems-Communications access for land mobiles (CALM)-Satellite networks, ISO Standard.
38. ITS Newsleter available at http://www.its.dot.gov/newsletter/PDF/Newsletter_august_V19.pdf, Aug 2013
39. Kandarpa, R., Chenzaie, M., Anderson, J., Marousek, J., Weil, T., Perry, F., Schworer, I., Beal, J., Anderson, C.: Final report: vehicle infrastructure integration proof of concept results and findings-Infrastructure, May 2009. Research and Innovative Technology Administration (RITA)-U.S. Department of, Transportation (2009).
40. Konstantinopoulou, L., Zwijnenberg, H., Fuchs, S., Bankosegger, D.: White paper: deployment challenges for cooperative systems. Available at http://www.cvisproject.org/download/Deliverables/Whitepaper_CooperativeSystemsDeployment.pdf
41. León-Coca, J.M., Reina, D.G., Toral, S.L., Barrero, F., Bessis, N.: Authentication systems using ID Cards over NFC links: the Spanish experience using DNIe. Proc. Comput. Sci. 2013 **21**, 91–98 (2013)
42. Miller, S., Rephlo, J., Armstrong, C., Jasper, K., Golembiewski, G.: National evaluation of the Safetrip-21 initiative: combined final report, March 2011, Science Applications International Corporation (SAIC) (2011).
43. PATH Website: Official web site of the PATH project, Website also available as http://www.path.berkeley.edu/Default.htm, Aug 2013
44. Piao, J., McDonald, M., Hounsell, N.: Cooperative vehicle-infrastructure systems for improving driver information services: an analysis of COOPERS test results, March 2012. Intell. Transp. Syst. (IET) 6(1), 9, 17 (2012).
45. Picture available at http://www.cvisproject.org/en/cvis_subprojects/technology/comm.htm
46. Picture available at http://www.safespot-eu.org/images/sub_projects.jpg
47. Picture available at http://www.umtri.umich.edu/content/SafetyPilot_map.JPG
48. Picture available at: http://local.iteris.com/itsarch/images/sausage.gif
49. Picture available at: http://www.iteris.com/itsarch/html/archlayers/transportationlayer.htm
50. PRECIOSA website available at: http://www.preciosa-project.org
51. Ries, E.: The Lean Startup: How Today's Entrepreneurs use Continuous Innovation to Create Radically Successful Businesses. Crown Business, New York (2011)
52. SAFESPOT website available at http://www.safespot-eu.org/
53. Safety Pilot Project Website available at http://safetypilot.umtri.umich.edu/
54. Schagrin, M.: Safety pilot final fact sheet, ITS Joint Program Office, Research and Innovative Technology Administration. Available at http://www.its.dot.gov/factsheets/pdf/SafetyPilot_final.pdf
55. Smart Cities by Hitachi website, available at http://www.hitachi.com/products/smartcity/index.html
56. Smarter Cities by IBM website available at http://www.ibm.com/smarterplanet/us/en/smarter_cities/overview/

57. Stahlmann, R., Festag, A., Tomatis, A., Radusch, I., Fischer, F.: Starting European field test for Car-2-X communication: The Drive C2X Framework. In: Proceedings of 18th ITS World Congress and Exhibition 2011.
58. The CAMP Vehicle Safety Communications Consortium: Vehicle safety communications project executive overview, Nov 2003, U.S. Department of Transportation-National Highway Traffic Safety Administration (2003).
59. The CAMP Vehicle safety communications consortium: vehicle safety communications project task 3 final report: identify intelligent vehicle safety applications enabled by DSRC, Mar 2005, U.S. Department of Transportation-National Highway Traffic Safety Administration (2005).
60. Web resource: PATH 2009 annual report. Also available as http://www.path.berkeley.edu/Data-Files/Annual_Reports/PATH-2009-Annual-Report.pdf, Aug 2013

Emerging Technologies in Health Information Systems: Genomics Driven Wellness Tracking and Management System (GO-WELL)

Timur Beyan and Yeşim Aydın Son

Abstract Today, with the technology-driven developments, healthcare systems and services are being radically transformed to become more effective and efficient. Omics technologies along with mobile sensors and monitoring systems are emerging disruptive technologies, which will provide us the opportunities of a paradigm shifting in medical theory, research and practice. Traditional methods are beginning to convert to a new personalized, predictive, preventive and participatory paradigm based on big data approaches. We anticipate that; next-generation health information systems will be constructed based on tracking all aspects of health status on 24/7, and returning evidence based recommendations to empower individuals. As an example of future personal health record (PHR) concept, GO-WELL is based on clinical envirogenomic knowledge base (CENG-KB) to engage patients for predictive care. In this chapter, we present the design principles of this system, after describing several concepts, including personalized medicine, omics revolution, incorporation of genomic data into medical decision processes, and the utilization of enviro-behavioural parameters for disease risk assessment.

1 Transforming Health in Information Era

Current healthcare systems are widely based on standalone or integrated information system infrastructures. *Health Information Systems* capture, store, share, transmit and manage data about health of the individuals or the transactions of the healthcare organizations. This concept includes integrated hospital and primary care information

T. Beyan (✉) · Y. Aydın Son
Informatics Institute, Department of Health Informatics, Middle East Technical University,
Ankara, Turkey
e-mail: tbeyan@yahoo.com

Y. Aydın Son
e-mail: yesim@metu.edu.tr

N. Bessis and C. Dobre (eds.), *Big Data and Internet of Things:*   315
A Roadmap for Smart Environments, Studies in Computational Intelligence 546,
DOI: 10.1007/978-3-319-05029-4_13, © Springer International Publishing Switzerland 2014

systems, clinical, laboratory, pharmacy, radiology and nuclear medicine information systems, patient administration, human resources, logistics and accounting management systems, and Picture Archiving and Communication Systems (PACS), etc. [71].

Regarding the capabilities of health information systems, three terms come into prominence; *Electronic Medical Record (EMR), Electronic Health Record (EHR)* and *Personal Health Record (PHR)* [28]. *EMR* is composed of clinical data repositories, clinical decision support systems, standard medical terminologies, computerized order entry, and documentation applications. Erroneously, EHR is sometimes used interchangeably with EMR, but essentially they are two different concepts. The *EHR* share an extraction of EMR among healthcare providers and other stakeholders [20]. *PHR* provides individuals to access their own medical data and engage in the healthcare [65].

Essentially, applications of information technologies in the health domain do not only consist of the health information systems. Today, healthcare services are continuously transforming under the influence of technological developments based on various intertwined concepts and applications. These are;

- Revolutionary paradigm shifting from traditional provider-centric to patient-centric personalized medicine,
- Commissioning of new health services based on information and communication technologies (mobile health systems, pervasive applications, environmental sensors, body area sensor networks etc.) and
- Development of evidence based healthcare systems with knowledge discovery capabilities driven by big data and knowledge infrastructure.

Today, volume, velocity and variety of data produced in the biomedical domain meet us the challenges and opportunities of big data appearing at all dimensions of healthcare systems. In 2012, healthcare data producing all over the world was 500 petabytes (10^{15} bytes), and it's estimated that will be 25,000 petabytes in 2020 [17]. Data volume and variety of EMR/EHR systems are exponentially increasing. It's argued that, if big data is used effectively, the US healthcare systems could make $300 billion in savings per year, falling costs by 8 % [45].

After the completion of Human Genome Project (HGP) in 2003; healthcare domain is immersed in biological data space such as genomic variations, gene expressions, proteomics, and metabolomics. By the end of 2011, genome sequencing capacity was estimated as 13 quadrillion bases per year on the worldwide. It's expected that, over the coming years, the National Cancer Institute (NCI) will analyse one million genomes (1,000 petabyte or 1 Exabyte) [45].

Today; researchers discover the new methods and approaches to incorporate big data into clinical systems. In a research study investigating genetic variations, 100,000 participant were involved, and totally 150 terabytes genotyping data were produced [42]. Furthermore, several Internet sources could be used to monitor of the population health conditions, e.g. Google Flu Trends to predict flu-related admissions and twitter updates to track epidemics [66].

In parallel, Internet connected devices containing embedded sensors and microprocessors Internet of Things (IOT) and mobile applications began to track and monitor personal clinical status (blood glucose level, heart and brain activity, sleep patterns, mood, etc.), behaviours (feeding, daily activity, calorie consumption, etc.) and environmental influences in a quantified manner [62]. The health data collected from machine-to-machine communications between these devices will be driving the big data challenge in healthcare [58].

In the context of next generation health information systems; evidence-based practice, seamless and efficient integration of clinical research outcomes into the clinical domain (translational research), omics driven personalized medicine, self-tracking and monitoring of the body and environment are the main emerging ideas. In the near future, all of these trends will be converged to provide effective and efficient health care services in learning health care systems.

In this chapter, a genomics driven wellness tracking and management system (GO-WELL) concept is proposed, which interprets genomic variation data, clinical, environmental, and behavioural observations to engage patients for predictive, personalized care. Before presenting the design principles of this system in prostate cancer, we briefly described several concepts about personalized medicine, area of omics, incorporation of genomic variation data into clinical processes, and the significance of environmental and behavioural parameters in disease mechanisms.

2 Omics Revolution and Personalized Medicine

Personalized medicine is a healthcare paradigm that aims to use individual's unique clinical, genomic, environmental, behavioural and sociocultural characteristics to predict disease susceptibility and assess risk, determine molecular characterization of disease for early diagnosis, tailor treatment regimens, and monitor prognosis. This emerging approach is based on new discoveries in bioinformatics i.e. *omics revolution* and provide us opportunities to provide precise, preventive, and effective medical care [12, 57].

2.1 Genomic Component of Diseases

Omics revolution, at first, emerged based on the genome sequencing research. Genome is the whole set of genetic material involved in a nucleated cell. Gene expression is a sequence of subcellular complex reactions aiming to convert inherited data (i.e. gene) into functional chemical molecules. Today, "ome" suffix is used to define various subcellular chemicals produced after these reactions. "Omics" are scientific research areas dealing with—omes using high-throughput screening techniques and producing big data [14, 24, 57] (Fig. 1).

Genomic DNA sequence is about 99.9 % identical between humans [14]. Compared with the reference sequence generated by the HGP, any individual's genome has about 3–4 million variants [13]. Genomic variations can range from single nucleotide

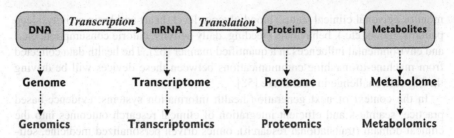

Fig. 1 Conceptual representation of gene expression, and corresponding -omes and -omics. mRNA (transcriptome) is synthesized from DNA strand in the cell nucleus (transcription). Then, in ribosomes, proteins are produced from amino acids using mRNA as a template (translation). Proteins are converted various metabolic products with enzymatic bioreactions

changes to gain or loss of whole chromosomes. In single nucleotide polymorphism (SNP), a single nucleotide in the genome sequence is different between individuals. Genomic variation can also be caused by insertion or deletion of nucleotides (indels) e.g. copy number variations [6].

SNPs are about 90 % of all the genomic variations. Although most of these are harmless, some of them have high values for disease risk assessment, medical diagnostics and pharmaceutical products [3, 13, 49]. Today, as sequencing and genotyping technologies gets cheaper and faster, a new line of companies emerged (e.g. 23andMe, deCODEme, Navigenics, Knome, Complete Genomics, etc.), who markets direct-to-consumer personal genomic services analysing SNPs to assess polygenic disease risks [8, 29]. With the advent of next generation sequencing (NGS) technologies, it's also possible to accomplish rapid and cheap whole genome sequencing (WGS). Researchers and clinicians expect that WGS will be one of the most important tools in personalized medicine age [7, 56, 72].

2.2 Environmental Component of Diseases

Since the HGP completed in 2003, the epidemiological researchers focused on to determine causative polymorphisms as genetic determinants of diseases [37]. With *Genome Wide Association Studies (GWAS)*, numerous variants have identified associated with diseases [5], but findings cannot explain the variability of diseases by only genetic polymorphism. Essentially, in 1980s, molecular epidemiologists discovered many biomarkers as reflection of the interaction between genetic and environmental factors. Currently, some authors propose that, in chronic diseases between 70 and 90 % of disease risks are due to environmental factors [51, 62]. Because modifying hereditary determinants of risk are not feasible except gene therapy, in order to target prevention efforts, it is critical to determine and evaluate changeable enviro-behavioural factors that interact with hereditary determinants to cause disease.

Therefore, to study and analysis of environmental factors, in a manner analogous to a GWAS, *Environmental Wide Association Studies (EWAS)* have started [5].

Today, a number of groups have undertaken efforts to determine the environmental causes of diseases. These factors concerning with mechanisms of human diseases can be categorized as socio-demographic parameters (age, ethnicity, race, gender, family health history), environmental causes (tobacco smoke, pollution, hazardous chemicals, occupational agents, microbial agents, radiation, etc.), behavioural factors (diet, physical activity, use of supplements, drugs, etc.), and internal environment of individual (ageing, body morphology, metabolism, hormones, microflora, inflammation, lipid peroxidation, oxidative stress, etc.) [51, 70]. In public health and clinical medicine, these factors can be classified as risk and protective factors according to the effect on disease mechanism and prognosis.

3 Genomic Data Integration into Medical Records

Today, in many countries, EMR/EHR systems are widely used to provide efficient and effective healthcare services. Integration of genomic data into medical practice via these systems may increase the performance and quality of the delivered healthcare services [16, 56, 73]. Genomic revolution challenges patients, clinicians, and healthcare information systems [55] and integration of genomic and clinical information into EMR/EHR is not routinely practiced except a few examples [31, 32].

3.1 Storing Genomic Data

WGS data contains about 3 billion base pairs and nucleated somatic human cells have two copies of genome sequence, which are about 3.2 GB. Collecting and sharing of personal raw genomic sequence exceeds the transmission and storage capacity in many healthcare organizations [35]. Due to the technical limitations, raw genomic data is stored outside of the EHR similar to PACS for medical images and clinical interpretation of data is preferable sent to the EHR database [23, 39, 59].

3.2 Clinicogenomic Information

In general, clinicians are not trained to understand and interpret the clinical importance of genomic variants [59]. Instead of presenting sequence data, incorporating meaningful and actionable clinicogenomic interpretation of variants into an EMR/EHR will be more effective [7, 38].

Today, the vast majority of variants has unknown significance and analysis of these can change over the course of a patient's lifetime. It would be completely impossible for individual clinicians to keep track of this stream of new information [59]. Therefore, to transform variant data into relevant clinical information, we need knowledge bases, which include clinicogenomic associations [72].

There are already some knowledge sources, e.g. Online Mendelian Inheritance of Man (OMIM), GeneReview, ClinVar (http://www.ncbi.nlm.nih.gov/clinvar), SNPedia (http://www.SNPedia.com), Human Gene Mutation Database (HGMD), Diagnostic Mutation Database (DMuDB), Clinical Genomics Database (CGD), Pharmacogenomics Knowledgebase (PharmGKB) (http://www.pharmgkb.org), disease and locus specific databases, etc. to provide clinicogenomic information related with human variations and disease.

Number of clinically relevant variant data and their changing clinicogenomic interpretations exceed the ability of unaided human mind. Exponentially increasing amount of scientific findings makes it impossible to stay up-to-date for physicians and, therefore, decision support systems, alerts and reminders become inevitable tools for utilizing clinicogenomic information [39, 59]. Essentially, in omics era, the healthcare systems need to shift from experience based practice to the system supported practice to make informed decisions in a timely manner [33].

In summary, to integrate clinicogenomic information into medical records, it is necessary to build a clinicogenomic knowledge base, an infrastructure which provides updated interpretation to physicians, and clinical decision support systems to use these data [3, 23].

3.3 Interoperability and Standards

To share data between different information systems, identifiers, terminology systems and messaging standards are needed. In genomic terminology, identifying gene symbols and identifiers are standardized by Human Gene Nomenclature Committee (HGNC), and variant nomenclature defined by the Human Genome Variation Society (HGVS). However, due to the complexity, HGSV nomenclature is not universally used by researchers. As a convention refSNP or "rs number, rs#" of dbSNP has been widely adopted and heavily referenced in the literature for identification of SNP variants [49, 63].

To integrate variants with clinical concepts and use clinicogenomic interpretation as part of decision support, we need to extend existing terminological systems (nomenclatures, classifications, coding systems, etc.) to involve genomic domain [32, 55, 67]. In the medical domain, there are well-established and generally accepted terminology standards.

Today, Logical Observation Identifiers Names and Codes (LOINC) has been extended to involve genomic variants. However, disease terminologies (Systematized Nomenclature of Medicine-Clinical Terms-SNOMED-CT, International Classification of Diseases-ICD, etc.) do not support the description of diseases at the genetic level efficiently. To construct the variant-drug and variant-surgical procedure relationships, we need standardized identification between databases and coding systems [67]. Several pharmacogenomic knowledge sources (PharmGKB, Drug-Bank) use the Anatomical Therapeutic Chemical (ATC) classification system. Clinical Bioinformatics Ontology (CBO) is a curated semantic network based on the combination of different clinical vocabularies (SNOMED-CT, LOINC) and National Center for Biotechnology Information (NCBI)'s bioinformatics resources (OMIM, GeneOntology) [30].

Messaging standards are required to share variant data and relevant clinical knowledge between health information systems. Health Level 7 (HL7) Clinical Genomics Work Group has published HL7 Version 2 implementation guide (Clinical Genomics; Fully LOINC-Qualified Genetic Variation Model) to report structured genotyping based test results and also include variants associated with disease and pharmacogenomic applications [67].

Partners Healthcare has developed the GeneInsight Suite for sharing the clinically relevant variant data based on HL7 standards, (GeneInsight Lab, GeneInsight Clinic and GeneInsight Network). GeneInsight Lab stores and manages genetic knowledge and enables to generate clinical reports. GeneInsight Network transmits genetic data and clinical interpretations or re-interpretations between labs and clinicians. GeneInsight Clinic presents genetic test results and subsequent interpretations from laboratories to the clinicians [3, 4, 39].

Today, some academic and commercial institutions are working to produce genomic information available within the EMR/EHR, including clinical decision support and clinicogenomic knowledge bases.

4 Sociodemographic Data for Health Records

Age, gender, ethnicity, and race are major socio-demographic factors affecting personal health status. Age is related with almost all medical conditions and often used to categorize patients for comparative studies. For example, prostate cancer is rarely seen in men younger than 40 years; the incidence rises rapidly with each decade thereafter [43]. In daily life, the terms of sex and gender are frequently used interchangeably, despite they have different meanings. Sex is defined as biological characteristics based on chromosomes, physiology, etc., while gender refers to the sociocultural construction of masculinity and femininity in a society [68]. This distinction is very important in some medical conditions. For example, in prostate cancer; as the male-to-female gender reassignment surgery generally does not involve prostatectomy, a female patient by gender can have prostate cancer [41].

Race is a socioeconomic construct of human variability based on differences in biological characteristics, physical appearance, social structures, and behaviour. This definition contains intertwined cultural and biological factors and sometimes used synonymously with ethnicity, ancestry, nationality, and culture. These are valuable predictors to assess disease risk [34]. For example, the highest incidence of prostate cancer occurs in African-American men, intermediate in whites, and is lowest in native Japanese [43].

5 Family Health History

Family health history (family history, family medical history,and family medical tree) is an aggregation of information about health status affecting a person and his/her family members. The scope of family members typically involves three generations of relatives by birth, person, his/her children and his/her siblings (parents, maternal and paternal grandparents, and maternal and paternal aunts and uncles) [2].

Family health history is accepted as the most efficient tool solving complex interactions between genes and environmental factors and assessing personal disease risk for a high number of disease e.g. arthritis, asthma, cancer, diabetes mellitus, hypertension, hypercholesterolemia, single-gene disorders (Mendelian inheritance), etc. [2, 21, 25].

Today, there are web-based tools to collect and assess family health [69]. Still the EMR/EHR is used to record and store family health history data in narrative format [32]. The American Health Information Community's (AHIC) Family Health History Multi-Stakeholder Workgroup proposed a structured data set for family health history within EMR/EHR [15, 21, 22].

6 Environmental Data for Health Conditions

6.1 Self-Tracking

Wearable electronics and multi-sensor platforms (e.g. smart watches, wristband sensors, wearable sensor patches, artificial reality-augmented glasses, brain computer interfaces, wearable body metric textiles), smartphone-based mobile applications, pre-programmed questionnaires can be used to collect personal physiological and psychological data [46, 62]. Currently most favourite mobile health applications are developed to track diet, exercises and weight management [19].

In addition to direct measurement tools, we can use several electronic databases to acquire personal environmental data. For example tracking by loyalty or credit cards, we can collect the personal food consumptions and consumer behaviour patterns [64].

To improve the measurement of critical environmental variables and standardize phenotypes and exposures, *PhenX (consensus measures for Phenotypes and eXposures) Toolkit* is developed as a web source, (https://www.phenxtoolkit.org) [26, 47]. In PhenX toolkit, for every measure, there are descriptions and protocols on how to collect data [60].

6.1.1 Nutrition and Dietary Supplements

Dietary intake contains all foods, beverages and sometimes dietary supplements consumed by the oral route [48, 53]. For assessing dietary intake, we need methods to collect nutritional data and food composition databases for the analysis of the data [37].

Today, there are software tools developed to collect and assess dietary characteristics of individuals [48]. *SuperTracker* (http://www.choosemyplate.gov/supertracker-tools/supertracker.html) is a free, web-based, and personalized tool to analyse diet and physical activity based on 2010 Dietary Guidelines for Americans [50]. Also, some commercial nutrient analysis software and databases are available, such as the Nutritionist Pro or NutriBase, Weight Watchers etc. [1].

Food composition databases which contain food descriptions, nutrients, dietary constituents, and portion weights have critical significance for dietary assessment methods [1, 40, 61]. The major food composition database in US is the *Nutrient Database for Standard Reference (NDSR)* (http://ndb.nal.usda.gov). *Food and Nutrient Database for Dietary Studies (FNDDS)* (http://www.ars.usda.gov/Services/docs. htm?docid=12089) is used to assess dietary intakes and has more than 7,000 foods and beverages and 65 food components limited brand names for foods with unique USDA food codes [44, 48]. Many developing countries use USDA databases to develop their own nutrition programs and guidance [1]. The European Union continues to improve the comparable European Food Information Resource Network (EuroFir) database (http://www.eurofir.org). The FAO of the United Nations and various national governments have developed food composition tables for many countries worldwide [61].

To assess essential nutrients and/or nutrient status, PhenX Nutrition, and Dietary Supplements Working Group choose 13 measures i.e. breast-feeding, dietary supplements use, total dietary intake, calcium, caffeine, sugar, dairy food, fiber, selenium, Vitamin D, fruits and vegetables intake, and percentage energy from fat [60].

Dietary supplements are products (other than tobacco) which are intended to supplement the diet including vitamins, minerals and herbal substances. In order to track personal consumption of supplements, *My Dietary Supplements (MyDS)* is developed by US Office of Dietary Supplements for consumers as a free mobile application (http://ods.od.nih.gov/HealthInformation/mobile/aboutmyds.aspx). The *Dietary Supplements Labels Database (DSLD)* includes the full label information from the entire dietary supplement products marketed in the US (http://www.dsld. nlm.nih.gov).

6.1.2 Physical Activity

Especially in chronic diseases, physical activity is recognized as an important variable for disease development and prognosis [27].

Quantitative measures of physical activity may be collected via wearable monitors, such as pedometers, accelerometers, and heart-rate monitors. Pedometers use pendulum mechanism that records steps. Heart-rate monitors typically measure electrical activity of the heart via a device worn on the chest in direct contact with the skin. Accelerometers record steps and estimate activity levels using chip sensors that determine the movement in different planes. Some accelerometers may allow wireless uploading of data onto websites that store and graph step and activity data [10].

Weight loss is an important indirect measurement parameter for physical activity. In some researches, it's proposed that modest weight loss correlated with increased activity levels [10].

6.2 Environmental Monitoring

Recently, there are growing number of sensors and applications to monitor several physical, chemical and biological agents e.g. temperature, barometric pressure, humidity, air quality, carbon monoxide, hydrogen sulphide, radiation levels, airborne contaminants etc. Some types of these sensors are used to record and share data in real-time [64].

Sensordrone (http://www.sensorcon.com) is a sensor platform which collects environmental data and transmits via Bluetooth to other connected devices including smartphones. Applications are designed to detect air quality, carbon monoxide levels, potential gas leaks, and body temperature etc.

Flexibity Internet Sensors (http://www.flexibity.com) is an Internet-connected sensor platform for wireless environmental monitoring.

Air Quality Egg (http://airqualityegg.wikispaces.com/AirQualityEgg) measures the air quality in the close environment, and share data in real-time [62].

The *Lapka* (https://mylapka.com) is a set of environmental sensors that plug into iPhone and can detect radiation, electromagnetic feedback, nitrates in raw foods, and temperature and humidity.

iGeigie is a portable Geiger counter using iPhone, which is developed after the Fukushima disaster in Japan (http://igeigie.com).

Today, energy consumption data (electricity and natural gas) is being tracked and stored electronically by utility companies for billing and these could be used to determine about exposure to electric and magnetic fields. Also, it's possible to track movements of individuals using mobile phones, and geolocation data can be used, for example, to help determine exposure e.g. air pollutants [64].

Some databases for identification, classification and disease-agent association are developed to collect and process data about hazardous exposures, such as ChemID-Plus, Comparative Toxicogenomics Database (CTD), Hazardous Substances Data Bank (HSDB), Environmental Health e-Maps etc. (http://toxnet.nlm.nih.gov).

7 Genomics Driven Wellness Tracking and Management System (GO-WELL)

General conceptual design of GO-WELL includes several components;

1. Collecting individual socio-demographic, genomic and environmental (external environment, internal environment and behavioural) data,
2. Processing these raw data to convert input parameters of rules in the clinical envirogenomic knowledge base (CENG-KB),

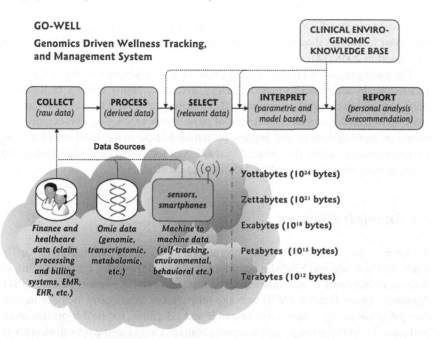

Fig. 2 General conceptual design of GO-WELL

3. Selecting and interpreting processed data to extract risk information based on CENG-KB and,
4. Reporting of individual disease risk degree regarding different perspectives of risk parameters (Fig. 2).

Most critical component of GO-WELL is CENG-KB. CENG-KB will contain clinical-genomic-environmental associations structured as IF-THEN rules and domain experts will develop, curate and enhance this knowledge base using scientific literature. CENG-KB will convert input parameters to clinical risk or protective information. Therefore, other components will be constructed according to characteristics of this knowledge base.

To provide more detailed information, we selected prostate cancer as an example medical condition and applied following steps to construct GO-WELL application:

1. Determination of knowledge sources,
2. Analysis of information types and models,
3. Extraction of input parameters,
4. Standardization of input identifiers and values to collect data,
5. Merging and value assignment to model free information,
6. Definition of model types and operators for model based information,
7. Construction of CENG-KB,
8. Designation of reporting interface.

Prostate cancer which is the second most common malignancy affecting men and fifth most frequently diagnosed cancer overall with an estimated 900,000 new cases diagnosed in 2008 [18].

The pathogenesis of prostate cancer could be partially explained by polygenetic factors [9]. In additional to genetic factors, age, race, family history, endogenous hormones, diseases, some environmental exposures and several behavioural features are proposed in literature as confounders of prostate cancer [43, 54]. This complicated nature of prostate cancer and burden on public health, make itself an ideal case for researching benefits of combined evaluation of individual genomic, clinical and enviro-behavioural data regarding personalized medicine.

7.1 Knowledge Sources

To acquire acceptable and accurate information based on medical literature, UpToDate, NCI web site and SNPedia are selected as basic references. *UpToDate* (http://www.uptodate.com) is an evidence-based clinical decision support resource [54]. *National Cancer Institute (NCI)* web site (http://www.cancer.gov) includes numerous publications on cancer care, research and education for health professionals and patients. *SNPedia* (http://www.snpedia.com) is a wiki sharing information about the effects of genomic variations. Clinically relevant risk and protective factors for prostate cancer are identified from these sources.

In addition to these knowledge sources, we studied several risk prediction and assessment tools for prostate cancer e.g. Prostate Cancer Prevention Trial (PCPT) risk calculator (http://deb.uthscsa.edu/URORiskCalc/Pages/uroriskcalc.jsp), European Randomized Study of Screening for Prostate Cancer (ERSPC) risk calculator (http://www.prostatecancer-riskcalculator.com), Sunnybrook risk calculator (http://sunnybrook.ca/content/?page=OCC_prostateCalc), etc. In these tools, clinical and laboratory parameters are used to extract predictions.

In the medical literature, there are various genomic assessment models and tools (SNP based multigene panels). An example of SNP driven prostate cancer risk assessment model is cumulative model. "Zheng, 2008" is a preliminary SNP based multigene panel example, which is based on the cumulative values of risky SNP counts and family health history [75]. Also, recently a predictive model for prostate cancer which includes genomic data (based on SNP associations) and health behaviours (body mass index-BMI, family health history, physical activity, smoking, drinking, lycopene intake, calories from fat, calcium intake) is proposed [74].

7.2 Analysis of Information

Two types of information model-free and model-based are extracted from the knowledge sources. Most of this information is model-free i.e. *independent associations*, whereas there are few model-based predictive tools. We categorized two types of model-free information as uniparameter and multiparameter independent associations. *Uniparameter independent associations* involve one input parameter but *multiparameter independent associations* have many. Independent associations are easily representable as simple IF-THEN statements. On the other hand, *model based information* can be accepted as collective impacts of risk parameters on a specific disease risk based on a predictive model. These types of associations are sophisticated and valuable tools regarding clinical practice. We can also construct such a type of model-based information unifying several IF-THEN statements within a risk assessment model.

7.3 Input Parameters

According to our knowledge sources, potential risk and protective factors for prostate cancer are listed in Table 1.

When we convert the narrative information to the structured IF-THEN rules, risk and protective factors becomes part of "antecedent" of rules. For every input parameter, we need to assign an input value. For some parameters (especially for environmental factors), time interval are also assigned.

7.4 Standardization of Input Data

Disease risk assessment is required continuous collection and evaluation of specific risk and protective factors. The relevant risk and protective factors are needed to be captured and stored. Data collection approaches and techniques are different according to the input parameters, and can be categorized mainly in three parts; individual entry (structured data entry or file uploading), collecting via automated data collection devices (e.g. barcode and QR (quick response) code readers, sensors, etc.), and transferring data from other systems (e.g. EMR, EHR etc.).

Sociodemographic data (age, ethnicity, and race) can be recorded permanently at once. Also, family health history (count of first degree male relatives with prostate cancer) can be recorded manually, but this data can change in time.

Personal genomics data file may contain a huge amount of variant data and it needs to be uploaded or transferred as a whole.

Data from enviro-behavioural sources can be gathered with different methods; besides manual typing, supplements and drugs data can be collected via automated devices. Drugs and procedural data can be transfer from EMR/EHR via web services. Unfortunately, most of the environmental and behavioural parameters are not recorded and stored in EMR/EHR in a structural way. Data about hazardous agents (if possible) can be collected via sensors. And also laboratory test results, previous diagnosis and anatomic measurements can be transferred from EMR/EHR via web services.

Table 1 Limited list of risk and protective factors for prostate cancer [43, 54]

Sociodemographic data

Age, family health history, ethnicity/race

Genomic data

SNPs   rs4880, rs16260, rs251177, rs629242, rs721048, rs1447295, rs1447295,
              rs1456315, rs1545985, rs1571801, rs1859962, rs2107301, rs2238135,
              rs2740574, rs2987983, rs4054823, rs4430796, rs5945572, rs6983267,
              rs7463708, rs7652331, rs10492519, rs13149290, rs16901979

Environmental sources

Nutrition and diet	Animal fat, fruits, legumes, yellow-orange and cruciferous vegetables, soy foods, dairy products, fatty fish, alcohol, coffee, green tea, modified citrus pectin, pomegranate
Supplements	Multivitamins, supplement containing products (vitamin E—with or without selenium, folic acid, zinc, calcium, vitamin D, retinoid), zyflamend
Drugs	5 alpha-reductase inhibitors, NSAIDs, statins, toremifene
Medical procedures	Vasectomy, barium enema, hip or pelvis x-rays, and external beam radiation therapy (EBRT) for rectal cancer
Tobacco use	Tobacco products

Environmental agents

Chemical agents	Agent orange, chlordecone, cadmium, docosahexaenoic acid, trans-fatty acids (TFA) 18:1 and 18:2, carotenoids, lycopene, calcium, folate, ethyl alcohol, selenium, vitamin E, folic acid, zinc, vitamin D, retinoids, zyflamend, finasteride, dutasteride, toremifene citrate
Physical agents	Ultraviolet light exposure, prostate radiation

Personal health status (internal environment)

Laboratory test results	Serum concentrations of testosterone derivatives, insulin-like growth factor-1 (IGF-1)
Diagnosis	Prostatitis, prostatic intraepithelial neoplasia
Anatomic measurements	High BMI (obesity)

After data collected, if required it is converted to other data formats while transferring through databases or using mathematical operators. For example, brand names of drugs (or supplements) are different from generic names. In a market, possibly, more than one drug (or supplement) can exist with same chemical ingredient(s). To combine these types of drugs, we need a database including both generic and product name.

To assess nutrition and nutrients, we need to use a food composition databases. Also, to calculate the effect of some type of parameters (especially for environmental agents); we should combine the effects in a time slot. At this point possibly we need to use fuzzy logic for inference or assign a confidence interval.

7.5 Merging and Value Assignment

In the literature, occasionally there are conflicting scientific conclusions. Effect type (risky, protective or normal), level of research evidence and impact of some

Table 2 Prostate cancer risk calculation by Zheng 2008 [75]

Model type	Model name	Total impact	Explanation (odds ratio)
Cumulative model	Zheng 2008	0	1.00 (by definition)
		1	1.50 (CI: 1.18–1.92); 1.62 (CI: 1.27–2.08)
		2	1.96 (CI: 1.54–2.49); 2.07 (CI: 1.62–2.64)
		3	2.21 (CI: 1.70–2.89); 2.71 (CI: 2.08–3.53)
		4	4.47 (CI: 2.93–6.80); 4.76 (CI: 3.31–6.84)
		5	4.47 (CI: 2.93–6.80); 9.46 (CI: 3.62–24.72)
		6	9.46 (CI: 3.62–24.72)

parameters might differ between studies. Therefore, these scientific findings should be evaluated, curated and merged by domain experts.

During this process, the clinical impact value and quality of evidence degree should be assigned to each type of independent associations. This is a critical process because a risk factor may have a weak effect and strong evidence, but another factor may have a strong effect but weak evidence. When reporting these risk (or protective) factors as a whole, all aspects of the data need to be presented regarding the dimensions of clinical significance [52].

In our knowledge sources, there are several different approaches and techniques to categorize these parameters i.e. UpToDate Grading Guide (http://www.uptodate.com/home/grading-guide), NCI Levels of Evidence for Cancer Screening and Prevention Studies (http://www.cancer.gov/cancertopics/pdq/screening/levels-of-evidence/HealthProfessional), and SNPedia concepts of magnitude (http://snpedia.com/index.php/Magnitude) and repute (http://snpedia.com/index.php/Repute). These grading methods should be compatible with each to be used as information sources in a single system.

7.6 Predictive Models

As explained above, model based information can be defined as a combination of several IF-THEN statements within a risk assessment model. For this type of information, it's required to assign impact values and operators for every input parameter to calculate the total personal risk value according to the model definition.

In "Zheng, 2008" model, existence of five different SNPs and family health history contributes to the total risk of the prostate cancer (calculated with odds ratio) for patients (Table 2). In this model, for example, if a patient has only one impact factor the risk of having prostate cancer increases by 1.5–1.62.

7.7 Design of CENG-KB

Collected and processed data interpreted and transformed to clinically relevant information based on CENG-KB. Content of CENG-KB must be extracted from the scientific literature by domain experts and continuously updated, and data collection must be re-interpreted after updating CENG-KB.

Model based and model free information is recorded in CENG-KB as structured IF-THEN rules. To constitute the rule definition table, we utilized a rule structure as below;

(Rule identifier)	IF	Input data	THEN	Medical data Model data Clinical significance data

In this representation, rules have a unique identifier assigned by system and values of inputs are previously defined. A clinical envirogenomic rules could be contained in one or more input parameters, value and time interval. Time interval is not an obligation and will be used to identify exposure period for some hazardous agents. Medical data category contains diagnosis code and name. Values of this data fields should be selected from a terminology standard (e.g. SNOMED, ICD, etc.). For model free (independent) associations, we have two choices i.e. risk and protective associations, and we assign magnitude of impact and quality of evidence degree as clinical significance degree. In the model based rules, we have to define the model type and name. Depending on the model type, we need some parameters (e.g. impact value of risk factors on the model) to calculate the complete value of risk.

7.8 Design of Assessment Report

After the input data is interpreted based on CENG-KB, there will be three types of assessment results i.e. parametric, multiparametric and model based. To present these results in a consistent, effective and non-redundant way, reporting is the critical last step.

In our approach, we present model based results at the top of the report. Model based results are simple and precise. To calculate these values, GO-WELL applications need to be developed based on the selected models. Then, independent associations are presented by categories and parameters. Multiparametric results are reported later with corresponding uniparametric explanations.

We can categorize enviro-behavioural parameters as three different categories i.e. source, agent and body (internal environment). Sources are the reservoir of the agents and exposure to a source depends on our behavioural and environmental characteristics such as nutrition, supplements, drugs etc. Same agents can be originated from different sources. For example, a person can take iron from nutrition, supplements, and drugs.

In the literature; sources, agents and their effect on the body are usually analysed as different layers and relationships between them are corrupted. For example, many references propose tomatoes as a protective factor for prostate cancer, but some of them analyse the possible agents (e.g. lycopene) at molecular level. Corruption between sources, agents and body in literature causes conflicts and redundancies in the final interpretation of these factors. Therefore, enviro-behavioural parameters are needed to be evaluated with respect to these three dimensions while known interrelationships between them are presented.

To visualize all independent uniparametric associations as a whole, we can use scatter graphs where the axes corresponds to categories (low, medium, and high) of impact and quality (Fig. 3).

7.9 An Example Case

Following case is an example of the GO-WELL workflow:

> Mr. John Doe is a white and 65 years old man. His father died 10 years ago from prostate cancer. He has rs4430796-AA, and rs16901979-AA SNPs in personal genomic file and weight = 81 kg and height = 180 cm. Regarding nutritional pattern, he consumes vegetable products 4 servings per day for 20 years and also more than 8 cup of coffee per day, for 15 years. Also he has taken about 120 mg/day zinc containing supplements between years for 10 years.

In our example, socio-demographic data (age, race, and family health history), some types of environmental (nutrition, supplements, and coffee consumption) and medical data (BMI, genomic variations) are tracked. In CENG-KB, there are many rules between the collected data and the prostate cancer. We accept that; there are nutritional databases to convert foods and meals to the food categories and nutrients and dietary supplements label database to convert commercial brands to generic chemicals. While these data is collected on time, we process and convert some of them to use as input for CENG-KB. For example, age derived from birth date; tomatoes and coffee consumptions acquired based on food composition database. And supplements ingredients captured based on dietary supplement database. BMI calculated with a formula; $(weight\text{-}kg)/(height\text{-}m)^2$.

We presented the final results according to report principles in (Fig. 4), and we exemplified all uniparametric environmental rules as a whole in (Fig. 3).

7.10 Development Stages of System

To realize GO-WELL concept, we begun to incrementally develop a system based on Turkish e-health information system. After a domain analysis, we developed a simple web application prototype using a rapid application development platform (Zoho Creator^{TM} and Zoho Reports^{TM}). In this preliminary system, we stored clinically relevant personal SNP data and some clinical and enviro-behavioural information, and generated individual recommendations based on some models and interpreting clinicogenomic associations. We accepted that, relevant data were acquired, processed and captured from original sources and stored in our database.

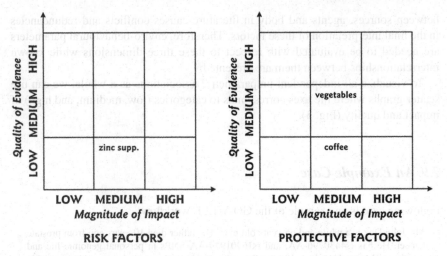

Fig. 3 A graphical visualization of complete environmental parameters for our example case

Then we focused to develop a flexible knowledge base. Currently, we are analysing some knowledge representation and management tools (e.g. BioXM, Protégé, Top-Braid, etc.) to integrate our application.

After that, we will continue to improve our system integrating with conventional and big data sources. At this phase, we will concentrate to overcome three different integration issues i.e. EMR data from national level EHR, enviro-behavioural tracking data from individual tracking and monitoring devices and personal genomic files from genomic laboratory. General architectural design of our system based on GO-WELL concept is in Fig. 5.

In Turkey, all personal medical data are collected and stored from caregiver organizations by National Health Information System of Turkey (NHIS-T). Current version of NHIS-T allows to transfer medical data from providers' (hospital and family practitioner) information systems to central servers via HL7 compliant web services. Pre-defined Minimum Health Data Sets (MHDS) are extracted from EMRs of healthcare providers, as aggregated clinical document elements i.e. transmission data sets or episodic EHRs.

Then this data sets are serialized into XML based on the HL7 Clinical Document Architecture Release Two (CDA-R2) to generate transmission schemas. Sending messages are stored in the central NHIS-T repositories [11, 36]. GO-WELL is a next-generation PHR concept and one of our final goal is to integrate our prototype application, GO-WELL and NHIS-T to capture actionable individual clinical data.

In the next step, we will integrate monitoring and tracking devices and collect personal enviro-behavioural raw data. We will use big data platforms to refine raw data and capture actionable information. This information will be sent for interpreting to the existing end-user web application.

Another phase will be integration of genomic laboratory system, NHIS-T and end user web application. Today, WGS is not a natural part of clinical processes. But,

RISK ASSESSMENT REPORT

Name, Surname : John Doe
Disease   : Prostate Cancer

MODEL BASED INTERPRETATION:

Model Type   : Cumulative Model
Model Name   : Zheng, 2008
Positive   :
Parameters

Parameter	Value
rs4430796-AA	Yes
rs6983267-AA	Yes
Family history	Yes

Total Risk Interpretation

Total impact	Odds ratio
3	2.21 (CI: 1.70-2.89); 2.71 (CI: 2.08-3.53)

INDEPENDENT ASSOCIATIONS:

SOCIODEMOGRAPHIC PARAMETERS:

Parameter	Value	Interpretation
Age	65	Prostate cancer rates increase exponentially after 50 years.
Race	White	The risk of developing and dying from prostate cancer is intermediate levels among whites.
Family history	Yes	To have first degree male relatives with prostate cancer is a risk factor.

GENOMIC VARIATIONS:

Variation	Interpretation
rs4430796-AA	A SNP in the TCF2 gene on chromosome 17q12, associated with increased risk for prostate cancer.
rs6983267-AA	A SNP on chromosome 8q24, associated with increased risk for several cancers, particularly prostate cancer.

ENVIRONMENTAL PARAMETERS:

Source driven interpretation

Parameter	Value	Time	Interpretation
Vegetables	4 servings/day	20 years	Lower prostate cancer risk (adjusted odds ratio 1.54).
Coffee consumption	8 cup of coffee/day	15 years	Lower risk of lethal prostate cancer (relative risk 0.44, 95% CI 0.22-0.75)
Zinc supplements	150 mg/day	10 years	Compared with nonusers, 2.29-fold increased risk of prostate cancer.
Physical activity	No data	No data	

Agent driven interpretation:

Parameter	Value	Time	Interpretation
Zinc	150 mg/day	10 years	Compared with nonusers, 2.29-fold increased risk of prostate cancer.

Effect driven interpretation

Parameter	Value	Time	Interpretation
BMI	32		Depending on physical activity and nutrition pattern.

Fig. 4 An example of individual risk assessment report

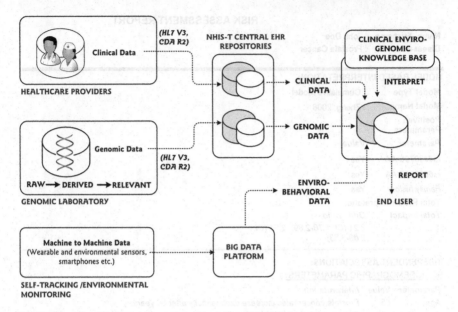

Fig. 5 General architecture of our system based on GO-WELL concept

authors proposed that, in the near future, personal genomic data will be an inevitable tools for clinical decision makers worldwide [7].

8 Conclusion and Future Researches

Health can be summarized as "a complete state of bio-psycho-social well-being", which means more than treatment of diseases and disorders. With the support of trending innovative technologies and challenging old paradigms, future healthcare systems would be closer to accomplish this idea. Technological shift will not only transform the concrete environment of health, but also will affect the mind, culture, and balance between individuals and professionals.

In this respect, we consider that next generation of health information systems will be constructed based on tracking and monitoring all aspects of individual health status in 24/7, and return evidence based recommendations to empower individuals. GO-WELL is an example of the next generation PHR concept and based on a clinical envirogenomic knowledge base to engage patients for predictive, personalized care. In this chapter, we present the design principles of this system, after describing several concepts including personalized medicine, omics revolution, incorporation of genomic data into clinical processes, and the utilization of environmental and behavioural parameters for disease risk assessment.

Currently, collecting personal environmental risk and protective parameters is feasible with existing technical opportunities. But significant point is to convert gathered raw data into clinically relevant format, such as a meal to its nutrient composition or a supplement product to its generic ingredients. Well established standards and identifiers are required to overcome this obstacle.

Nowadays, EMR/EHR systems are spreading, and personal clinical data accumulates in health information systems. Development and utilization of PHR is encouraged. As the cost and time required for genotyping is decreasing the direct-to consumer personal genomics sector is growing up. Self-tracking and environmental monitoring devices are being used more frequently each day. All of these improvements are complicating the data structure of the health domain. Empowered patients want variety of new services from the healthcare providers and traditional decision tools are insufficient to provide the desired predictive, preventive and personalized care. GO-WELL aims to integrate these three areas (clinical status, genomic and enviro-behavioural components) to infer personal preventive recommendations. In this approach, we aim to use traditional and new data sources together with a minimalistic and reductive approach. In our architecture, we aim to extract actionable data from huge data stacks and use to infer recommendations based on a comprehensive knowledge base.

As we emphasized before, manually curated and accredited knowledge base is the most important component of the GO-WELL. Based on this knowledge base, collected risk data will gain a predictive meaning and improvements in clinical sciences will be reflected to individuals by the reinterpretation of collected data. At this point, we need additional and improved prediction tools based on genomic and environmental parameters. To make it easier to extract and manually curate existing references for domain experts, we aim to develop a knowledge repository integrating some knowledge bases with semantic technologies and adding some automatic evaluation techniques.

The primary purpose of GO-WELL is to provide true and actionable information for patients and their family practitioners. GO-WELL will return evidence based recommendations to the individuals processing data and make them responsible about their preferences and consequences. Presentation and visualization of the interpreted results is another important part of GO-WELL. We need to produce effective approaches and techniques visualizing our multidimensional and scattered data for users.

The proposed model of GO-WELL is partially static and need improvements. In the near future, population based research will be performed based on real and synchronous data, and similarity metrics will be used to calculate possible personal disease risks. This will be required to use big data analytics for predictive and preventive health monitoring.

Omics area represent various type of data sources beyond genomic data and in the near future these omics data will be added to the clinical practices e.g. transcriptomics, proteomics, metabolomics, and epigenomics. Also, systems medicine is an emerging approach and possibly will increase the effectiveness of risk prediction strategies.

In addition, we aim to enhance GO-WELL, by integrating data warehouses for research. With this capability, integrated genomic and environmental data groups can also be used for clinical research. We would extract the meaningful relationship patterns via this system and by using these patterns we could calculate risks of groups who have similar characteristics e.g. family members or communities.

Eventually, big data driven integrated systems in healthcare domain will have critical significance in the dawning of the personalized medicine era. So, design and the development of innovative systems and new approaches are immediately required for this conversion.

References

1. Ahuja, J.K., Moshfegh, A.J., Holden, J.M., et al.: USDA food and nutrient databases provide the infrastructure for food and nutrition research, policy, and practice. J. Nutr. **143**(2), 241S–249S (2013)
2. Alspach, J.G.: The importance of family health history: your patients' and your own. Crit. Care Nurse **31**(1), 10–15 (2011)
3. Aronson, S.J., Clark, E.H., Varugheese, M., et al.: Communicating new knowledge on previously reported genetic variants. Genet. Med. (2012). doi:10.1038/gim.2012.19
4. Aronson, S.J., Clark, E.H., Babb, L.J., et al.: GeneInsight Suite: a platform to support laboratory and provider use of DNA-based genetic testing. Hum. Mutat. **32**, 532–536 (2011)
5. Balshaw, D.M., Kwok, R.K.: Innovative methods for improving measures of the personal environment. Am. J. Prev. Med. **42**, 558–559 (2012)
6. Barnes, M.R.: Genetic variation analysis for biomedical researchers: a primer. In: Barnes, M.R., Breen, G. (eds.) Genetic Variation: Methods and Protocols. Humana Press (2010) (Methods in Molecular Biology)
7. Berg, J.S., Khoury, M.J., Evans, J.P.: Deploying whole genome sequencing in clinical practice and public health: Meeting the challenge one bin at a time. Genet. Med. **13**, 499–504 (2011)
8. Bloss, C.S., Schork, N.J., Topol, E.J.: Effect of direct-to-consumer genome-wide profiling to assess disease risk. N. Engl. J. Med. **364**, 34–524 (2011)
9. Boyd, L.K., Mao, X., Lu, Y.L.: The complexity of prostate cancer: genomic alterations and heterogeneity. Nat. Rev. Urol. **11**, 64–652 (2012)
10. Van Camp, C.M., Hayes, L.B.: Assessing and increasing physical activity. J. Appl. Behav. Anal. **45**, 871–875 (2012)
11. Dogac, A., Yuksel, M., Avcı, A., et al.: Electronic health record interoperability as realized in the Turkish health information system. Methods Inf. Med. **50**, 9–140 (2011)
12. Downing, G.J.: Key aspects of health system change on the path to personalized medicine. Transl. Res. **154**, 272–276 (2009)
13. Drmanac, R.: The ultimate genetic test. Science **336**, 1110–1112 (2012)
14. Dziuda, D.M.: Data Mining for Genomics and Proteomics, Analysis of Gene and Protein Expression Data. Wiley, New York (2010)
15. Feero, W.G., Bigley, M.B., Brinner, K.M.: New standards and enhanced utility for family health history information in the electronic health record: an update from the American health information community's family health history multi-stakeholder workgroup. J. Am. Med. Inform. Assoc. **15**, 723–728 (2008)
16. Feero, W.G.: Genomics, health care, and society. N. Engl. J. Med. **365**, 1033–1041 (2011)
17. Feldman, B., Martin, E.M., Skotnes, T.: Big Data in Healthcare, Hype and Hope, Dr. Bonnie 360° (2012)
18. Ferlay, J., Shin, H.R., Bray, F., et al.: GLOBOCAN 2008 v1.2, Cancer incidence, mortality and prevalence worldwide. In: IARC CancerBase No. 10. (2008). http://globocan.iarc.fr/factsheet. asp. Accessed October 2013
19. Fox, S., Duggan, M.: Mobile health 2012, pew internet & American life project (2012). http://www.pewinternet.org/Reports/2012/Mobile-Health.aspx. Accessed Mar 2013

20. Garets, D., Davis, M.: Electronic medical records vs. electronic health records: yes, there is a difference. HIMSS Analytics (2006). http://www.himssanalytics.org/docs/WP_EMR_EHR. pdf. Accessed Jun 2013
21. Ginsburg, G.S., Willard, H.F.: Genomic and personalized medicine: foundations and applications. Transl. Res. **154**, 87–277 (2009)
22. Glaser, J., Henley, D.E., Downing, G., et al.: Advancing personalized health care through health information technology: an update from the American Health Information Community's Personalized Health Care Workgroup. J. Am. Med. Inform. Assoc. **15**, 6–391 (2008)
23. Green, R.C., Rehm, H.L., Kohane, I.S.: Clinical genome sequencing. In: Ginsburg, G.S., Willard, H.F. (eds.) Genomic and Personalized Medicine, 2nd edn. Academic Press, New York (2013)
24. Gubb, E., Matthiesen, R.: Introduction to Omics. In: Matthiesen, R. (ed.) Bioinformatics Methods in Clinical Research, Methods in Molecular Biology, Humana Press, Totowa (2010)
25. Guttmacher, A.E., Collins, F.S., Carmona, R.H.: The family history-more important than ever. N. Engl. J. Med. **351**, 2333–2336 (2004)
26. Hamilton, C.M., Strader, L.C., Pratt, J.G., et al.: The PhenX Toolkit: get the most from your measures. Am. J. Epidemiol. **174**, 253–260 (2011)
27. Haskell, W.L., Troiano, R.P., Hammond, J.A., et al.: Physical activity and physical fitness: standardizing assessment with the PhenX Toolkit. Am. J. Prev. Med. **42**, 92–486 (2012)
28. Hayrinen, K., Sarantoa, K., Nykanen, P.: Definition, structure, content, use and impacts of electronic health records: A review of the research literature. Int. J. Med. Inform. **77**, 291–304 (2008)
29. Helgason, A., Stefánsson, K.: The past, present, and future of direct-to-consumer genetic tests. Dialogues Clin. Neurosci. **12**, 61–68 (2010)
30. Hoffman, M., Arnoldi, C., Chuang, I.: The clinical bioinformatics ontology: a curated semantic network utilizing RefSeq information. Pac. Symp. Biocomput. **10**, 139–150 (2005)
31. Hoffman, M.A.: The genome-enabled electronic medical record. J. Biomed. Inform. **40**, 44–46 (2007)
32. Hoffman, M.A., Williams, M.S.: Electronic medical records and personalized medicine. Hum. Genet. **130**, 33–39 (2011)
33. IOM (Institute of Medicine).: Evidence-Based Medicine and the Changing Nature of Healthcare: 2007 IOM Annual Meeting Summary. The National Academies Press, Washington (2008)
34. IOM (Institute of Medicine).: Race, Ethnicity, and Language Data: Standardization for Health Care Quality Improvement. The National Academies Press, Washington (2009)
35. Kahn, S.D.: On the future of genomic data. Science **331**, 728–729 (2011)
36. Kose, I., Akpınar, N., Gürel, M., et al.: Turkey's national health information system (NHIS). In: Proceedings of the eChallenges Conference, Stockholm, s.n., pp. 170–177 (2008)
37. Lioy, P.J., Rappaport, S.M.: Exposure science and the exposome: an opportunity for coherence in the environmental health sciences. Environ. Health Perspect. **119**, A466–467 (2011)
38. Marian, A.J.: Medical DNA sequencing. Curr. Opin. Cardiol. **26**, 175–180 (2011)
39. Masys, D.R., Jarvik, G.P., Abernethy, N.F., et al.: Technical desiderata for the integration of genomic data into electronic health records. J. Biomed. Inform. **45**, 419–422 (2012)
40. Merchant, A.T., Dehghan, M.: Food composition database development for between country comparisons. Nutr. J. **5**, 2 (2006)
41. Miksad, R.A., Bubley, G., Church, P., et al.: Prostate cancer in a transgender woman 41 years after initiation of feminization. JAMA **296**, 2316–2317 (2006)
42. Miliard, M.: IBM helps Coriell Institute keep cool. Healthcare IT News (2011). Available at http://www.healthcareitnews.com/news/ibm-helps-coriell-institute-keep-cool. Accessed Oct 2013
43. National Cancer Institute.: Risk Factors for Prostate Cancer Development (2013). Available at http://www.cancer.gov/cancertopics/pdq/prevention/prostate/healthprofessional/page3. Accessed Oct 2013
44. Ng, S.W., Popkin, B.M.: Monitoring foods and nutrients sold and consumed in the United States: dynamics and challenges. J. Acad. Nutr. Diet. **112**, 41–45 (2012)

45. O'Driscoll, A., Daugelaite, J., Sleator, R.D.: 'Big data', Hadoop and cloud computing in genomics. J. Biomed. Inform. **46**, 81–774 (2013)
46. Paddock, C.: Self-tracking rools help you stay healthy. Medical News Today (2013). http://www.medicalnewstoday.com/articles/254902.php. Accessed Jun 2013
47. Pan, H., Tryka, K.A., Vreeman, D.J., et al.: Using PhenX measures to identify opportunities for cross-study analysis. Hum. Mutat. **33**, 849–857 (2012)
48. Pennington, J.A., Stumbo, P.J., Murphy, S.P., et al.: Food composition data: the foundation of dietetic practice and research. J. Am. Diet. Assoc. **107**, 2105–2113 (2007)
49. Poo, D.C., Cai, S., Mah, J.T.: UASIS: universal automatic SNP identification system. BMC Genomics **12**(Suppl 3), S9 (2011)
50. Post, R.C., Herrup, M., Chang, S., et al.: Getting plates in shape using SuperTracker. J. Acad. Nutr. Diet. **112**, 354–358 (2012)
51. Rappaport, S.M., Smith, M.T.: Epidemiology, environment and disease risks. Science **330**, 460–461 (2010)
52. Riegelman, R.K.: Public health 101: healthy people-healthy populations Jones & Bartlett Learning (2010).
53. Rutishauser, I.H.: Dietary intake measurements. Public Health Nutr. **8**(7A), 1100–11007 (2005)
54. Sartor, A.O.: Risk factors for prostate cancer, In: UpToDate. Vogelzang, N., Lee, R., Richie, J.P.: (ed.) (2013). http://www.uptodate.com/contents/risk-factors-for-prostate-cancer. Accessed: March 2013
55. Sax, U., Schmidt, S.: Integration of genomic data in electronic health records, opportunities and dilemmas. Methods Inf. Med. **44**, 546–550 (2005)
56. Scheuner, M.T., de Vries, H., Kim, B., et al.: Are electronic health records ready for genomic medicine? Genet. Med. **11**, 510–517 (2009)
57. Schneider, M.V., Orchard, S.: Omics technologies, data and bioinformatics principles. In: Mayer, B. (ed.) Bioinformatics for Omics Data. HumanaPress (Methods and Protocols) (2011)
58. Schultz, T.: Turning healthcare challenges into big data opportunities: a use-case review across the pharmaceutical development lifecycle. Bull. Assoc. Inf. Sci. Technol. **39**, 34–40 (2013)
59. Starren, J., Williams, M.S., Bottinger, E.P.: Crossing the omic chasm: a time for omic ancillary systems. JAMA **309**, 1237–1238 (2013)
60. Stover, P.J., Harlan, W.R., Hammond, J.A., et al.: PhenX: a toolkit for interdisciplinary genetics research. Curr. Opin. Lipidol. **21**, 136–140 (2010)
61. Stumbo, P.J., Weiss, R., Newman, J.W., et al.: Web-enabled and improved software tools and data are needed to measure nutrient intakes and physical activity for personalized health research. J. Nutr. **140**, 2104–2115 (2010)
62. Swan, M.: Sensor mania! the internet of things, wearable computing, objective metrics, and the quantified self 2.0. J. Sens. Actuator. Netw. **1**, 217–253 (2012)
63. Thomas, P.E., Klinger, R., Furlong, L.I., et al.: Challenges in the association of human single nucleotide polymorphism mentions with unique database identifiers. BMC Bioinf. **12**(Suppl 4), S4 (2011)
64. Van Tongeren, M., Cherrie, J.W.: An integrated approach to the exposome. Environ. Health Perspect. **120**, A103–104 (2012)
65. Tran, B.Q., Gonzales, P.: Standards and guidelines for personal health records in the United States: finding consensus in a rapidly evolving and divided environment. J. Health Med. Inf. (2012). doi:10.4172/2157-7420
66. Transforming Health Care through Big Data, Institute for Health Technology Transformation (IHT2), New York, 2013. https://iht2bigdata2013.questionpro.com. Accessed Oct 2013
67. Ullman-Cullere, M.H., Mathew, J.P.: Emerging landscape of genomics in the electronic health record for personalized medicine. Hum. Mutat. **32**, 512–516 (2011)
68. Verdonk, P., Klinge, I.: Mainstreaming sex and gender analysis in public health genomics. Gend. Med. **9**, 402–410 (2012)
69. Weitzel, J.N., Blazer, K.R., MacDonald, D.J., et al.: Genetics, genomics, and cancer risk assessment, state of the art and future directions in the era of personalized medicine. CA Cancer J. Clin. **61**, 327–359 (2011)

70. Wild, C.P.: The exposome: from concept to utility. Int. J. Epidemiol. **41**, 24–32 (2012)
71. Winter, A., Reinhold, H., Ammenwerth, E.: Health Information Systems, Architectures and Strategies, 2nd edn. Springer, London (2011)
72. Wright, C., Burton, H., Hall, A., et al.: The implications of whole genome sequencing for health in the UK. PHG Foundation (2011)
73. Yao, L., Zhang, Y., Li, Y., et al.: Electronic health records, implications for drug discovery. Drug Discov. Today **16**, 13–14 (2011)
74. Yücebaş, C., Aydın Son, Y.: A prostate cancer model build by a novel SVM-ID3 hybrid feature selection method using both genotyping and phenotype data from dbGaP, PLOS ONE, (2013). doi:10.1371/journal.pone.0091404
75. Zheng, S.L., Sun, J., Wiklund, F., et al.: Cumulative association of five genetic variants with prostate cancer. N. Engl. J. Med. **358**, 910–919 (2008)

Part III
Advanced Applications and Future Trends

Sustainability Data and Analytics in Cloud-Based M2M Systems

Hong-Linh Truong and Schahram Dustdar

Abstract Recently, cloud computing technologies have been employed for large-scale machine-to-machine (M2M) systems, as they could potentially offer better solutions for managing monitoring data of IoTs (Internet of Things) and supporting rich sets of IoT analytics applications for different stakeholders. However, there exist complex relationships between monitored objects, monitoring data, analytics features, and stakeholders that require us to develop efficient ways to handle these complex relationships to support different business and data analytics processes in large-scale M2M systems. In this chapter, we analyze potential stakeholders and their complex relationships to data and analytics applications in M2M systems for sustainability governance. Based on that we present techniques for supporting M2M data and process integration, including linking and managing monitored objects, sustainability monitoring data and analytics applications, for different stakeholders who are interested in dealing with large-scale monitoring data in M2M environments. We present a cloud-based data analytics system for sustainability governance that includes a Platform-as-a-Service and an analytics framework. We also illustrate our prototype based on a real-world cloud system for facility monitoring and analytics.

1 Introduction

Consider complex buildings with thousands of monitoring sensors that monitor hundreds of objects. In such buildings, three main types of data will be collected and integrated: (1) data about building elements (e.g., floors, equipment, and electricity systems), (2) data about sensor configuration and status (e.g., sensor configuration

H.-L. Truong (✉) · S. Dustdar
Distributed Systems Group, Vienna University of Technology, Vienna, Austria
e-mail: truong@dsg.tuwien.ac.at

S. Dustdar
e-mail: dustdar@dsg.tuwien.ac.at

N. Bessis and C. Dobre (eds.), *Big Data and Internet of Things:*
A Roadmap for Smart Environments, Studies in Computational Intelligence 546,
DOI: 10.1007/978-3-319-05029-4_14, © Springer International Publishing Switzerland 2014

parameters, sensor location and sensor data type), and (3) monitoring and analysis data [e.g., monitoring status of building elements, energy consumption data, and Greenhouse Gas (GHG) data]. These rich types of data have different lifetime, constraints, and usefulness and have different relevancy to different stakeholders. These types of data play a crucial role in understanding, realizing and optimizing business opportunities centering around IoTs (Internet of Things) in smart environments. Therefore, to foster various data analytics for multiple stakeholders, we need to have efficient ways to manage not only data themselves but also complex relationships among them and analytics applications and stakeholders.

Our work is focused on supporting different types of analytics for understanding complex sustainability measurements (e.g., electricity consumption and GHG calculation) and maintaining M2M environments [e.g., monitoring failure of chillers and optimizing the operation of HVAC (heating, ventilation, and air conditioning) systems]. Such data analytics are crucial for different operation and maintenance processes in sustainability governance of buildings that are required by different stakeholders [1], such as building operators, equipment manufacturers, and auditors. Research effort so far has been concentrated on techniques for low level data management, such as sensor integration, data storage, and data query mechanisms, but has neglected the complexity and diversity of stakeholders and their interests to business and data analytics processes. Therefore, sustainability data and corresponding analytics applications are not well managed to support multiple stakeholders. For supporting data analytics required by multiple stakeholders, we cannot simply focus on single types of data but we have to make use of the interconnected relationships among different types of data monitored and gathered from many types of objects in the same environment. This requires us to also address the linked data problems in M2M environments as well as diversity of analytic forms.

In this chapter, we analyze stakeholders and their requirements for data analytics in cloud-based M2M environments with a focus on solutions for Platform-as-a-Service (PaaS) for integrating IoT data and processes in sustainability governance using cloud technologies. Our goal is to develop techniques for linking monitored objects, monitoring data, and applications to stakeholder's needs and to provide a cloud-based data analytics system with a Platform-as-a-Service (PaaS) and supporting tools for stakeholders to access data and perform data analytics of their own interest. This chapter contributes (1) a detailed analysis of stakeholders in cloud-based sustainability governance in smart buildings and their requirements for data analytics, (2) techniques for linking, enriching and managing data for sustainability analysis in M2M environments, and (3) a data analytics system including a PaaS and an analytics framework for managing relationships between stakeholders, monitored objects, monitoring data and applications. Our techniques can be adopted by cloud-based M2M platform providers. To illustrate the usefulness of our work, we present our prototype atop a real-world cloud system for facility monitoring and analytics.

This chapter is an extension of [2]. We have substantially revised and extended stakeholders and requirements analysis, data linked models and platform design as well as introduced new techniques to manage complex data for different types

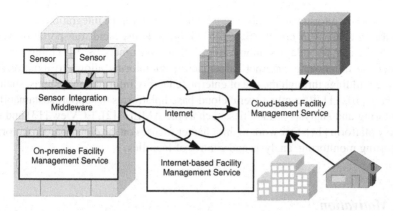

Fig. 1 Typical model of building's IoT monitoring

of analytics and provided detailed techniques inside our system. The rest of this chapter is organized as follows: Sect. 2 presents background, motivation and related work. In Sect. 3 we analyze stakeholders and their requirements for data analytics in cloud-based sustainability governance platforms. We present techniques for enriching, linking and managing M2M data in sustainability governance in Sect. 4. Section 5 describes our cloud-based data analytics system for sustainability governance. Section 6 presents our prototype and experiments. We conclude the chapter and outline our future work in Sect. 7.

2 Background, Motivation, and Related Work

2.1 Background

In an end-to-end M2M system several *monitoring sensors* will perform the monitoring and measurement of facility elements and their surrounding environments, such as equipment, air quality, and meter usages. These elements and environments are *monitored objects* whose status and operations will be monitored, analyzed and controlled to meet stakeholder's business requirements. Depending on different situations and configurations, data captured from sensors can be considered significant for being aggregated/pre-processed in *Sensor Integration Middleware* (also called *M2M gateway*). The significant data is then propagated to storage, analysis, and monitoring services which can be hosted on premise or on cloud-based infrastructure.

Figure 1 describes typical elements in such above-mentioned systems. Depending on systems, data analytics can be performed on premise, over the Internet with or without cloud computing systems. Today, several frameworks and middleware have been provided to support capturing and storing monitoring data and to perform data analytics on-premise or over the cloud [3]. Examples of systems that support on-premise monitoring and analysis are for homes [4], for the integration of different monitoring sensors to provide data for buildings, houses and transportation vehicles [5–7], and for relaying monitoring data to consumers [8, 9]. Going beyond on-premise monitoring, Internet-based facility monitoring allows to monitor and analyze buildings through the use of enterprise facility information systems, such as shown in [10, 11]. Recently, several cloud-based platforms to support sustainability monitoring and analysis of facilities, such as the AMEE [12], ECView [13] and the Galaxy platform [14]. Our work in this chapter focuses on the cloud-based platforms supporting monitoring, analysis and control of facilities.

2.2 Motivation

Our work is motivated by complex, diverse and mass-customized needs of data analytics for sustainability governance from different stakeholders that a cloud-based M2M provider must support. First, we need to identify main types of stakeholders and which analytics applications, processes and constraints are associated with these stakeholders. Generally, different stakeholders will need different processes, applications and access controls that handle, utilize and apply to different types of data. Currently, it is not very clear how to manage such complex associations in clouds, given the fact that each stakeholder might need only a part of monitoring data of a given monitored object (e.g., a chiller) as well as a stakeholder might need to access the same type of monitoring data across different facilities based on business needs and contracts [3]. This is very different from (open) e-science data or sensor Web platforms in which usually one type of stakeholders (e.g., scientists) tends to access large-scale datasets of similar data types (e.g., satellite images).

Second, we want to support different types of analytics of sustainability measurements that involve multiple types of data. However, data composition for such analyses remains challenging partially due to the lack of techniques for discovering the right data associated with monitored objects. For example, being able to obtain measurements about indoor air quality (IAQ) and costs would allow us to control and adapt the operation of several devices and systems to make sure that the air quality is suitable for specific contexts. Explained in a study of the Environmental Protection Agency (EPA) in [15], several types of building structure data and monitoring data are needed, e.g., floor area, windows, and HVAC monitoring data. But finding all relevant data for such analyses requires us to manage complex, interdependent relationships between monitoring data, monitored objects, applications and stakeholders. This is different from the low-level data management for monitoring data [16] or analytics tasks managements, such as in [17, 18].

Third, data analytics in M2M-based sustainability governance are usually developed and performed by domain experts, who lack IT skills but require different supporting environments for developing and executing data analytics. Their analytics applications are diverse in terms of program execution models (e.g., sequential application, batch jobs, workflows, Hadoop, and automatic lightweighted applications), execution environments and programming languages (e.g., R, Java, and Python). From the business perspective, stakeholders also require different forms of support, e.g., programmable data analytics APIs for service integrator or platform/application developer, Software-as-a-Service (SaaS) for the end-user, and automatic lightweight applications for OEM (Original Equipment Manufacturer) and equipment operators. Thus, providing PaaS will also hide IT complexity to enable domain experts to develop and test their different applications for specific monitored objects in their domains, fostering different models for data analytics in M2M-based sustainability governance.

2.3 Related Work

Recently, several cloud-based platforms to support the monitoring of energy consumption have been introduced, such as Tendril,[1] AlertMe [19] and xively [20]. Furthermore, there are systems supporting sustainability governance for buildings, such as the Galaxy platform [14]. While these systems can manage different types of data and provide different applications to analyze the data, most of them let the user to manage the complex relationships between monitoring data and monitored objects and do not provide a generic framework for managing these relationships.

Several stakeholder analyses have been studied for sustainability technologies [21], but we are not aware of stakeholder analysis for sustainability governance of facilities in clouds. In [22] the authors discuss about a cloud system for ubiquitous cities. However, stakeholders and application and data controls have not been discussed.

Generic data management techniques for sensor data, such as [23], discuss general security, privacy, and provenance, but not the complex relationships between data, application and stakeholders. Investigation of cloud computing for storing and processing sensor data has been conducted recently, such as techniques to access sensor data stored in their cloud using HBase [24] and using NoSQL [25]. However, they do not deal with complex relationships between monitoring data and sustainability governance features for facilities.

Several general cloud services have been developed, such as CA AppLogic [26], Appistry [27], Google App Engine [28], Microsoft Azure [29] and Amazon DynamoDB [30]. and Parabon Frontier [31]. They offer APIs for accessing data but do not provide support for specifying and managing complex relationships between monitoring data and monitored objects. Thus when using them for storing monitoring data, we need to develop models for managing relationships among data, monitored objects, stakeholders and applications. Several works have been focused on data analytics in cloud and grid for e-science data [32] but they do not address facility monitoring data.

---

[1] http://www.tendrilinc.com

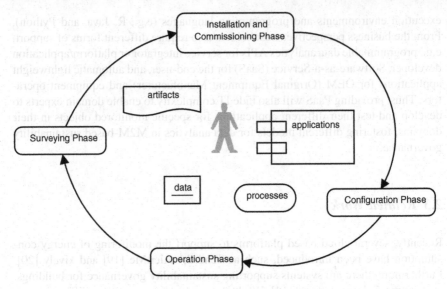

Fig. 2 Phases in establishing and operating facility monitoring and analysis

3 Stakeholders and Requirements of M2M Data Analytics

3.1 M2M-based Sustainability Governance

Several phases exist in the establishment and operation of sustainability monitoring and analysis of a facility, as shown in Fig. 2. In *Surveying Phase*, *Installation and Commissioning Phase* and *Configuration Phase*, few stakeholders are involved, such as the cloud facility management provider and facility owner. However, in the operation phase, several stakeholders involved in different tasks require different combining sets of data, artifacts, applications and processes [3]. Overall, these tasks center around the following types of *monitored objects*:

- *Facilities*: including facility elements, such as buildings, floors, and rooms,
- *Facility equipment and systems*: including equipment and systems used to operate facilities, e.g., electricity systems, freezers, compressors, chillers and fans, and
- *IT monitoring systems*: including sensors, gateways, networks and platforms used to gather monitoring data about facilities, and facility equipment and systems.

There are diverse types of and large amount of monitoring data collected for these monitored objects. Such data can be processed and manipulated by various types of *governance features* implemented as *applications* in a cloud-based sustainability governance platform, as described in the following:

- Near real-time *MonitoringApp*: are used for online monitoring and controlling facilities, such as alarm monitoring and near-real time energy consumption, thus handling monitoring data on the fly and possibly controlling monitored objects at runtime.

- Offline *AnalysisApp*: are used for analyzing sustainability measurements collected via a period of time, such as monthly energy consumption and GHG, for facilities. Typically, this kind of applications will analyze vast data available in the platform. These applications can be utilized by near real-time *MonitoringApp* in order to decide suitable actions at runtime.

Both *MonitoringApp* and *AnalysisApp* can carry out their functionality centralized, e.g., having all their tasks executed in a single server, or distributed, e.g., having multiple monitoring and analysis tasks executed in the cloud servers and M2M gateways. They might implement a simple or a complex process of several tasks. Sustainability governance features used by stakeholders will handle monitoring data under *data constraints*. Such constraints specify different conditions about, e.g., which types of monitoring data, how many data streams, and how many monitored objects. Typically, conditions are associated with the business models agreed between stakeholders and the provider of the cloud-based sustainability governance platform.

3.2 Stakeholders and Sustainability Data

Sustainability governance is not only in the interest of stakeholders that own and/or operate facilities but also other stakeholders who can benefit from data sharing and analytics functions of sustainability governance platforms, such as regulators, community users and developers (see [3] for possible stakeholders). Therefore, understanding stakeholders will enable the modeling and management of complex relationships between stakeholders, monitored objects, monitoring data and applications. For a given type of stakeholders, different roles can exist. For example, the equipment manufacturer and the maintainer can utilize data to perform (1) individual equipment maintenance, (2) electricity system maintenance, or (3) mechanical system maintenance in different ways.

To illustrate stakeholders in sustainability governance, let us consider 5 facilities in Dubai shown in Table 1. Overall, there exist several stakeholders, roles and types of data. Even with only mechanical, electrical, and plumbing (MEP) systems, in these buildings, we have around 60 types of monitored objects and each monitored object has several data streams indicating types of monitoring data. From our analysis of stakeholders and existing platforms in [3], we see the need to enable the customization of sustainability governance feature provisioning and to recommend suitable data streams and applications based on complex relationships between stakeholders, monitored objects, monitoring data and applications. With respect to service provisioning and customization, for a large-scale facility, there could be various stakeholders interested in the same monitored object, but different types of monitoring data, as well as there could be a stakeholder who allows to access to multiple monitored objects. This requires a fine-grained management of applications,

Table 1 Example of facilities and their data streams and stakeholders

Name	Data streams	Nr. stakeholders	Nr. roles	Nr. users
JAFZA	2,656	4	9	15
CARRIER AL FUTTAIM MOSQUE	66	4	9	9
CARRIER BIN HINDI TOWER	98	4	9	10
GREEN BUILDING	657	4	9	17
G2 INTERNATIONAL	155	4	9	15

monitored objects, and monitoring data for each stakeholder. A real-world sustainability governance platform in the cloud will probably serve for hundred thousands of buildings. The number of data streams (for monitoring objects) in each building is large and applications are diverse, making the management of their relationships complex.

3.3 Characteristics of Sustainability Monitoring Data and Analytics Applications

Sustainability monitoring data typically are associated with multiple types of monitored objects which affect each other. A particular interesting characteristic is that these types of data can be monitored and analyzed in isolation or in combination. For example, monitoring data of a chiller can be analyzed alone by the manufacturer or the maintenance operator to understand the operation of the chiller, while analyzed in combined with facility building sensing temperature and outdoor temperature in order to understand how to tune the indoor air quality and comfort. Generally, we cannot assume that different types of monitoring data will always be analyzed together because a specific stakeholder (e.g., chiller manufacturer) and applications for the stakeholder require only specific types of data (e.g., only chiller monitoring data). This requirement is associated with business models (e.g., payment) and compliance issues. Thus, the access to data must be controlled.

Due to very large types of monitoring data and stakeholders, diverse types of analytics exist. First, different monitored objects require different monitoring and analysis applications, and, even for the same category of monitored objects (e.g., chillers) each manufacturer can have a different analysis model for his/her type of objects. Second, diverse types of users exist, including normal users using SaaS and scientists/developers/domain experts writing complex sustainability monitoring and analysis applications/jobs. Third, diverse types of execution platforms exist due to the diversity of experts and analysis models for sustainability governance. Therefore, although different types of monitoring data can be managed by a single platform, analysis applications will not follow a single model. They will be developed using different languages (e.g., Java, R, Matlab and Python), and execution models (e.g., sequential program and workflows), depending stakeholders and their analytics requirements.

4 M2M Data and Process Integration for Data Analytics

4.1 M2M Data Fragments

One of the most important issues in managing a large number of sensors is how to link them to monitored objects. In many systems, this link is identified only in the deployment and configuration phase, in which sensor identifier is mapped to monitored object identifiers, by the people who install, configure, and manage the sensors. However, in complex facilities, monitored objects are not atomic, they can include other monitored objects, or linked together to provide a virtual monitored object. Furthermore, links from sensor monitoring to monitored objects are not established in a single process but multiple ones, involving multiple stakeholders. This requires us to manage the change of linking data over time.

In our study, several of data collection processes, except in the Operation Phase, are carried out by different teams and currently implemented with software support and manual paper sheets without integration, leading to the lack of links between monitoring data to monitored objects and dependencies among different types of data. To link monitoring data with monitored object information to support data analytics, we consider the following situations:

- Situation 1—rich monitored object information is available in a well-defined specification (e.g., newly design building) and can be supplied by the facility owner. In this case, it is possible to develop a correlation between monitoring data and monitored object information. For example, in most of the above-mentioned phases, we just need to capture information about sensors and map sensors to monitored objects using their identifiers.
- Situation 2—facility information is not available in a well-defined specification (e.g., old building), facility information is not completed, and/or facility information can only be provided by the end-user at a high level (e.g., in cases of home owners). In this case, it is possible to enrich monitoring data using user-specific facility information, but, in addition to information about sensors, the above-mentioned processes must also capture monitored object structures and dependencies.
- Situation 3—facility information is not available (e.g., the facility owner does not want to reveal the facility structure, yet still wants to enjoy complex analysis) but it is possible to annotate sensor data with certain metadata about monitored objects and locations. In this case, enriching monitoring data is performed at the sensor side.

Table 2 Main elements required for sustainability data analytics

Type	Concept	Description
Monitored objects in large-scale facilities	Building	Describe the whole building
	Floor	Represent a floor in a building
	Component	Represent various types of components of a system
	System	Represent different MEP systems inherent in buildings
Information about sensors and monitoring data	Monitoring sensor	Describes sensors used to monitor objects
	Monitoring data type	Describes the type of monitoring data provided by monitoring sensors
	Data stream	Describes a time series of monitoring data entries
	Data item	Describes single monitoring data entries

- Situation 4—facility information is not available and sensor monitoring data cannot be annotated with information about facility elements. In this case, automatic correlation techniques could be possible solutions.

In our work, we link monitoring data and sensor management using an external information service. Building information will be provided by the building owner by uploading their building information based on existing specifications. This way is suitable for Situation 1. When the building information is not well-defined and user-specific, then the building information service can provide interfaces for the user to specify his/her building information. In this case, suitable for Situation 2, standards or specific building information can be used. In overall, building information and monitored objects will be mapped to monitoring sensors using identifiers via integrated processes carried out through different phases.

4.2 Linked Data Model

Several specifications have been developed, such as the IFC classes [33], to cover building structure data. From the concept of existing specifications, main data concepts that can be used to enrich monitoring data for analytics are physical containment objects (e.g., building and building floor) and physical MEP (e.g., equipment and components). On the other hand, from the facility monitoring, we have several types of metadata (e.g., sensor configuration and description) besides a large amount of near-realtime monitoring data. Table 2 explains possible elements from building structure and monitoring data that should be linked.

To manage complex relationships among data, we have developed a linked data model, shown in Fig. 3. With this model, before the Operation Phase, most monitored objects and their dependencies, such as *Building*, *System*, *Component*, and *Floor*, and most entities related to sensors and their configurations, such as *MonitoringSensor*, *SensorConfiguration*, and *SensorModel*, are collected. Especially, *MonitoringSensor* will be used to link monitored objects to monitoring data types. Information about sensors and monitored objects can be obtained from the repository of sensors and monitored objects provided to multiple stakeholders by cloud facility management providers. During the Operation Phase, monitoring sensors will be executed, thus data items in sensor data streams will be stored separately from other types of data but they can be linked by using identifiers.

We utilize existing gateways which provide different monitoring data streams. We use data type identifiers, monitored object identifiers as well as data stream identifiers for identifying data. Let *dataURI* be the unique identifier of a type of monitoring data, *dataStreamURI* be the identifier of a data stream in a gateway, *dataTypeURI* be the identifier of data type, and *monitoredObjectURI* be the identifier of a monitored object. Overall, a *dataID* is a combination of *dataStreamURI* and (*dataTypeURI*, *monitoredObjectURI*). Using *dataTypeURI* and *monitoredObject URI* we will able to obtain metadata about types of monitoring data and monitored objects, while utilizing *dataStreamURI* we can obtain monitoring data. For example, to indicate the low suction pressure of a chiller, we can use either http://pcccl/dataStream/stream124 or (http://pcccl/dataType/LowSuctionPressure, http://pcccl/monitoredObject/chiller123).

4.3 Mixed Data Management

Since we have multiple types of monitored objects and different types of data and analytics, we do not expect that a single data representation, such as a traditional SQL-based model or a NoSQL-based model, would be suitable for all types of monitored objects. Consider that:

1. monitoring data and facility data are mainly gathered, managed and owned by the M2M service provider
2. each analytics stakeholder utilizing only limited, based on contracts, sets of these data for his/her analytics, and
3. the there types of monitoring data, facility data and analytics results have different structures, scales, volumes and access needs.

We utilize a mixed configuration of relational/graph data models and NoSQL data models to store, manage and link these types of data.

Figure 4 presents the conceptual model of our data management.[2] The data described by the linked data model in Sect. 4.2 have complex but structured

---

[2] Detailed design and implementation of these conceptual models are out of the scope of this chapter.

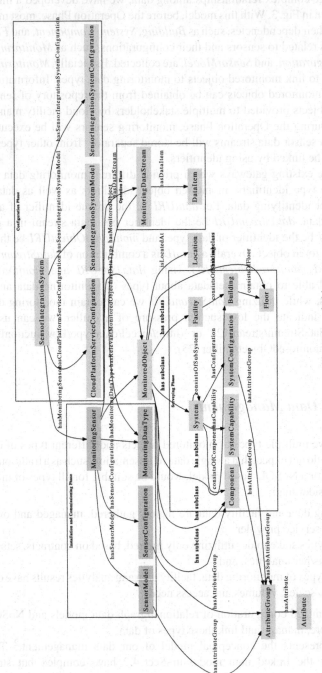

Fig. 3 Simplified linked data model for information about sensors and monitored objects

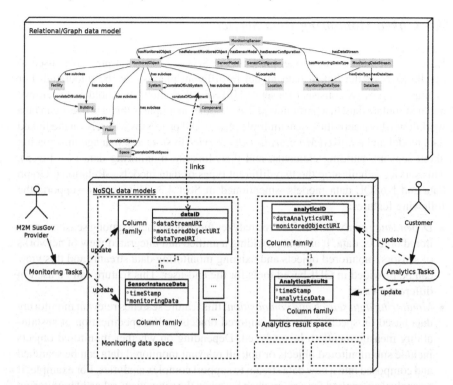

Fig. 4 Combining different types of data models for managing monitoring data, facility data and analytics results

relationships among their elements. Therefore, we utilize a relational/graph data model to manage them. This data model is selected also based on the rationale that the linked data are not changed often and their volume is much smaller than monitoring data which evolve over the time. On the other hand, monitoring data from sensors are voluminous and change rapidly over the time. They are mainly time series data and require elastic, scalable infrastructure to manage. Therefore, we utilize NoSQL data models to manage the monitoring data. The analytics results have different structures, depending on the types of analytics, as well as they are also associated with different unstructured documents about the analytics, e.g., documentation on how the analytics are carried out. Therefore, we also utilize NoSQL models which allow flexible structures of analytics and documents to be managed.

In our model, through Monitoring Tasks, the M2M provider manages monitoring data in NoSQL DaaS using different spaces, each is configured for a facility or a group of connected facilities. Within a space for monitoring data, we manage sensor data and link sensor information to facility data using dataID column family. Through Analytics Tasks, the results of analytics for each stakeholder will be managed in a specific analytics result space for that stakeholder. To support different operations, the NoSQL DaaS will provide different APIs for storing and accessing spaces.

4.4 Linked Monitoring Data Service

Based on the linked data model presented in Sect. 4.2, different processes, used to handle parts of the linked data model in different phases, can be integrated. For example, a process consists of a set of UI forms that allow stakeholders to retrieve and manipulate data in a particular phase. Another example is that a process can be a workflow whose activities span multiple phases. All processes will rely on the linked data model and a *Linked Monitoring Data Service* to store and manage information about sensors, building elements, and their links to monitoring data (see Fig. 5). This service is built atop the two different types of data models—Relational/Graph data and NoSQL data models—mentioned in Sect. 4.3. This service supports the following features:

- *Dependency analysis*: this feature analyzes the dependencies among sensors, monitored objects, data. It aims at providing a unified and integrated view of networks of sensors, monitored objects and existing monitoring data streams and this view can be provided in GUI-based tools for the end-user. This feature can be used in different phases.
- *Monitoring data search and composition*: this feature searches relevant monitoring data based on specification of monitored objects and/or specification of sustainability measurements to be analyzed. Depending on whether monitored objects include sub monitored objects or not, all relevant monitored data can be searched and composed into a new data stream to support complex analytics. For example, if an analysis is applied for a room, then based on the room view, relevant monitoring data will be discovered, whereas if an analysis is applied to energy consumption, then relevant data for energy analysis will be searched.
- *Quality of monitoring data analysis*: this feature utilizes linked data in order to determine the quality of monitoring data and the influence of the quality of monitoring data on the detection of faults, context-aware applications, and sustainability analysis. The quality of monitoring data is strongly dependent on, e.g., sensor model and configuration, monitored object structure, and gateway configuration, thus having all of these data linked will foster the evaluation of quality of monitoring data [34].

5 Cloud Services for Sustainability Data Analytics

5.1 Overview

Figure 6 presents core cloud services for sustainability data analytics which integrate stakeholder information, monitored objects, monitoring data and applications. At the core of the system is the *SusGovPaaS* which is a platform-as-a-service for conducting sustainability analytics. *SusGovPaaS* provides features for analytics and management via service APIs. Based on these features, *SusGovSaaS* will offer governance

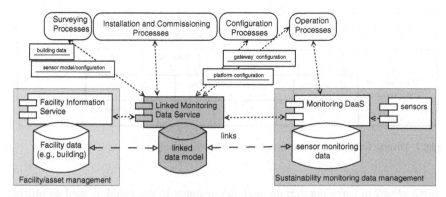

Fig. 5 Integrating processes for managing linked data

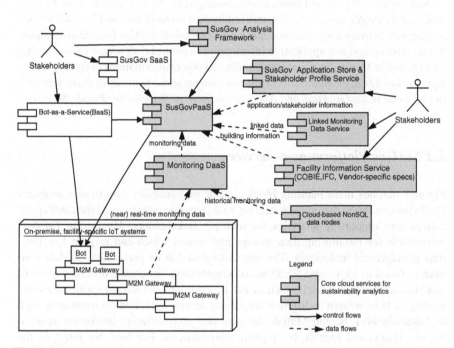

Fig. 6 Conceptual architecture of the cloud-based data analytics system for sustainability governance

analysis for stakeholders using the SaaS model. A *SusGov Analysis Framework* aiming at sustainability analysis expert that exploits *SusGov PaaS* is provided. Furthermore, to enable the development and provisioning of automated, light-weighted monitoring applications (called Bots) [35], *Bot-as-a-Service* (*BaaS*) which supports the development and management of Bots is also integrated. *BaaS* could deploy its Bots in *SusGov PaaS* and *M2M Gateways* and these Bots can utilize

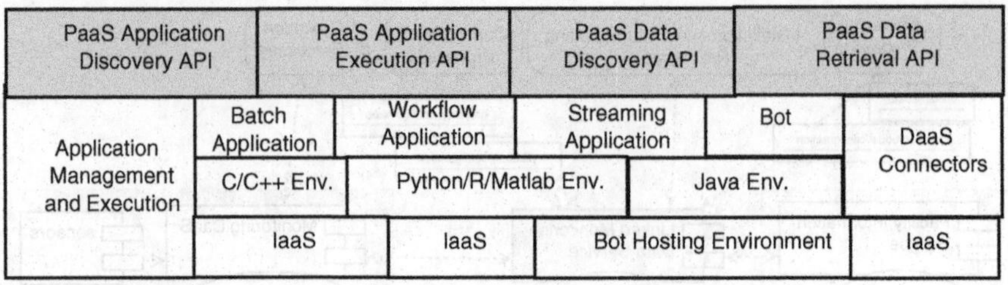

PaaS Application Discovery API		PaaS Application Execution API		PaaS Data Discovery API		PaaS Data Retrieval API	
Application Management and Execution	Batch Application	Workflow Application		Streaming Application	Bot	DaaS Connectors	
	C/C++ Env.	Python/R/Matlab Env.			Java Env.		
	IaaS	IaaS		Bot Hosting Environment		IaaS	

Fig. 7 Diverse forms of applications and execution environments

SusGovPaaS to carry out certain analytics features in the cloud as well as utilize *M2M Gateways* to perform analytics and monitoring activities in specific gateways.

Stakeholder profiles and applications, managed by *SusGov Application Store & Stakeholder Profile Service* will be utilized by *SusGovSaaS* and *SusGovPaaS* to support service provisioning, customization, and execution. *SusGovPaaS* utilizes stakeholder information and application information to control data access to monitoring data (stored in *MonitoringDaaS*) and monitored object information (stored in *Facility Information Service*). In the following, we discuss some features in these services, in particular in *SusGovPaaS* and *Application Store & Stakeholder Profile Service*.

5.2 SusGov Platform-as-a-Service

Figure 7 outlines main building blocks in our sustainability governance analytics PaaS (SusGovPaaS) that aims at dealing with dynamic properties of data and applications and supporting analytics for multiple stakeholders. *Monitoring DaaS* is responsible for monitoring data storage and access which can rely on low level data management techniques. The access to data can be performed via data connectors (such as SQL- and REST-based connectors). To enable different types of analytics application models, such as batch, workflow and stream applications and intelligent bots, written in different languages, several execution environments, such as based on Java, R, and Matlab, are provided atop different Infrastructure-as-a-Service (IaaS) and specific bot hosting environments. For bots, we integrate the Bot-as-a-Platform [35]. To expose capabilities of applications and monitoring data for different stakeholders, several APIs are provided for discovering, retrieving and accessing/executing data and applications.

While the execution environment platform is not in the focus of this chapter, we should note that sustainability applications in sustainability governance are very different that in contemporary data analytics platforms, such as in e-science, due to the fact that applications are needed and required for different stakeholders. Applications can be simple sequential programs that examine only a single type of chiller data or can be complex workflows that optimize air quality based on, e.g., temperature, chiller, and electricity consumption information.

5.3 Describing and Managing SusGov Applications

In our work, the types of applications mentioned in Sect. 3.1 are encapsulated in the so-called *sustainability governance applications* (*SusGovApp*). Managing *SusGov-Apps* is a complex task due to (1) a large number of types and instances of applications exist, and (2) control parameters and input data for applications are complex. Thus, an efficient management of these applications is crucial for complex sustainability governance analysis and service provisioning and configuration recommendation.

We develop a model that can be used to describe *SusGovApps*. Figure 8 presents a simplified conceptual model for *SusGovApps*. Overall, information about a *Sus-GovApp* will include (1) category of the application (SusGovAppCat egory), (2) metadata about the application (SusGovAppDescription), (3) concrete monitoring data (MonitoringDataIdentification), (4) abstract supporting monitoring data types, (TypeOfMonitoringData), and (5) output presentation (OutputPresentation). Detailed information about execution environments can be specified in application description, whereas with concrete monitoring data types, we can constrain types of data accessed by the application. Many SusGovApps can be managed by an application store (SusGovAppStore). This model can manage applications which are executed at gateways.

Listing 1 presents a simplified example of the description for an application analysis named Demand_CP4E which is used to analyze energy consumption.

Listing 1 Example of application descriptions

```
<owl:NamedIndividual rdf:about="http://pcccl/susgovapp#
    DemandCP4E">
  <rdf:type rdf:resource="http://pcccl/susgovapp#
    SusGovApp"/>
  <rdfs:label xml:lang="en">DemandCP4E</rdfs:label>
  <appCategory rdf:resource="http://pcccl/susgovapp#
    AnalysisApp"/>
  <appDescription rdf:resource="http://pcccl/susgovapp#
    DemandCP4EDescription"/>
  <monitoringData rdf:resource="http://pcccl/susgovapp#
    Stream66273"/>
</owl:NamedIndividual>
```

To implement *SusGovApp*, we consider two possibilities: *SusGovApp* will access data from gateways in buildings and *SusGovApp* access data directly from *Monitoring-DaaS* in data centers. In both cases, *SusGovApp* will use certain data connectors, however, we consider these connectors are internal part. In our model, we consider applications as black boxes and these applications can be exposed via a generic interface. For example, instead of defining a detailed logic of a complex analysis application as a workflow and managing the workflow, the workflow can be considered as an input parameter of the application which is implemented as a batch job by invoking a workflow engine to execute the workflow. Depending the implementa-

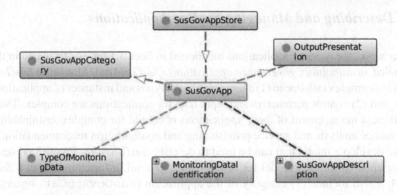

Fig. 8 Description of SusGovApp

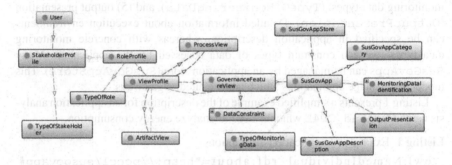

Fig. 9 Linking stakeholders, monitored objects, monitoring data and governance features

tion, the internal logic of a *SusGovApp* can be simple java implementation, complex workflow (for extracting, transforming and loading data) or complex event processing (e.g., for monitoring). In our work, we consider *SusGovApp* can be developed using different methods and frameworks, such as the R framework, Java, Matlab, and complex scientific workflows, due to the diversity and complexity of sustainability measurements. Therefore, a `TypeOfApp` is used to support the identification of possible execution environments.

5.4 Linking Types of Data, Applications and Stakeholders

To link data, applications and stakeholders together, we have developed a conceptual model for managing stakeholders and their views on sustainability governance features and data. Figure 9 describes our model for managing stakeholder in sustainability governance. A stakeholder is described by `StakeholderProfile`. Each stakeholder will be associated with different roles, described by `RoleProfile`. `ViewProfile`, used to specify what a role can be done, will be associated with one

Table 3 Examples of SusGovPaaS APIs

APIs	Description
`discover/monitoredobjects/`	Discover all monitored objects
`discover/monitoredobjects/` `{monitoredobjecturi}/...`	Discover a monitored object based on its uid using `monitoredobjecturi` and other information
`discover/apps/`	Discover all available applications based on some metadata
`discover/apps/{appid}/...`	Discover a specific application indicated by `appid` and the application's metadata
`execute/apps/{appid}/...`	Execute applications
`discover/data/{monitored` `objecturi}/{uid}`	Discover all data related to the monitored object (indicated by `monitoredobject`) and/or its data indicated by `uid`
`download/{category}/{uid}...`	Download monitoring data (`category=data`), applications (`category=apps`) and analytics reports (`category=reports`) by using their uid

or multiple roles. `ViewProfile` will identify which processes should be used for a particular role, using `ProcessView`, and which governance features are allowed, specified by `GovernanceFeatureView`. Governance features will be characterized via possible applications to be used, described by `SusGovApp` and possible constraints on data, described by `DataConstraints`. Both `SusGovApp` and `DataConstraint` are linked to monitoring data they apply, specified by `MonitoringDataIdentification`. The data that the stakeholder can use will be identified via identifiers of monitoring data and data constraints. Each stakeholder can have different permissions for different types of applications and corresponding monitoring data. The above-mentioned model is described in RDF and used to provision sustainability governance analytics features.

5.5 PaaS APIs and Analytics Framework

To foster the utilization and integration of features of SusGov PaaS, we provide different APIs for different consumers to control, execute and manage analytics. Table 3 presents some examples of (simplified) APIs in our PaaS. In general, we provide APIs for (1) discovering monitored objects, analytics and monitoring applications, and data, (2) executing analytics applications, and (3) managing and downloading applications and analytics reports.

To support advanced users to develop analytics applications and conduct sustainability analysis, we develop a *SusGov Analysis Framework* offering rich GUIs and exploiting RESTful APIs for *SusGovPaaS* (e.g., shown in Table 3) to discover and access data and applications as well as to execute applications in our SusGovPaaS.

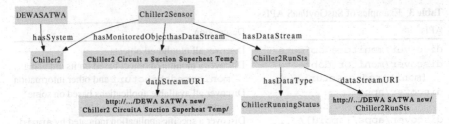

Fig. 10 An exemplified graph for DEWASATWA building and one chiller

6 Illustrating Experiments

In the prototype, (time series) monitoring data are obtained from gateways at the building site; gateways are based on Niagara AX³ as part of the Galaxy platform [14]. The data is in XML but it does not include metadata about monitored objects which are known by engineers who perform the setup and configuration of sensors and gateways. Therefore, *dataTypeURI*, *monitoredObjectURI* and *dataStreamURI* are constructed by extracting metadata about sensors in combination with configuration information. In particular, the configuration information, manually collected via different processes and described in our linked data model, allows the specification of dependencies among different monitored objects. For example, configuration can specify which sensors are associated with which chillers in which buildings. Figure 10 presents an example of a graph describing how monitoring data is linked to sensors and buildings using RDF. Based on dependencies, we could define data constraints for specific stakeholders as well as reason relevant types of monitoring data for specific analyses of monitored objects.

We utilize Jersey—and implementation of JAX-RS for RESTful Web Services—and Weblogic 10.3 for developing our SusGovPaaS. *Application Store & Stakeholder Profile Service* are RESTful Web services that store their information under RDF/XML format and use Allegro Graph.⁴ We use Cassandra as NoSQL data nodes which are controlled and accessed by/via the *MonitoringDaaS* as a REST-based Web service.

Figure 11 shows a snapshot of our *SusGov Analysis Framework*. First of all utilizing monitored objects/data discovery APIs, we can show the relationships among monitored objects that a user can access via a dependency graph (in the top-left window of Fig. 11). This graph allows the user to examine the influence among monitored objects, e.g., which monitored objects can contribute to the analysis of a room. For each monitored object, the user can also see which types of monitoring data he/she can access (in the bottom-left window of Fig. 11). Such access controls are controlled by the cloud providers or stakeholders who have the right to control the monitored object. Similarly, the user can also discover existing applications (in the top-middle window—Application Discovery—of Fig. 11). Based on

³ http://www.niagaraax.com/

⁴ http://www.franz.com/agraph/allegrograph

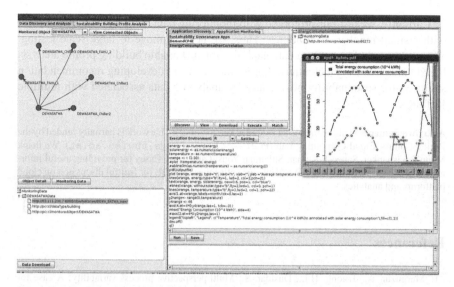

Fig. 11 Example of SusGov analysis framework

application description and monitoring data discovered, the user can select applications and ask our SusGovPaaS to execute them. Monitoring data, applications and reports from analyses can be downloaded back to the user working place, subject to the access control. For example, the chart in Fig. 11 shows an analysis result from a R-based application. By downloading monitoring data and applications, the user can also execute the applications in his/her local execution environment (e.g., shown in the right-middle window—Execution Environment—in Fig. 11).

7 Conclusions and Future Work

In this chapter, we analyze stakeholders, their relationships to sustainability monitoring data, analysis applications and monitored objects. We have developed models for managing these relationships. Based on that we design our SusGovPaaS with APIs for supporting data and application discovery, and analysis application management and execution that can be used by various stakeholders and automated applications. We have illustrated that by linking monitoring data streams, monitored objects, applications and stakeholders, we can manage their complex relationships, thus facilitating complex sustainability governance monitoring and analysis features within cloud-based M2M systems. This also hides complexity by not showing low-level data streams and supports customized features in the SaaS model for sustainability governance. Overall, our techniques aim at supporting cloud-based M2M platform providers to solve complex issues within their cloud systems.

Currently, we test our prototype with monitoring data from the Galaxy platform with a small number of buildings. Thus, we need to move to large-scale testing. We are currently focusing on data analytics for sustainable building profile analysis based on data mining techniques. Furthermore, we are developing recommendation solutions for sustainability governance by analyzing data associated with multiple facilities.

Acknowledgments This chapter is an extended version of [2]. This work is partially funded by the Pacific Control Cloud Computing Lab (PC3L—pc31.infosys.tuwien.ac.at). We thank Manu Ravishankar, Saneesh Kumar, Sulaiman Yousuf and Terry Casey for providing useful information and datasets of facilities. We thank our colleagues in PC3L for fruitful discussion on cloud platforms and analytics.

References

1. Murguzur, A., Truong, H.L., Dustdar, S.: Multi-perspective process variability: A case for smart green buildings (short paper). In: 6th IEEE International Conference on Service-Oriented Computing and Applications (SOCA 2013) (2013)
2. Truong, H.L., Dustdar, S.: M2M platform-as-a-service for sustainability governance. In: IEEE SOCA, pp. 1–4 (2012)
3. Truong, H.L., Dustdar, S.: A survey on cloud-based sustainability governance systems. Int. J. Web Inf. Syst. **8**(3), 278–295 (2012)
4. Acker, R., Massoth, M.: Secure ubiquitous house and facility control solution. In: Proceedings of the 5th International Conference on Internet and Web Applications and Services, ICIW '10, Washington, DC, USA, pp. 262–267. IEEE Computer Society (2010)
5. Tompros, S., Mouratidis, N., Draaijer, M., Foglar, A., Hrasnica, H.: Enabling applicability of energy saving applications on the appliances of the home environment. Netw. Mag. Glob. Internetwkg. **23**, 8–16 (2009)
6. Krishnamurthy, S., Anson, O., Sapir, L., Glezer, C., Rois, M., Shub, I., Schloeder, K.: Automation of facility management processes using machine-to-machine technologies. In: Proceedings of the 1st International Conference on the Internet of Things, IOT'08, pp. 68–86. Springer, Berlin, Heidelberg (2008)
7. Choi, J., Shin, D., Shin, D.: Research and implementation of the context-aware middleware for controlling home appliances. In: International Conference on Consumer Electronics, ICCE 2005, pp. 161–162. Digest of Technical Papers (2005)
8. Broering, A., Foerster, T., Jirka, S., Priess, C.: Sensor bus: An intermediary layer for linking geosensors and the sensor web. In: Proceedings of the 1st International Conference and Exhibition on Computing for Geospatial Research & #38; Application, COM.Geo '10, New York, pp. 12:1–12:8. ACM (2010)
9. Consortium, B.: Deliverable d2.2: End-to-end platform specification beywatch data model (annex). http://www.beywatch.eu/docs/D22_Annex.pdf. June 2010
10. Granderson, J., Piette, M., Ghatikar, G.: Building energy information systems: User case studies. Energ. Effi. **4**, 17–30 (2011). doi:10.1007/s12053-010-9084-4
11. Swords, B., Coyle, E., Norton, B.: An enterprise energy-information system. Appl. Energ. **85**(1), 61–69 (2008)
12. AMEE: http://www.amee.com/. Last Accessed 7 Feb 2012
13. Thyagarajan, G., Sarangan, V., Sivasubramaniam, A., Suriyanarayanan, R., Chitra, P.: Managing carbon footprint of buildings with ecview. Computer **99**(PrePrints) (2010)
14. Pacific Controls: Galaxy. http://pacificcontrols.net/products/galaxy.html (2011). Last Accessed 7 Feb 2012

15. EPA: Energy cost and iaq performance of ventilation systems and controls—executive summary. Technical Report EPA-4-2-S-01-001, United States Environmental Protection Agency (2000)
16. Bulut, A., Singh, A.: A unified framework for monitoring data streams in real time. In: Proceedings of 21st International Conference on Data Engineering, ICDE 2005, pp. 44–55, Apr 2005
17. Hauder, M., Gil, Y., Liu, Y.: A framework for efficient data analytics through automatic configuration and customization of scientific workflows. In: eScience, pp. 379–386. IEEE Computer Society (2011)
18. Ekanayake, J., Pallickara, S., Fox, G.: Mapreduce for data intensive scientific analyses. In: Proceedings of the 4th IEEE International Conference on eScience, Washington, DC, USA, pp. 277–284. IEEE Computer Society (2008)
19. AlertMe: http://www.alertme.com. Last Accessed 17 Feb 2012
20. xively: https://xively.com/. Last Accessed 27 Aug 2013
21. Matsuura, M., Suzuki, T., Shiroyama, H.: Stakeholder assessment for the introduction of sustainable energy and environmental technologies in Japan. In: IEEE International Symposium on Technology and Society, ISTAS '09, pp. 1–9, May 2009
22. Yun, C.H., Han, H., Jung, H.S., Yeom, H.Y., Lee, Y.W.: Intelligent management of remote facilities through a ubiquitous cloud middleware. In: IEEE International Conference on Cloud Computing, CLOUD '09, pp. 65–71, Sept 2009
23. Balazinska, M., Deshpande, A., Franklin, M.J., Gibbons, P.B., Gray, J., Hansen, M., Liebhold, M., Nath, S., Szalay, A., Tao, V.: Data management in the worldwide sensor web. IEEE Pervasive Comput. **6**, 30–40 (2007)
24. Rolewicz, I., Catasta, M., Jeung, H., Miklós, Z., Aberer, K.: Building a front end for a sensor data cloud. In: Proceedings of the International Conference on Computational Science and its Applications—Volume Part III, ICCSA'11, pp. 566–581. Springer, Berlin, Heidelberg (2011)
25. Li, T., Liu, Y., Tian, Y., Shen, S., Mao, W.: A storage solution for massive iot data based on nosql. In: GreenCom, pp. 50–57. IEEE (2012)
26. CA AppLogic: http://www.ca.com/us/cloud-platform.aspx (2013). Last Accessed 23 Dec 2013
27. Appistry Platform: http://www.appistry.com/products (2013). Last Accessed 23 Dec 2013
28. Google App Engine: http://code.google.com/appengine/ (2013). Last Accessed 23 Dec 2013
29. Microsoft Azure Services Platform: http://www.microsoft.com/azure/default.mspx (2013). Last Accessed 23 Dec 2013
30. Amazon: Amazon DynamoDB. http://aws.amazon.com/dynamodb/. Last Accessed 17 Feb 2012
31. Parabon Frontier: http://www.parabon.com/ (2013). Last Accessed 23 Dec 2013
32. Grossman, R.L., Gu, Y., Mambretti, J., Sabala, M., Szalay, A., White, K.: An overview of the open science data cloud. In: Proceedings of the 19th ACM International Symposium on High Performance Distributed Computing, HPDC '10, New York, NY, USA, pp. 377–384. ACM (2010)
33. BuildingSMART International Limited: Ifc4. http://www.buildingsmart-tech.org/specifications/ifc-releases/ifc4-release/ifc4-release-summary. Last Accessed 15 Nov 2013
34. Truong, H.L., Dustdar, S.: On evaluating and publishing data concerns for data as a service. In: APSCC, pp. 363–370. IEEE Computer Society (2010)
35. Truong, H.L., Phung, P.H., Dustdar, S.: Governing bot-as-a-service in sustainability platforms—issues and approaches. Proc. CS **10**, 561–568 (2012)

Social Networking Analysis

Kevin Curran and Niamh Curran

Abstract A social network indicates relationships between people or organisations and how they are connected through social familiarities. The concept of social network provides a powerful model for social structure, and that a number of important formal methods of social network analysis can be perceived. Social network analysis can be used in studies of kinship structure, social mobility, science citations, contacts among members of nonstandard groups, corporate power, international trade exploitation, class structure and many other areas. A social network structure is made up of nodes and ties. There may be few or many nodes in the networks or one or more different types of relations between the nodes. Building a useful understanding of a social network is to sketch a pattern of social relationships, kinships, community structure, interlocking dictatorships and so forth for analysis.

1 Introduction

Communication is and has always been vital to the growth and the development of human society. An individual's attitudes opinions and behaviours can only be characterised in a group or community [5]. Social networking is not an exact science, it can be described as a means of discovering the method in which problems are solved, how individuals achieve goals and how businesses and operations are run. In network theory, social networks are discussed in terms of node and ties (see Fig. 1). Nodes are individual actors and ties are relationships within networks. The social capital of individual nodes/actors can be measured through social network diagrams, as can measures of determination of the usefulness of the network to the actors individually [4]. The shape of a social network helps determine a network's

K. Curran (✉) · N. Curran
School of Computing and Intelligent Systems, Faculty of Computing and Engineering,
University of Ulster, Northern Ireland, UK
e-mail: kj.curran@ulster.ac.uk

N. Bessis and C. Dobre (eds.), *Big Data and Internet of Things:*   367
A Roadmap for Smart Environments, Studies in Computational Intelligence 546,
DOI: 10.1007/978-3-319-05029-4_15, © Springer International Publishing Switzerland 2014

Fig. 1 Social networking
diagram

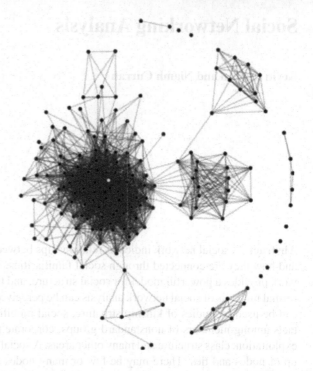

usefulness to its individuals. Smaller, tighter networks can be less useful to their
members than networks with lots of weak ties to individuals outside the main network.
More open networks, with many weak ties and social connections, are more likely
to introduce new ideas and opportunities to their members than closed networks
with many redundant ties. In other words, a group of friends who only do things
with each other already share the same knowledge and opportunities. A group of
individuals with connections to other social worlds is likely to have access to a wider
range of information. It is better for individual success to have connections to a
variety of networks rather than many connections within a single network. Similarly,
individuals can exercise influence or act as brokers within their social networks
by bridging two networks that are not directly linked essentially filling structural
holes [1].

Resulting graphs from node/tie diagrams can be complex. Social Networks operate
on many different levels from families up to nations, and play a critical role in
determining the way problems are solved, organisations are run and the degree in
which people succeed in achieving their goals. Below is an example of a social
network diagram, the node with the highest betweenness centrality (Betweenness—
The extent to which a node lies between other nodes in the network. This measure
takes into account the connectivity of the node's neighbours, giving a higher value
for nodes which bridge clusters. The measure reflects the number of people who a
person is connecting indirectly through their direct links, and Centrality—measure

giving a rough indication of the social power of a node based on how well they "connect" the network) is marked in yellow [11].

A few analytic tendencies distinguish social network analysis. There is no assumption that groups are the building blocks of society: the approach is open to studying less-bounded social systems, from nonlocal communities to links among websites. Rather than treating individuals (persons, organizations, states) as discrete units of analysis, it focuses on how the structure of ties affects individuals and their relationships. In contrast to analyses that assume that socialization into norms determines behaviour, network analysis looks to see the extent to which the structure and composition of ties affect norms.

In fact, long before it became the commercialised and significant entertainment juggernaut that it is today, social networking was nothing more than a theory. However, this theory of social networking stems back as far as the late 1800s as numerous sociologists were able to outline its basic principles [6]. German sociologist, Ferdinand Tönnies was a major contributor to sociological theory and it was him who initially highlighted that social groups exist by containing individuals which are linked together through shared beliefs and values. By the turn of the twentieth century, another major German sociologist, Georg Simmel became the first scholar to think appropriately in social network terms. Simmel produced a series of essays that pinpointed the nature of network size. He further displayed an understanding of social networking with his writings as he highlighted that social interaction existed within loosely-knit networks as opposed to groups [9]. The next real significant growth of social networking didn't really commence until the 1930s when three main social networking traditions emerged. The first tradition to emerge was pioneered by Jacob Levy Moreno, who was recognised as one of the leading social scientists. Moreno began the systematic recording and analysis of social interaction in smaller groups such as work groups and classrooms. The second tradition was founded by a Harvard group which began to focus specifically on interpersonal relations at work. The third tradition originated from Alfred Radcliffe-Brown, an English social anthropologist. Radcliffe-Brown strongly urged the systematic studies of networks; 'Social Network Analysis' was born. However, SNA did not advance further until the 1950s. This was when social network analysis was developed through the kinship studies of Elizabeth Bott who studied at the University of Manchester in England. It was here that the University's group of anthropologists began a series of investigations of community networks in regions such as Africa and India [11]. This research set the trend as more universities began similar investigations and studies as time progressed. During the 1960s a group of students at Harvard University began work to unite the different tracks and traditions already associated with social networking. Additional research was carried out in universities such as the University of California, Irvine and the University of Toronto. The latter contained a sociology group that emerged in the 1970s. The research undertook by this group argued that viewing the world in terms of social networks provided a greater analytical advantage. This view is also supported by Wasserman, S. and K. Faust, 1994, in their Social Network Analysis writings in the Cambridge University Press as they explain the extent of which SNA provides analytical advantage; *"The unit of analysis in network analysis is not the individual,*

but an entity consisting of a collection of individuals and the linkages among them".
In recent times, social networking theories have been put aside as social networking
has transferred to social media such as social networking sites like Facebook and
MySpace. Although nowadays social networking is seen as more of an entertainment
package, its roots stem back to the theoretical studies of sociologists such as Tönnies
and Simmel as well as the progression of Social Network Analysis [7].

2 Social Networking

Social groups can exist as personal and direct social ties that either link individuals
who share values and beliefs or impersonal, formal, and instrumental social links.
Durkheim gave a non-individualistic explanation of social facts arguing that social
phenomena arise when interacting individuals constitute a reality that can no longer
be accounted for in terms of the properties of individual actors. He distinguished
between a traditional society—"mechanical solidarity"—which succeeds if individual differences are lessened, and the modern society that develops out of support
between differentiated individuals with independent roles. Social network analysis
has emerged as a key technique in modern sociology, and has also gained a following
in anthropology; biology, communication studies, economics, geography, information science, organizational studies, social psychology, and sociolinguistics, and has
become a popular topic of speculation and study.

- Anthropology—is the study of humanity. It has origins in the natural sciences, the
  humanities, and the social sciences.
- Biology—is a natural science concerned with the study of life and living organisms, including their structure, function, growth, origin, evolution, distribution,
  and taxonomy.
- Communication Studies—is an academic field that deals with processes of communication, commonly defined as the sharing of symbols over distances in space
  and time. Hence, communication studies encompasses a wide range of topics and
  contexts ranging from face-to-face conversation to speeches to mass media outlets
  such as television broadcasting.
- Economics—is the social science that analyzes the production, distribution, and
  consumption of goods and services.
- Geography—is the science that deals with the study of the Earth and its lands,
  features, inhabitants, and phenomena.
- Information Science—is an interdisciplinary science primarily concerned with the
  analysis, collection, classification, manipulation, storage, retrieval and dissemination of information.
- Organizational Studies—encompass the systematic study and careful application
  of knowledge about how people act within organizations.

- Social Psychology—is the study of the relations between people and groups; or how situational factors affect the thoughts, feelings, and/or behavior of an individual.
- Sociolinguistics—is the study of the effect of any and all aspects of society, including cultural norms, expectations, and context, on the way language is used, and the effects of language use on society.

It has now moved from being a suggestive metaphor to an analytic approach to a pattern with its own theoretical statements, methods social network analysis software and researchers. The potential for computer networking to facilitate new forms of computer-mediated social interaction was suggested early on. Efforts to support social networks via computer-mediated communication were made in many early online services, including Usenet, Arpanet, listserv, and bulletin board services (BBS). Many prototypical features of social networking sites were also present in online services such as America Online, Prodigy, and CompuServe. Early social networking on the World Wide Web began in the form of generalized online communities such as TheGlobe.com (1995), Geocities (1994) and Tripod.com (1995). Many of these early communities focused on bringing people together to interact with each other through chat rooms, and encouraged users to share personal information and ideas via personal web pages by providing easy-to-use publishing tools and free or inexpensive web space. Some communities—such as Classmates.com took a different approach by simply having people link to each other via email addresses. In the late 1990s, user profiles became a central feature of social networking sites, allowing users to compile lists of "friends" and search for other users with similar interests.

New social networking methods were developed by the end of the 1990s, and many sites began to develop more advanced features for users to find and manage friends. This newer generation of social networking sites began to flourish with the emergence of Makeoutclub in 2000, followed by Friendster in 2002, and soon became part of the Internet mainstream. Friendster was followed by MySpace and LinkedIn a year later, and finally Bebo and Facebook in 2004. Attesting to the rapid increase in social networking sites' popularity, by 2005, MySpace was reportedly getting more page views than Google. Facebook launched in 2004, has since become the largest social networking site in the world. Today, it is estimated that there are now over 200 active sites using a wide variety of social networking models. Web based social networking services make it possible to connect people who share interests and activities across political, economic, and geographic borders. Through e-mail and instant messaging, online communities are created where a gift economy and mutual unselfishness are encouraged through collaboration. Information is particularly suited to gift economy, as information is a non-rival good and can be gifted at practically no cost. Facebook and other social networking tools is increasingly the object of scholarly research. Scholars in many fields have begun to investigate the impact of social networking sites, investigating how such sites may play into issues of identity, privacy, social capital, youth culture, and education Several websites are beginning to tap into the power of the social networking model for humanity. Such models provide a means for connecting otherwise fragmented industries and

Fig. 2 Matrix of group
relationships

	Paul	Laura	Aidan	Sarah
Paul	---	1	0	1
Laura	1	---	1	0
Aidan	0	1	---	0
Sarah	1	0	0	---

small organizations without the resources to reach a broader audience with interested users. Social networks are providing a different way for individuals to communicate digitally. These communities of hypertexts allow for the sharing of information and ideas, an old concept placed in a digital environment. The amount of information needed to describe even the smallest of social networks can be quite big. As the information can be in large quantities, this can make managing and manipulating the date to show patterns of social structure quite complicated. Tools from mathematics are used to help all of the tasks of social network methods. Matrices are useful for recording information such as the calculation of indexes describing networks. An example of a simple matrix is shown in Fig. 2.

The above matrix shows the structure of a close friendship in a group of four people: Paul, Laura, Aidan and Sarah. It describes a pattern of linking ties with a point-to-point matrix where the rows represent choices by each actor. We put a "1" if a person likes another, and a "0" if they don't. One reason for using mathematical and graphical techniques in social network analysis is to represent the descriptions of networks efficiently and more economically. This also enables us to use computers to store and manipulate the information quickly and more accurately than we can by hand. The use of computers in social networks to show mathematical representations is also very important because if you had a huge amount of data that need manipulated, it could take years to do by hand, whereas it can be done by a computer in a few minutes. Formal Methods are used to represent data because matrices and graphs summarise and present a lot of information quickly and easily and allow us to apply computers in analysing data. This helps because most of the work is repetitive but requires accuracy and that is exactly the sort of thing computers do well. Matrices and graphs have rules and conventions. These help us communicate better but sometimes the rules and conventions of the language of the graphs and mathematics themselves lead us to see things in our data that might not occurred for us to look for if we described it ourselves.

The most popular social networking site is facebook.com. It currently has over 500 million active users world-wide and is a free use service. Through facebook.com you can search for friends, browse a news feed consisting of recent status updates, photo uploads, future events, videos, links posted by other users that are 'friends', learn more about people you see every day, comment and post on other people's walls, join 'Clubs', keep in contact with friends at other schools/colleges, share photos privately or publicly, etc. These social networks start out by an initial set of founders sending out a message inviting members of their own personal networks to join the site. The new members then repeat this process, growing the total number of members and links in the network. These sites then offer different features like viewable profiles, chat, etc. Social connections can also be used for business connections. Blended networking is an approach to social networking that combines both offline elements (face-to-face events) and online elements. Social computing is the use of social software, which is based on creating or recreating social conversations and social contexts online through the use of software and technology. An example of social computing is the use of email for maintaining social relationships. The current metrics for social network analysis are as follows:

- Bridge—An edge is said to be a bridge if deleting it would cause its endpoints to lie in different components of a graph.
- Centrality—This measure gives a rough indication of the social power of a node based on how well they "connect" the network. "Betweenness", "Closeness", and "Degree" are all measures of centrality.
- Betweenness—The extent to which a node lies between other nodes in the network. This measure takes into account the connectivity of the node's neighbors, giving a higher value for nodes which bridge clusters. The measure reflects the number of people who a person is connecting indirectly through their direct links.
- Closeness—The degree an individual is near all other individuals in a network (directly or indirectly). It reflects the ability to access information through the "grapevine" of network members. Thus, closeness is the inverse of the sum of the shortest distances between each individual and every other person in the network. The shortest path may also be known as the "geodesic distance".
- Centralization—The difference between the number of links for each node divided by maximum possible sum of differences. A centralized network will have many of its links dispersed around one or a few nodes, while a decentralized network is one in which there is little variation between the number of links each node possesses.
- Clustering Coefficient—A measure of the likelihood that two associates of a node are associates. A higher clustering coefficient indicates a greater 'cliquishness'.
- Density—The degree a respondent's ties know one another/proportion of ties among an individual's nominees. Network or global-level density is the proportion of ties in a network relative to the total number possible (sparse vs. dense networks).
- Degree—The count of the number of ties to other actors in the network.
- Cohesion—The degree to which actors are connected directly to each other by cohesive bonds. Groups are identified as 'cliques' if every individual is directly tied to every other individual, 'social circles' if there is less stringency of direct contact, which is imprecise, or as structurally cohesive blocks if precision is wanted.
- Eigenvector centrality—A measure of the importance of a node in a network. It assigns relative scores to all nodes in the network based on the principle that connections to nodes having a high score contribute more to the score of the node in question.

- Prestige—In a directed graph prestige is the term used to describe a node's centrality. "Degree Prestige", "Proximity Prestige", and "Status Prestige" are all measures of Prestige.
- Reach—The degree any member of a network can reach other members of the network.
- Structural hole—Static holes that can be strategically filled by connecting one or more links to link together other points. Linked to ideas of social capital: if you link to two people who are not linked you can control their communication.

Another type of social network is a sexual network, which is defined by the sexual relationships within a set of individuals. They can be formally studied using the mathematics of graph theory [2]. Epidemiological studies (scientific study of factors affecting the health and illness of individuals and populations) have researched into sexual networks, and have discovered that the statistical properties of sexual networks are crucial to the spread of sexually-transmitted diseases (STDs). Social network analysis has been used to help understand how patterns of human contact aid or inhibit the spread of diseases such as HIV in a population. The evolution of social networks can sometimes be modelled by the use of agent based models, providing insight into the interplay between communication rules, rumour spreading and social structure. Social Contract is a political theory that explains the basis and purpose of the state and of human rights. Within a society, all its members are assumed to agree to the terms and conditions of the social contract by their choice to stay within the society without violating the contract.

The Social Safety Net is a term used to describe a collection of services provided by the state, e.g. (welfare, homeless shelters, etc.). They help prevent anyone from falling into poverty beyond a certain level. An example of how the safety-net works would be in the case of a single mother unable to work. In many western world countries, she will automatically receive benefits to the support the child so the child will have a better chance at becoming a successful member of society. On large social networking services, there have been growing concerns about users giving out too much personal information and the threat of sexual predators. Users of these services also need to be aware of data theft or viruses. However, large services, such as MySpace and Netlog, often work with law enforcement to try to prevent such incidents. There is a perceived privacy threat in relation to placing too much personal information in the hands of large corporations or governmental bodies, allowing a profile to be produced on an individual's behavior on which decisions, harmful to an individual, may be taken. Privacy on social networking sites can be undermined by many factors. For example, users may disclose personal information, sites may not take adequate steps to protect user privacy, and third parties frequently use information posted on social networks for a variety of purposes. For the internet generation, social networking sites have become the preferred forum for social interactions. However, because such forums are relatively easy to access, posted content can be reviewed by anyone with an interest in the users' personal information. There is also an issue over the control of data or information that was edited or removed by the user may in fact be retained and/or passed to third parties. This danger was highlighted when the controversial social networking site Quechup harvested e-mail addresses from users' e-mail accounts for use in a spamming operation. Interpersonal communication has been a growing issue as more and more people have turned to social networking as a means of communication. Beniger [3] describes how mass media has gradually replaced interpersonal communication as a socializing force. Further, social networking sites have become popular sites for youth culture to explore themselves, relationships, and share cultural artifacts. Many social networks also provide an online environment for people to communicate and exchange personal information for dating purposes. Objectives can vary from looking for a one time date, to a short-term relationship to a long-term relationship. Most of these social networks, just like online dating services, require users to give out certain pieces of information. This usually includes a user's age, gender, location, interests, and perhaps a picture. Releasing very personal information is usually discouraged. This allows other users to search or be searched by some sort of criteria, but at the same time people can maintain a degree of anonymity similar to most online dating services. Online dating sites are similar to social networks in the sense that users create profiles to meet and communicate with others, but their activities on such sites are for the sole purpose of finding a person of interest to date. However, an important difference between social networks and online dating services is the fact that online dating sites usually require a fee, where social networks are free. This difference is one of the reasons the online dating industry is seeing a massive decrease in revenue due to many users opting to use social networking services instead. Many popular online dating services such as Match.com, Yahoo Personals, and eHarmony.com are seeing a decrease in users, where social networks like MySpace and Facebook are experiencing an increase in users.

3 Network Analysis

There are two basic kinds of network analysis, reflecting two different kinds of data: ego network analysis, and complete network analysis. Ego network analysis can be done in the context of traditional surveys. Each respondent is asked about the people they interact with, and about the relationships among those people. Since respondents might be sampled randomly from a large population, in ego network analysis it is unlikely that the respondents will know anyone in common, and no attempt is made to link up the networks. Typically, the analysis of ego networks involves assessing the quality of a person's networks (size, diversity, average income, etc.) or relating attributes of ego with attributes of their alters (homophily). Ego network analysis is convenient because it can be used in conjunction with random sampling, which enables classical statistical techniques to be used to test hypotheses.

Complete network analysis is where you try to obtain all the relationships among a set of respondents, such as all the friendships among employees of a given company. Most of the rhetoric surrounding network analysis is based on complete networks.

Fig. 3 Ego network

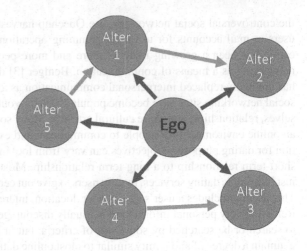

Techniques such as subgroup analysis, equivalence analysis and measures like centrality all require complete networks (Fig. 3).

There are numerous terms used to describe connections within social networks. Homophily is the extent to which actors make ties with similar versus dissimilar others. The similarity is defined by factors such as gender, race, age or other salient characteristic [8]. Multiplexity is the number of content-forms contained in a tie. For example, two people who are friends and work together would have a multiplexity of two. It can also be seen as simply the strength of a relationship. Network closure is the measure of the completeness of relational triads while propinquity is the tendency for actors to have more ties with others who are close geographically.

Of course, visual representation of social networks can be helpful in order to understand the network data and relay the result of the analysis. Most software presents data through visual nodes and ties attributing colours, size and other advanced properties to nodes. For complex information such as large social graphs with varying attributes, these visual representations can be a powerful means to convey complex information. Likewise, collaboration graphs can be used to show human relationships—both negative and positive. For instance, a positive edge between two nodes denotes a positive relationship (i.e. friendship) and a negative edge between two nodes denotes a negative relationship (i.e. hatred).

In the real world, social network analysis can be used in counter-intelligence and law enforcement activities. There are a number of hacking tools also known as information gathering tools online which allow one to build up detailed analysis of an individual's social graph. One such free tool is FOCA (Fingerprinting Organizations with Collected Archives). FOCA can do much more than build social graphs. In fact, it has recently come to the attention of the world that the National Security Agency (NSA) has been using its huge collections of data to create sophisticated social network graphs of some Americans' social connections that can identify their associates, their locations at certain times, their travelling companions and other personal information, according to newly disclosed documents and interviews with officials [10]. This started around 2010. After initial maps of the social network are complete, we believe that analysis is then performed to determine the structure of the network and determine, for example, the leaders within the network. This allows military or law enforcement assets to launch capture-or-kill decapitation attacks on the high-value targets in leadership positions to disrupt the functioning of the network.

4 Conclusion

Social Networks are social structures made up of nodes and ties; they indicate the relationships between individuals or organisations and how they are connected through social familiarities. They are very useful for visualising patterns. They operate on many levels and play an important role in solving problems, on how organisations are run and they helps individuals succeed in achieving their targets and goals. In today's society social networks allow two people in different locations to interact with each other socially (e.g. chat, viewable photos, etc.) over a network. They are also important for the Social Safety Net because this is helping the society with the likes of the homeless, unemployment. Group politics relate to 'In-Groups' and 'Out-Groups' as each competes with each other. The use of mathematical and graphical techniques in social network analysis is important to represent the descriptions of networks compactly and more efficiently. Social Networking is all around us and so there is always going to be friends and casual acquaintances both within the sub-groups and outside it. These status types link all sub-groups together as well as the internal structure of a group. Hence there are direct and in-direct connections to link everyone together within social circle websites like facebook.com.

References

1. Anderson, C.: People power. Wired magazine. http://www.wired.com/wired/archive/14.07/people.html (2006)
2. Arnold, N., Paulus, T.: Using a social networking site for experiential learning: appropriating, lurking, modeling and community building. Internet High. Educ. **13**, 188–196 (2010)
3. Beniger, J.: The Control Revolution: Technological and Economic Origins of the Information Society. Harvard University Press, Cambridge (1987)
4. Bowers-Campbell, J.: Cyber pokes: motivational anecdote for developmental college readers. J. colle. reading lear. **39**(1), 74–87 (2008)
5. Conole, G., Culver, J.: The design of Cloudworks: applying social networking practice to foster the exchange of learning and teaching ideas and designs. Comput. Educ. **54**, 679–692 (2010)
6. Mason, R., Rennie, F.: E-learning and Social Networking Handbook: Resources for Higher Education, 1st edn. New York, Routledge (2008)
7. Mazer, J.P., Murphy, R.E., Simonds, C.J.: I'll see you on 'Facebook': the effects of computer mediated teacher self-disclosure on student motivation, affective learning, and classroom climate. Commun. Educ. **56**(1), 1–17 (2007)

8. McPherson, N., Smith-Lovin, L., Cook, J.: Birds of a feather: homophily in social networks. Ann. Rev. Sociol. **27**, 415–444 (2001)
9. Prensky, M.: Digital natives, digital immigrants. On the horizon **9**(5), 1–6 (2001)
10. Risen, J., Poitras, L.: N.S.A. gathers data on social connections of U.S. citizens. N.Y. Times **28** (2013)
11. Scott, J.: Social Network Analysis—A Handbook, 2nd edn. SAGE Publications, Beverly Hills (2000)

Leveraging Social Media and IoT to Bootstrap Smart Environments

David N. Crowley, Edward Curry and John G. Breslin

Abstract As we move towards an era of Smart Environments, mixed technological and social solutions must be examined to continue to allow users some control over their environment. Realisations of Smart Environments such as Smart Cities and Smart Buildings bring the promise of an intelligently managed space that maximises the requirements of the user while minimising resources. Our approach is to create lightweight Cyber Physical Social Systems that aim to include building occupants within the control loop to allow them some control over their environment. We motivate the need for citizen actuation in Building Management Systems due to the high cost of actuation systems. We define the concept of citizen actuation and outline an experiment that shows a reduction in average energy usage of 26 %. We outline a use case for citizen actuation in the Energy Management domain, propose architecture (a Cyber-Physical Social System) built on previous work in Energy Management with Twitter integration, use of Complex Event Processing (CEP), and discuss future research in this domain.

1 Introduction

Realisations of Smart Environments such as Smart Cities and Smart Buildings bring the promise of an intelligently managed space that maximises the requirement of the users (e.g. citizen commute experience, building occupant comfort) while minimising

D. N. Crowley (✉) · E. Curry
DERI, NUI, Galway, Ireland
e-mail: david.crowley@deri.org

E. Curry
e-mail: ed.curry@deri.org

J. G. Breslin
Electrical and Electronic Engineering, NUI, Galway, Ireland
e-mail: john.breslin@nuigalway.ie

N. Bessis and C. Dobre (eds.), *Big Data and Internet of Things:*
A Roadmap for Smart Environments, Studies in Computational Intelligence 546,
DOI: 10.1007/978-3-319-05029-4_16, © Springer International Publishing Switzerland 2014

resources required (e.g. transportation costs, energy costs, pollution, etc.). Smart Environments are often taken as the norm in research but in fact the majority of current environments have very little support for sensing or actuation and are far removed from being intelligent environments. The existence of Sensor Actuator Networks (SANs) is a key requirement for the delivery of Smart Environments; however retrofitting SANs into existing buildings is costly, disruptive in business situations, and a time consuming process with large scale rollouts being a medium to long-term vision. Deploying sensors in buildings is relatively cheap and time effective, sensors that measure energy usage can be a cost-effective way of monitoring cost in a building.

The Economist in 2010 in relation to the Internet of Things (IOT) said, "Everything will become a sensor—and humans may be the best of all" [27]. We envision by combining these already deployed sensors in buildings and through employing systems extensively utilised on the Social Web such as microblog services that the Social Web will act as an enabling layer, the "social glue" between the Cyber-Physical System (CPS), government agencies, and the community. Whether that community is a city, town, a group of interested citizens, or a group formed around a social object, these groups when connected through social connections or objects [40] can be envisaged as systems and as a whole as a System of Systems (SoS). In the near-term, we need an alternative approach if we are to realise Smart Environments. Our alternate approach builds on an Internet of Things (IOT) architecture, Linked Data, and Social Media and coupled with our key concept of citizen actuation to create a Cyber-Physical Social System (CPSS) to reduce energy costs and aims to keep humans "in the loop". To test the concept of citizen actuation we conducted an experiment to examine if our hypothesis i.e. including citizen actuators in an energy management use case would help lower energy usage and during the test period energy usage was lowered by 26 % on average.

2 Smart Environments

Smart environments are physical worlds that are interwoven with sensors, actuators, displays and computational elements, embedded seamlessly into everyday objects and connected through a continuous network [38] or a small world where different kinds of smart device are continuously working to make inhabitants' lives more comfortable [11] where a smart device is an electronic device, generally connected to the Internet or other networks through WiFi, 3G, or other protocols usually with a display and now commonly with touch or voice activated controls. As people, things, and the world gets more interconnected the vision of smart environments has moved into reality and the interlinking of these physical worlds as Weiser describes them allows for the creation of larger systems or System of Systems which integrates systems into complex systems that offer better functionality and performance than simply the sum of the constituent systems [28]. In the next two sections, we will examine two separate smart environments that, while different, encapsulate the challenges within this research area. We will then discuss the challenges in smart environments.

2.1 Smart Buildings

Defining what Smart Buildings are is inherently difficult as definitions rely on the concept of Building Automation Systems (BAS). BAS is an example of a control system that controls building elements or other systems such as electrical systems, fire, security, heating, and air conditioning. Snoonian states that for BAS to be effective, any automation system must enable all these mechanical and electrical systems to work from a single building control point [36]. The vision of a Smart Building is of one that optimises its internal environment for the comfort and usability of its occupants while minimising the resources required to run and maintain the building. Within the context of a smart office building the objective would be to optimise the operation of the building to provide the ideal working conditions to increase staff productivity (e.g. internal lighting, temperature, room CO_2 levels, etc.) while minimising operational costs (e.g. energy consumption, water consumption, CO_2 emission etc.). The heart of a Smart Building is the Building Management System (BMS) and the Building Automation System (BAS) that provide advanced capabilities for the control and management of the building.

These systems rely heavily on the use of sensors and actuation to monitor and control operations (air-conditioning, ventilation, heating, lighting, etc.) within a building. While deploying sensors in buildings is relatively cheap and time effective. Sensors that measure energy usage can be a cost-effective way of monitoring energy costs in a building. The cost of full management systems is often prohibitive for Small and Medium Enterprises (SMEs) as retrofitting existing buildings is costly and can disrupt business. While BMS and BAS systems are becoming more popular within new building construction, most buildings are not currently equipped with sophisticated building management of building automation systems. The opportunity to reduce energy consumption in these building will require the retrofit of such systems at significant cost and time. An alternative lower-cost solution is needed; we believe that citizen actuation can offer this benefit without the need for high-cost installation of building automation equipment.

2.2 Smart Cities

Caragliu et al believe a city to be smart when investments in human and social capital and traditional (transport) and modern Information and Communications Technology (ICT) infrastructure fuel sustainable economic growth and a high quality of life, with a wise management of natural resources, through participatory governance [9]. This is a very holistic view of a Smart City where services are integrated and humans are involved through participatory governance. Deploying urban sensor networks, while costly, is a realistic goal and some cities and countries are already investing heavily in smart energy grids, traffic monitoring sensors, weather stations, and parking sensors to help manage the city. Research projects such as Smart Santander[1] are trying to realise the dream of a technologically smart city. While this technologically is a step forward it does not include any social data from inhabitants of the city. Concurrently there is also research looking at mobile social reporting applications that allow citizens to report on issues within their local environment [13] and IBM's Smarter Cities research aims to incorporate social elements.[2] In our research, we aim to allow people (building users or people living in an area) the opportunity to both report but also to be active in fixing the issues in their environment and this has led to the creation of the concept of citizen actuators.

2.3 Challenges

Integration of services and data from multiple sources from control systems (traffic, electrical, and emergency services), sensor networks, and social data is a serious challenge for large-scale smart environments. Interoperability between these systems is huge problem, aggregating data from proprietary software, legacy systems, and new systems can have a high cost. Linked Data discussed in Sect. 4 describes in detail an approach for data management by aggregating and linking heterogeneous data from various sources and transforming them to Linked Data. Sensor data can be represented in many formats like SensorML, Observations, and Measurements (OM) from the Open Geospatial Consortium (OGC), and more recently the W3C Semantic Sensor Network (SSN) ontology. The SSN ontology merges the concepts from SensorML (sensor focused), the OGC OM (observation focused), and system models. It develops a general representation of sensors and relies on upper-level ontologies to define the domain, and an operation model that describes the implementation of the measurement. The representation of a sensor in the ontology links together what it measures (the domain), the physical sensor (the grounding) and its functions and processing (the models) [10]. In this work, Linked Data principles were adhered to and play an important role in integrating data from over ten legacy systems, sensor data, and social data.

Improving energy performance, especially through changing the way an organisation operates, requires a number of practical steps, which will include the need for a systematic and holistic approach for information gathering and analysis. Creating a holistic view of energy information for an organisation is not a straightforward task; the complexities of real-world organisations mean they use energy in many different ways. Energy Intelligence platforms need to support four key requirements:

- Holistic energy consumption
- Multi-level energy analysis
- Business context Energy Consumption

---

[1] http://www.smartsantander.eu/

[2] http://www.ibm.com/smarterplanet/ie/en/smarter_cities/overview/index.html

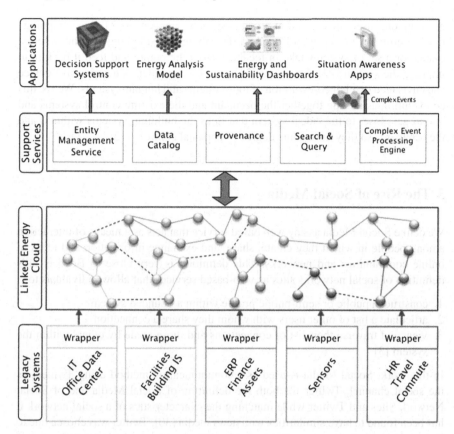

Fig. 1 Linked dataspace for energy inteligence

- Energy situational awareness

Implementation of these goals is described in [16] uses Linked Data to allow connection of multiple data sources including human resources systems, sensor networks, facility management, and finance systems to create a Linked Dataspace for Energy Inteligence as shown in Fig. 1. Our approach concentrated on energy situational awareness to allow building occupants to become actively involved in energy management and to play an important role in lowering energy consumption. A key element in improving building control systems in existing structures is installing SANs and while embedding sensors into an environment can be relatively cost-effective, the cost of installing actuation systems can be prohibitive (for example installing automated windows generally requires the replacement of the existing windows). In our experiment, we examined lowering energy usage within a building environment that had energy usage sensors but no form of automated actuation. Using IoT technologies and leveraging social media platforms we aim to bootstrap smart environments by developing light-weight CPSS that encourage (building or environment) users to become part of the system. The concept of "bootstrapping" in computing originated with Engelbart describing bootstrapping systems where a system or a set of systems take in feedback from the output and feed it back in to improve the system [22]. Engelbart [21] saw the computer as a supportive device and envisaged explicit co-evolution of tool-systems and human-systems. We see this co-evolution as linking together the occupant and the existing control systems and aim to get closer to the "ideal system that links the building, systems within it and the occupants so they have some degree of personal control" [32].

3 The Rise of Social Media

We define Social Media as any web based service that acts as a means of interaction among people in which they create, share, and exchange information and ideas in online communities and networks. This definition is very close to Boyd et al.'s definition of social networks sites as web-based services that allow individuals to

1. construct a public or semi-public profile within a bounded system
2. articulate a list of other users with whom they share a connection
3. view and traverse their list of connections and those made by others within the system [3]

In this work, Social Media is used as a communication method using Twitter[3] as the social channel. Twitter fits both the definition of Social Media and of Social Network Sites and Twitter while matching the characteristics of a social network it has been found to have similarities with news media [30]. Twitter was chosen due to its high use in the community chosen for the experiment described later. Other forms of communication were considered during the design process such as Facebook,[4] internal email, and other social media platforms. Internal email was considered too formal a mode of communication as this email is used for college communication and research. Facebook's communication mechanism did not have the directness needed beyond participant mentioning. Other networks did not have the user penetration in the participant community.

In the past ten years, we have seen the growth of online social networks and an explosion of user-generated content on the Web, in particular published from mobile devices. For example, a popular microblogging platform Twitter was founded in 2006 with an extremely fast growing user base with 175 million users[5] by October 2010 and about 340m posts processed per day as of March 2012 with 140m active users.[6] The ease of posting to services like Twitter while attaching data such as

---

[3] http://twitter.com/

[4] https://www.facebook.com/

[5] http://www.pcmag.com/article2/0,2817,2371826,00.asp

[6] http://blog.twitter.com/2012/03/twitter-turns-six.html

pictures, videos, and links significantly contributes to the growth in the volume of user-generated content.

In this work, it is a uni-directional communication between the building occupant and the system but in future work we would like to allow users to post photos issues or problems [7, 13] which would hopefully engage users more with the system. Social media platforms could also offer building-users the ability to view the energy consumption of the building and offer comment if the consumption is unexpectedly high through dashboards and social media platforms. Knowledge can also be shared by gauging people's experience with an environment and their reaction to changes to that environment.

3.1 Citizen Sensing

The concept of citizen sensing [25, 35] classed as opportunistic sensing where people report on issues or events in their surroundings and this information is then analysed to try to create insights into these events. In participatory sensing [8], the person "opts in" to send data concerning some task or request, for example eBird [37] (a real-time, online checklist program, eBird has revolutionized the way that the birding community reports and accesses information about birds[7]). In Sheth's citizen sensing, the people themselves can be seen as acting in a similar manner to physical sensors, but what is being sensed must typically be derived from the texts of their status updates, photo captions, or microblog posts [35]. Sheth defines the role of these citizen sensors as "humans as citizens on the ubiquitous Web, acting as sensors and sharing their observations and views using mobile devices and Web 2.0 services" [35]. Citizen sensing and crowdsourcing have been applied to a large number of use cases as described in [29]. Research has examined concepts of a "sensor tweeting standard" and tweet metadata [15, 20] to embed structured metadata into Tweets. As discussed previously the goal of this work is to allow both citizen sensing and actuation in smart environments.

3.2 Citizen Actuation

This notion of "human in the loop" sensing has led to the creation of the concept of citizen actuation [14], where people can report on their surroundings but when these reports are combined with other data sources then actionable requests can be constructed and sent to users. While citizen sensors [33, 35] only sense and report on their surroundings, citizen actuators can sense and act. The concept of citizen actuation comes from the need to close the loop started by citizen's reporting about events in their surrounding environment. While citizen sensing examines collecting

---

[7] http://ebird.org/content/ebird/about/

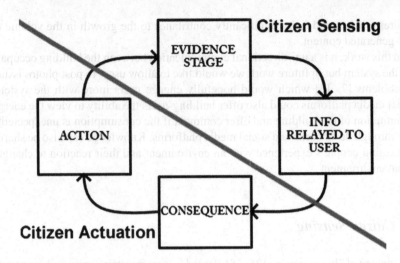

Fig. 2 Feedback loop diagram indicating where citizen sensing and citizen actuation occur

updates and extracting meaningful information, citizen actuation aims to make these reports an actionable item.

A good example of this is FixMyStreet, this application allows users to report (sense) issues with their locality, and this report is referred to the corresponding local government body. With citizen actuation (depending on the issue reported), the local people that reported the issue could fix the problem (e.g. by collecting litter in their housing estate, painting over graffiti on walls etc.).

Figure 2 displays how the citizen sensing and citizen actuation elements take place and how in conjunction they form a feedback loop. We define citizen actuation as the activation of a human being as the mechanism by which a control system acts upon the environment. Control systems often incorporate feedback loops, a feedback loop can be visualised as in Fig. 2. Feedback loops can be split into four stages, the data acquisition or evidence stage is the first. This stage collects the data and processes it for presenting to the user. The second stage relays the information to the user with richer context. This can be through visual representations like graphs, signs, or even warnings. A good example of this is a speed sign that measures a car's current speed displaying it to the driver in comparison to the speed limit. The third stage is consequence, which shows the gain from what the user has reported. The final stage is action, where the user completes an action or makes a choice then this action/choice has a reaction and the feedback loop can restart [24]. By encouraging citizen sensors to interact with their environment, we aim to allow for the creation of a feedback loop where people's actions will feed back into the loop.

Feedback has been examined as a method of lowering energy consumption in multiple studies [7, 12, 23, 34]. These studies vary in success from zero reduction in energy consumption to some reporting a saving of 20%, usual reported savings are between 5 and 12% [23] in household studies. The feedback methods used range

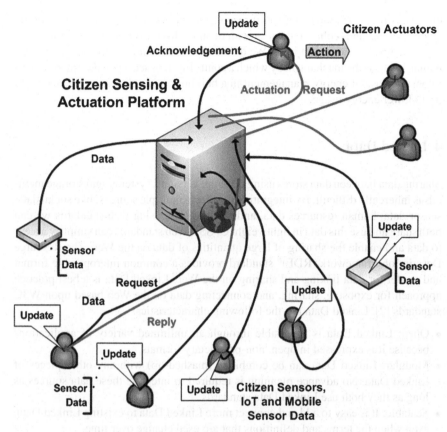

Fig. 3 Citizen actuators and citizen sensors within a cyber-physical social system

from feedback incorporated into an electricity bill to electronic feedback through a web interface or a smart meter. In designing our experiment discussed in Sect. 5 we examined the literature on feedback as our experiment can be seen as a kind of active or directed feedback using social media as the delivery channel and smart meters and similar devices can be seen as passive or undirected feedback. Our feedback can be described as active or directed as all the participants were individually messaged through social media.

Figure 3 shows a Cyber-Physical Social System including how updates flow from citizen sensors, IoT objects, and mobile sensor data. These updates may contain sensor data, aggregated sensor data from multiple sensors or sensor stations, or citizen sensor updates with or without annotated sensor data attached. Updates are posted to a citizen sensing and actuation platform (evidence stage of the feedback loop) that processes or aggregates and displays them to the users for the information relay stage of the feedback loop. The user then can choose to complete an action or in the case of our specific system send an actuation request to the user. The system then waits for an acknowledgment; if received it returns to the evidence stage of the feedback loop, and replies to the user with thanks. This citizen actuation request and response flow is shown in Fig. 5. We define citizen actuation as the activation of a human being as the mechanism by which a control system acts upon the environment. In our use case of energy management in a building, we deploy citizen actuators to try to lower energy usage.

4 Linked Data

Sharing data between data stores using different versions, systems, and storage methods is inherently difficult. By integrating data from multiple sources, like social data, sensor data, human resources data, and building data using web standards we can both share and use this data in lightweight CPSSs. Web standards can simplify access to data and enable the sharing of large quantities of data on the Web. The Resource Description Framework (RDF)[8] standard provides a common interoperable format and model for data linking and sharing on the Web. Linked Data is a best practice approach for exposing, sharing, and connecting data on the Web based upon W3C standards [2]. Linked Data has the following characteristics:

- Open: Linked Data is accessible through an unlimited variety of applications because it is expressed in open, non-proprietary formats.
- Modular: Linked Data can be combined (mashed-up) with any other pieces of Linked Data. No advance planning is required to integrate these data sources as long as they both use Linked Data standards.
- Scalable: It is easy to add and connect more Linked Data to existing Linked Data, even when the terms and definitions that are used change over time.

We propose that RDF and Linked Data provide an appropriate technology platform to enable the sharing of cross-domain information relevant to the operation of a building. We propose that as we move data to the cloud, Linked Data technology offers a viable medium for the sharing and reuse of data across silos. Whether it is the integration of multiple energy management systems, billing systems, building management systems, or spreadsheets, Linked Data offers a method of exposing, sharing, and connecting data in a manner that is reusable and not a one-off integration solution. Linked Data's characteristics that enable this sharing of cross domain data is derived from Linked Data publishing practices [1]:

- Use URIs as names for things: the use of Uniform Resource Identifiers (URI) (similar to URLs) to identify things such as a person, a place, a product, a organization, a building, a room, an event or even concepts such as risk exposure or net profit, simplifies reuse and integration of data.
- Use HTTP URIs so that people can look up those names: URIs are used to retrieve data about objects using standard web protocols. For a person this could be their

---

[8] http://www.w3.org/RDF/

organization and job classification, for an event this may be its location, time, and attendance, for a product this may be its specification, availability, price, etc.

- When someone looks up a URI, provide useful information using the standards: when someone looks up (dereferences) a URI to retrieve data, they are provided with information using a standardized format. Ideally in Semantic Web standards such as RDF.
- Including links to other relevant URIs so that people can discover more things: retrieved data may link to other data sources, thus creating a data network e.g. data about a product may link to all the components it is made of, which may link to their supplier.

This integration of data from different sources allows integration of social data, IoT data, and building management data to be easily reused in building energy management. This linking of social data and sensor data has been proposed by Breslin et al. previously [4] and can be achieved by linking lightweight social data models like Semantically Interlinked Online Communities (SIOC) [5], Friend of a Friend (FOAF) [6], and Semantic Sensor Network (SSN) ontology [10].

5 Use Case

In order to study and visualise the effect of citizen actuators we chose to set up an experiment in the Digital Enterprise Research Institute.[9] (DERI) part of National University of Ireland, Galway[10] DERI is a research institute with about 130 members divided into research streams, research units, and administrative staff. DERI consists of about 20 organisational units. Generally, unit members are co-located in offices, wings, or floors. For this experiment, we selected one area, the North wing located on the First Floor as highlighted in Fig. 4 to monitor energy usage patterns and to build a model of energy usage over time. This wing of the building was selected as it contains two smaller meeting rooms and one larger conference room and these rooms are principally used in normal business hours (9am to 6pm). In our experiment, we examined the data from 6pm to 9pm, as this would allow us to track energy usage out of office hours and model energy usage to detect abnormal usage.

Individuals' seating location and unit membership details are stored in a graph database using RDF, a standard model for data interchange on the Web [19]. This information is stored as per Linked Data principles [1] and is stored with other data relevant to the Linked Energy Intelligence Dataspace [18] and Sustainable DERI Platform [17]. Using this Linked Data representation of members' seating location and the booking schedule for the meeting rooms and conference room, we can both analyse when the meeting rooms are not in use and which individual is normally in close proximity. By modelling energy usage and data on room usage, abnormal usage outside of times when meetings are not scheduled can be monitored. A Complex

---

[9] http://www.deri.ie/

[10] http://www.nuigalway.ie/

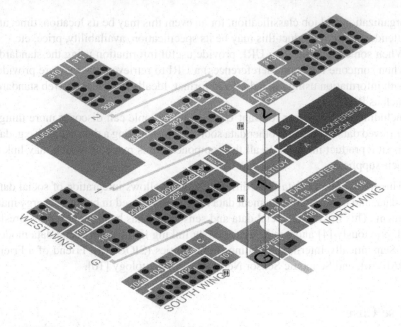

Fig. 4 DERI floorplan—meeting rooms highlighted in *red*

Event Processing (CEP) system can then be used to process real-time information streams (energy usage sensors) to detect situations of interest (abnormal energy usage) with the use of event patterns. When a pattern is matched, a new higher-level event is generated and can be utilised in further processing, be visualised in a dashboard or sent to other processing tools. By utilising, the Semantic Sensor Network ontology [10] to represent the sensors and sensor readings, the readings can be implemented with Linked Data principles in mind and this information can be used by a CEP engine.

For this experiment, a CEP engine as described in [26] detected abnormal energy usage from sensor data as set by a threshold from modelling energy usage over time. This combined with temporal data and information from the room management system allows the system to process the data and create a higher level event, in this case the event can be described as abnormal energy use for both the time of day and room status (booked for a meeting or not). This event then sends a Twitter message to an appropriate user to request the user to check on the issue. This is what we have called the actuation request. Twitter was used due to its available API and relevant heavy usage within the institute. Other platforms were investigated for usage but none had the same usage or availability on devices. Twitter is also highly suited to this experiment because of its ease of posting and lack of barriers to entry. The request took this form:

> Hi @username could you check if lights/air-con were left on in the 1st Floor North Wing please? And turn them off if possible. Thanks

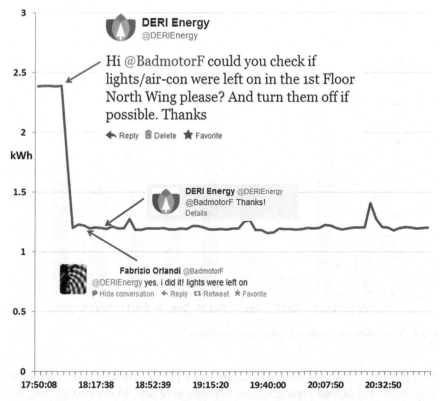

Fig. 5 Actuation request–request–ackowledgment and task completed–Thanks

The CEP engine then waits for a response from the user and if the user completes the request and replies to the system then the system checks the energy usage and replies to the user with thanks. Figure 5 displays an example of an actuation request, acknowledgment from a user, and the energy usage.

We designed an experiment that included citizen actuators in a Cyber-Physical Social System to test our hypothesis that citizen actuation could help to lower energy usage. For our experiment we collected data over a thirty two week period from November 2012 to August 2013. A control period of thirty-two weeks without any abnormal interference was chosen. This period was chosen as it included the start/end of months to try to cancel any abnormal usage events out; events like project meetings, proposal deadlines, or end of financial reporting periods when meeting rooms would be more heavily used. Weekend data was removed from the experiment, as it would be have been impossible to reconcile this data with data that included actuation requests because at weekends most (if not all) actuation requests would not be completed, as the users would not be on site. Fifteen volunteers were selected for the experiment and for each request; one volunteer was chosen at random to receive the request.

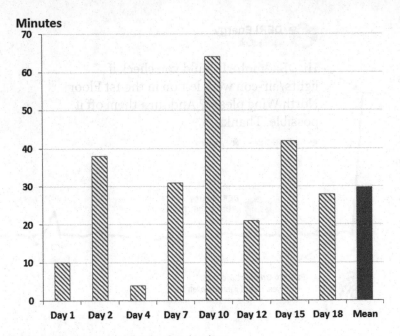

Fig. 6 Time taken for actuation to take place in minutes

Overall, seven volunteers were sent requests over the four-week period used for the experiment and in total eight actuation requests were sent. In the next section, we will display and discuss the results from the experiment.

6 Results

The experiment ran from Monday to Friday over a four-week period (twenty working days excluding weekends). During the twenty-day experimental period, eight actuation requests were sent to seven randomly chosen volunteers and in each case, the volunteer completed the requested action. Figure 6 shows the time to complete the actuation request but does not show data for the days when no actuation request was sent. Actuation requests were sent on eight days and actuation was completed on each of those eight days. These days when actuation requests were sent are marked with a* in Table 1, this table also displays the average energy usage in kilowatt hours[11] (kWh).

The time taken for the actuation to take place varies greatly from a minimum of five minutes to a maximum of sixty-four minutes. From this data, days with actuation have higher averages than days without actuation, this is due to the fact that days

---

[11] http://en.wikipedia.org/wiki/Kilowatt_hour

Table 1 Average kWh by day energy usage

Day	Day 1*	Day 2*	Day 3	Day 4*	Day 5	Day 6	Day 7*	Day 8	Day 9	Day 10*
kWh	1.5359	1.5727	1.2344	1.5984	1.4041	1.2059	1.6658	1.2031	1.3067	1.6484

Day	Day 12*	Day 13	Day 14	Day 15*	Day 16	Day 17	Day 18*	Day 19	Day 20	Avg Day
kWh	1.5901	1.2934	1.2540	1.5731	1.2854	1.4013	1.6251	1.4393	1.4205	1.93

*Denotes a day where Actuation took place. Avg Day is the daily average energy usage of the control period

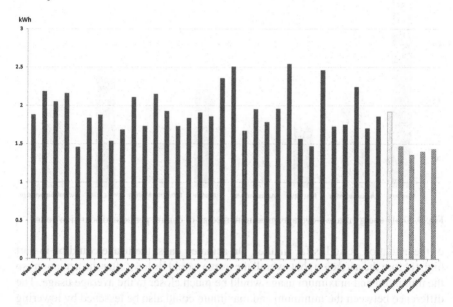

Fig. 7 Average weekly energy usage

with actuation have some time period with higher energy usage than days without actuation (days where everything has been turned off). However, when compared to the average of the control period these days have lower energy consumption. The average time between request and acknowledgment was just under thirty minutes (29.75) as shown in Fig. 6. In this experiment, re-routing of the request was not implemented as a decision was made to only examine the data and chance of success when one request was issued. The success rate was 100 % in this experiment when only one request was sent. This will be discussed further in Sect. 7.

Figure 7 shows the average energy usage (kWh) over each of the thirty-two weeks of the control period, the average of those thirty-two weeks and the average energy usage (kWh) in the four weeks with actuation. The actuation weeks' energy usage is lower than all but one of the control weeks (week 5) and the actuation weeks' average usage is 0.503 kWh lower than the average of the thirty-two control weeks. Figure 8 displays the greater variance between the minimum and maximum energy usage during the control period, while the variance in the minimum and maximum

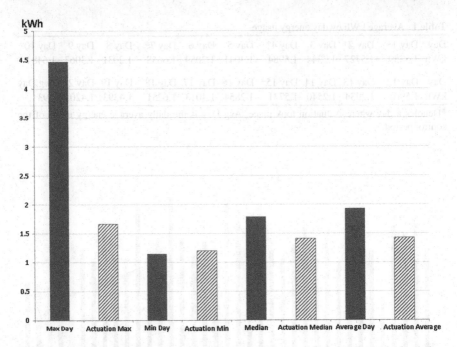

Fig. 8 Daily energy usage—average, max, min, median of control period and actuation period

of the actuation period is far smaller. On actuated days, the usage could be in part high (before the actuation) but overall the usage would be low (after the actuation) so the minimum and maximum usage would be much closer to the average usage. The difference between the minimum and maximum could also be lessened by lowering the time between the actuation request and the person completing the action.

Figure 9 shows energy usage of three days, the dotted line represents a sample day taken when every device was turned off manually and checked periodically to get a low baseline for energy usage (this was done before the control period). The dashed line shows the average daily usage from the control period. The solid line shows the actuation day graph from Fig. 5 that shows the energy usage before and after the actuation took place.

Overall, the results show that the energy usage on average declined compared to the control weeks during the weeks that had active participation (experimental weeks) from users which were the weeks these users received actuation requests and completed the actions of turning off electrical components. This saving on average was 0.503 kWh which when compared to the average energy used in the control weeks of 1.93 kWh; this equates to a decrease of energy usage by 26 %. Each actuation week's energy usage was equal to or lower in value to the lowest control week apart from one week (which compared to the other control weeks is considerably lower). This is quite a large drop in energy usage during a time when energy usage should be generally lower as the rooms are not in use. In the next section we will discuss these results and then we will discuss related future work.

kWh

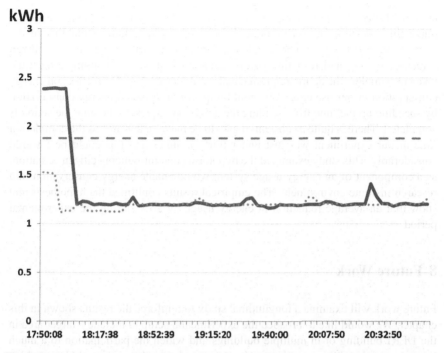

Fig. 9 Energy usage comparison—*dashed line* is average energy usage from control period, the *solid line* is energy usage with actuation, and the *dotted line* is a day when devices were turned off manually

7 Conclusion

Our initial hypothesis was that enabling citizen actuators to interact with their environment in an energy management use-case would help lower power usage and lower costs. This hypothesis would be supported if the energy usage data from when citizen actuators were enabled would decrease significantly from the control period. By lowering the average energy usage by 0.503 kWh the hypothesis is confirmed. A 26 % drop in energy usage is quite a considerable reduction and if transposed to financial savings over a year the benefit would be significant. Further research would be needed to show if this is possible in our experimental setup on an institute level or in a larger enterprise environment. Furthermore, other factors that could have been integrated into this experiment, like better education for users or signage encouraging people to turn off electrical apparatus, which has proved successful in other social studies [31].

In this experiment, the actuation request success rate is 100 % which may or not always be achievable as people in an enterprise could be travelling/on holidays or not checking their social media accounts. This is why routing and rerouting of requests is an important aspect of future work in this area. Deciding which person is the most likely (according to some optimisation criteria) to complete the request is an interesting question. In an enterprise this criteria could include availability of the person accessed from an internal calendar system. The selection process of citizen actuator, routing, and rerouting of the actuation request could improve the time taken to complete the action requested and would have a beneficial effect on energy conservation in our use case. This could improve the gains in energy conservation by speeding up the time from sending the actuation request to the time the action is completed. There is quite a substantial gain to be made in this area as the completion time in our experiment was just under thirty minutes and this could be lowered considerably. This study examined the hypothesis that introducing citizen actuation, as a component of an energy usage system would enable energy conservation in a research institute environment. The empirical results confirmed the hypothesis and show that an average reduction in energy usage by 26 % during the experimental period.

8 Future Work

Future work will examine a longitudinal study to reinforce the results shown in this chapter. We would like to broaden the experimental setup to include a larger area in the DERI building or in multiple buildings and widen the participation to a much larger group. In addition to using energy usage sensors the goal would be to widen the scope of the research with the inclusion of other sensors like motion, light, or heat would also create a clearer picture of energy usage and occupancy. Welsh [39] describes issues with sensor networks for scientific experiment especially in creating dense networks especially in built up areas or pre-built buildings so in addition to stationary sensors the presence of mobile device sensors could be utilised to improve the data gained drawing on previous work in [15]. The integration of a reporting system to allow building occupants to post about issues in their environment is also being examined for furture work. In separate but related work we have looked at implementing game elements in non-game applications (called Gamification) [13], this could also be examined to see if this can engage users in longer studies and improve gains. Routing of requests to users that best fit to the requirement of the task is another area of future work—where the initial user to receive the actuation request is chosen by examining multiple selection criteria and if this user does not complete the action then the task will be rerouted to the next most suitable candidate. The choosing of the best candidate might also include other features such as data collected on personal fitness tracking devices/services like Fitbit[12], Nike+[13] or mobile phone applications. Fitness and wellness of employees is a concern for enterprise and a person's step count for that day could be taken into account as one of the criteria for choosing the best fit for the task (e.g. the person with the lowest step count).

---

[12] http://www.fitbit.com/
[13] http://nikeplus.nike.com/plus/

Furthermore, other interesting avenues of research would be creating formal models for Citizen Actuation and generalising the concept to other areas. This could lead to further examination of the cost of Smart Environments with actuation versus the cost of implementing CPSS to fulfil the needed actuation tasks. With generalisation of Citizen Actuation, the definition of Smart Environments could be redefined to include the symbiotic relationship between man and machine. As the Internet of Things is expanding to become the Internet of Everything (IoE), the human element has become included in the design of systems and as Cicso describe the IoE as the bringing together of people, process, data, and things to make networked connections more relevant and valuable.[14] It is through taking advantage of both data from the Social Web and combining this with sensor data that the next generation of applications can be developed and researched.

References

1. Berners-Lee, T.: Linked Data-Design Issues. http://www.w3.org/DesignIssues/LinkedData.html (2006)
2. Bizer, C., Heath, T., Berners-Lee, T.: Linked data—the story so far. Int. J. Semant. Web Inf. Syst. **5**, 1–22 (2009). doi:10.4018/jswis.2009081901
3. Boyd, D., Ellison, N.B.: Social network sites: definition, history, and scholarship. J. Comput. Mediated Commun. **13**(1), 210–230 (2007)
4. Breslin, J., Decker, S., Hauswirth, M., Hynes, G., Le-Phuoc, D., Passant, A., Polleres, A., Rabsch, C., Reynolds, V.: Integrating social networks and sensor networks. In: W3C Workshop on the Future of Social Networking. Barcelona (2009)
5. Breslin, J.G., Harth, A., Bojars, U., Decker, S.: Towards semantically-interlinked online communities. In: The Semantic Web: Research and Applications, pp. 500–514. Springer, Berlin (2005)
6. Brickley, D., Miller, L.: FOAF vocabulary specification 0.98. Namespace document. http://xmlns.com/foaf/spec/ (2010)
7. Bull, R., Irvine, K.N., Fleming, P., Rieser, M.: Are people the problem or the solution? A critical look at the rise of the smart/intelligent building and the role of ict enabled engagement. In: European Council for an Energy Efficient Economy, ECEEE. ECEEE, Toulon/Hyres, France (2013)
8. Campbell, A.T., Eisenman, S.B., Lane, N.D., Miluzzo, E., Peterson, R.A., Lu, H.L.H., Zheng, X.Z.X., Musolesi, M., Fodor, K., Ahn, G.S.A.G.S.: The rise of people-centric sensing. IEEE Internet Comput. **12**(4), 12–21 (2008). doi:10.1109/MIC.2008.90
9. Caragliu, A., Del Bo, C., Nijkamp, P.: Smart cities in Europe. J. Urban Technol. **18**(2), 65–82 (2011)
10. Compton, M., Barnaghi, P., Bermudez, L., García-Castro, R., Corcho, O., Cox, S., Graybeal, J., Hauswirth, M., Henson, C., Herzog, A., Huang, V., Janowicz, K., David Kelsey, W., Le Phuoc, D., Lefort, L., Leggieri, M., Neuhaus, H., Nikolov, A., Page, K., Passant, A., Sheth, A., Taylor, K.: The SSN Ontology of the W3C semantic sensor network incubator group. Web Semant. Sci. Serv. Agents World Wide Web **17**, 25–32 (2012). doi:10.1016/j.websem.2012.05.003
11. Cook, D., Das, S.: Smart Environments: Technology, Protocols and Applications. Wiley, London (2004)

---

[14] http://www.cisco.com/web/about/ac79/innov/IoE.html

12. Costanza, E., Ramchurn, S.D., Jennings, N.R.: Understanding domestic energy consumption through interactive visualisation: a field study. In: Proceedings of the 2012 ACM Conference on Ubiquitous Computing, pp. 216–225. ACM, Pittsburgh, PA (2012)

13. Crowley, D.N., Corcoran, P., Young, K., Breslin, J.: Gamification of citizen sensing through mobile social reporting. In: IEEE International Games Innovation Conference 2012 (IGIC), pp. 1–5. Rochester, NY (2012)

14. Crowley, D.N., Curry, E., Breslin, J.G.: Closing the loop-from citizen sensing to citizen actuation. In: 7th IEEE International on Digital Ecosystems and Technologies. Menlo Park, CA (2013)

15. Crowley, D.N., Passant, A., Breslin, J.G.: Short paper: annotating microblog posts with sensor data for emergency reporting applications. In: 4th International Workshop on Semantic Sensor Networks 2011 (SSN11), pp. 84–89. Bonn, Germany (2011)

16. Curry, E.: System of systems information interoperability using a linked dataspace. In: IEEE 7th International Conference on System of Systems Engineering (SOSE 2012), pp. 101–106. Genova, Italy (2012)

17. Curry, E., Hasan, S., ul Hassan, U., Herstand, M., O'Riain, S.: An entity-centric approach to green information systems. In: Proceedings of the 19th European Conference on Information Systems (ECIS 2011). Helsinki, Finland (2011)

18. Curry, E., Hasan, S., O'Riain, S.: Enterprise energy management using a linked dataspace for energy intelligence. In: Second IFIP Conference on Sustainable Internet and ICT for Sustainability. Pisa, Italy (2012)

19. Curry, E., ODonnell, J., Corry, E., Hasan, S., Keane, M., ORiain, S.: Linking building data in the cloud: integrating cross-domain building data using linked data. Adv. Eng. Inform. **27**(2), 206–219 (2013)

20. Demirbas, M., Bayir, M.A., Akcora, C.G., Yilmaz, Y.S., Ferhatosmanoglu, H.: Crowd-sourced sensing and collaboration using twitter. In: IEEE International Symposium on a World of Wireless Mobile and Multimedia Networks (WoWMoM), pp. 1–9. IEEE, Sydney, Australia (2010)

21. Engelbart, D.C.: Toward augmenting the human intellect and boosting our collective IQ. Commun. ACM **38**(8), 30–32 (1995). doi:10.1145/208344.208352

22. Engelbart, D.C.: Augment, bootstrap communities, the Web: what next? In: CHI 98 Cconference Summary on Human Factors in Computing Systems, CHI '98, pp. 15–16. ACM, New York, NY, USA. doi:10.1145/286498.286506 (1998)

23. Fischer, C.: Feedback on household electricity consumption: a tool for saving energy? Energ. Effi. **1**(1), 79–104 (2008)

24. Goetz, T.: Harnessing the power of feedback loops. http://www.wired.com/magazine/2011/06/ff_feedbackloop/all/1 (2011)

25. Goodchild, M.: Citizens as sensors: the world of volunteered geography. GeoJournal **69**(4), 1–15 (2007)

26. Hasan, S., Curry, E., Banduk, M., O'Riain, S.: Toward situation awareness for the semantic sensor Web: complex event processing with dynamic linked data enrichment. In: 4th International Workshop on Semantic Sensor Networks 2011 (SSN11), pp. 69–81. Bonn, Germany (2011)

27. Herring, M.: A sea of sensors. http://www.economist.com/node/17388356 (2010)

28. Jamshidi, M.: System-of-systems engineering-a definition. In: IEEE International Conference on Systems, Man, and Cybernetics (SMC). Hawaii, USA (2005)

29. Boulos Kamel, M.N., Resch, B., Crowley, D.N., Breslin, J.G., Sohn, G., Burtner, R., Pike, W.A., Jezierski, E., Chuang, K.Y.S.: Crowdsourcing, citizen sensing and sensor seb technologies for public and environmental health surveillance and crisis management: trends, OGC standards and application examples. Int. J. Health Geographics **10**(1), 67 (2011)

30. Kwak, H., Lee, C.: What is Twitter, a social network or a news media? In: Proceedings of the 19th International Conference on World Wide Web, WWW '10, pp. 591–600. New York, NY (2010)

31. Lindenberg, S., Steg, L.: Normative, gain and hedonic goal frames guiding environmental behavior. J. Soc. Issues **63**(1), 117–137 (2007). doi:10.1111/j.1540-4560.2007.00499.x
32. Parsons, D.J., Chatterton, J.C., Clements-Croome, D., Elmualim, A., Darby, H., Yearly, T., Davies, G.: Carbon brainprint case study: intelligent buildings. Techincal report, Centre for Environmental Risks and Futures. Cranfield University (2011)
33. Sakaki, T., Okazaki, M., Matsuo, Y.: Earthquake shakes twitter users: real-time event detection by social sensors. In: Proceedings of the 19th International Conference on World Wide Web. WWW'10, pp. 851–860. ACM, New York, NY (2010)
34. Shaw, D.: Smart, social energy: can software change our energy habits? http://www.bbc.co.uk/news/technology-20173641
35. Sheth, A.: Citizen sensing, social signals, and enriching human experience. IEEE Internet Comput. **13**(4), 87–92 (2009)
36. Snoonian, D.: Smart buildings. IEEE Spectr. **40**(8), 18–23 (2003)
37. Sullivan, B., Wood, C., Iliff, M., Bonney, R., Fink, D., Kelling, S.: eBird: a citizen-based bird observation network in the biological sciences. Biol. Conserv. **142**(10), 2282–2292 (2009)
38. Weiser, M., Gold, R., Brown, J.: The origins of ubiquitous computing research at PARC in the late 1980s. IBM Syst. J. **38**(4), 693–696 (1999)
39. Welsh, M.: Sensor networks for the sciences. Commun. ACM **53**(11), 36 (2010). doi:10.1145/1839676.1839690
40. Zengestrom, J.: Why some social network services work and others don't—or: the case for object-centered sociality. http://www.zengestrom.com/blog/2005/04/why-some-social-network-services-work-and-others-dont-or-the-case-for-object-centered-sociality.html (2005)

Four-Layer Architecture for Product Traceability in Logistic Applications

Jose Ivan San Jose, Roberto Zangroniz, Juan Jose de Dios
and Jose Manuel Pastor

Abstract In this chapter, we describe our work on the design of an auto-managed system for the tracking and location of products in transportation routes, called Transportation Monitoring System (TMSystem). Manufacturers, retailers and customers require tracking of goods in production and distribution lines. Automatic Vehicle Location (AVL) Systems are being introduced in many cities around the world. These systems are aimed for cost reduction purposes, and also provide optimization of time and resources. Companies usually control the quality of their production during the manufacturing phase, but products can also be controlled along the distribution and transportation phases, before they are delivered to customers. When controlling all the phases, including also the location of the transport, a large amount of data (Big Data) will be generated, and should be processed in order to get useful information, and so all the resources in the company are optimized. A four-layer system is proposed in order to provide an efficient solution for the Real-Time Monitoring (RTM) of goods. Two Web Services are proposed, Location Web Service and Check Order Web Service, so that customers can easily access information about the shipment of their orders. Finally, a Web Application is developed to access those Web Services.

J. I. San Jose (✉) · R. Zangroniz · J. J. de Dios · J. M. Pastor
Institute of Audiovisual Technologies (http://itav.uclm.es), University of Castilla-La Mancha,
Cuenca, Spain
e-mail: JoseIvan.SanJose@uclm.es

R. Zangroniz
e-mail: Roberto.Zangroniz@uclm.es

J. J. de Dios
e-mail: JuanJose.deDios@uclm.es

J. M. Pastor
e-mail: JoseManuel.Pastor@uclm.es

N. Bessis and C. Dobre (eds.), *Big Data and Internet of Things:*   401
A Roadmap for Smart Environments, Studies in Computational Intelligence 546,
DOI: 10.1007/978-3-319-05029-4_17, © Springer International Publishing Switzerland 2014

1 Introduction

1.1 Background

In the dynamic and ever changing world we are living in, companies, which are able to deliver products in a faster and better way, will be more likely to success. Tracking of each unit of product should be performed during the manufacturing and distribution phases, and the exact location and the environmental conditions of each product can be determined at any time.

This is a critical issue, in particular, when the process involves handling of special loads. The specific properties of this type of products, such as fragile or perishable goods and whose quality needs to be preserved, mostly requires that the productive process is carried out in a location where all the units can be completely controlled.

Automatic Vehicle Location (AVL) [1] provides real-time location information for any mobile assets upon which it is installed. AVL is used for different tracking purposes, especially for those related to tracking one vehicle or a fleet of vehicles. Tracking system technology [2] was made possible by the integration of several navigational technologies, as Global Positioning System (GPS) [3], Geographic Information System (GIS) [4] and General Packet Radio Service (GPRS) [5].

The traceability and the tracking of products can be implemented in an economic way by means of RFID (Radio Frequency Identification) technology [6, 7]. It can benefit consumers through improved product availability, speed of service, and quality insurance. The application of RFID technology [8] helps businesses improve supply chain efficiency, generating a high added value by improving the distribution process and optimizing the available resources.

The Electronic Product Code (EPC) [9] is a unique identification code that is generally thought of as the next generation of the standard bar code [10]. It is stored on an RFID tag and assigns a unique global number to each product. Therefore EPC can be associated to specific product information, such as manufacturing date, origin and destination.

By means of this technology, a major reduction in the time required for processing of orders is achieved, and the errors in the ordering system are also minimized. Products are received from the manufacturers, for example, in pallets or containers and in the picking tasks non-homogeneous loads are obtained by combining different products as required [11].

This improvement in the traceability and tracking of goods is not only available for companies, but for the customers too. In this way, the customer is able to know the real-time status of its product after its delivery, as well as its geographical location and other parameters. Besides, it is easier to allocate the responsibilities when they are damaged or altered during the distribution phases, when real-time monitoring of products is provided to the customer.

The traceability of any kind of product is monitored through any Smart Object [12, 13] using sensors and it is located on real time on a map with a GPS device.

These devices generate a large amount of data, which will be stored in a database and will be filtered in order to obtain the useful information and discard un-useful data.

The technologies and architectures known as Big Data were developed to store, process and manage big volumes of data as quick as possible, or even in real time. Big Data will be considered a complete system for working with large amounts of data and combine them with existing databases.

Companies usually control the quality of their production during the manufacturing phase, but products can also be controlled along the distribution and transportation phases, before they are delivered to customers. When controlling all the phases, including also the control of the location of the transport and the information generated by the sensors of the smart object in our system, a large amount of data (Big Data) will be generated, and should be processed in order to filter the useful information.

All the data collected both from the sensors and the data sent by the GPS device, the transportation can be faster and more efficient, and the products will be completely controlled till they are delivered. In order to transmit all the data, a smart object has been designed for sending and storing data during the whole process of transportation. This information can be queried at any time through a Web page or via an app installed on any mobile device.

1.2 Motivations

Most companies need their products to be monitored during production, storage and distribution phases. Also, real-time fleet location is needed for many companies. Sometimes, this monitoring does not only include its location and real-time working, but also the capability to get information about the environmental conditions or physical situation of the products.

Customers can also use the system to collect some relevant information about the products they buy, and check if they have correctly received their orders before opening them.

Our goal is to implement an RFID-based development deployed at four different layers, which has been called Transportation Monitoring System (TMSystem). An RFID system which is able to support passive RFID tags and Wireless Sensor Network (WSN) nodes, and managed by a middleware capable of properly processing all the generated data used for tracking each unit of product from its manufacturing until its final location, including transportation and also installation.

We will also provide a solution to two well-known problems in the logistics chain by implementing the two Web Services that we are proposed: first, the real-time location of the transport and second, the checking of products at any time.

1.3 Technologies

RFID is one of the most innovative technologies and promises important benefits to customers and businesses in object location and identification. RFID is a generic term that is used to describe a system that transmits the identity (unique serial number) of any kind of object or person wirelessly using radio waves [14]. To capture these data, active and passive tags can be used.

RFID passive tags are used as they are a good solution for tracking products along the distribution process and they are used in the first layer of our proposed architecture.

RFID passive tags do not need maintenance and they are more widespread than their active counterparts, so the required handling equipment is generally available in buildings, such as warehouses. However, it is possible to use RFID active tags or specific applications, such as sensors, if required.

A WSN [15] provides RFID devices with important applications such as remote environmental monitoring and target tracking. The sensors are equipped with wireless interfaces with which they can use to communicate with each other to build a whole network. Usually, a WSN consists of several sensor nodes working together for monitoring a region to obtain data about the environment.

ZigBee [16] is a standard that defines a set of communication protocols for low data-rate short-range wireless networking. ZigBee-based wireless devices operate in 868, 915 MHz and 2, 4 GHz frequency bands, at a maximum data rate of 250 Kbps. This protocol is mainly targeted for battery-powered applications where low data rate, low cost, and long battery life are their main requirements. Compared to other network technologies [17], ZigBee has some features that make it a more suitable choice for sensor and control networks.

IEEE 802.15.4/ZigbeeWireless Network is used for the communication between different WSN nodes in the second layer of our proposed architecture. Sometimes, the ZigBee network cannot be directly connected to a computer. In these cases, the PAN coordinator is connected to a GPRS module that can communicate over big distances in real time. By using this module, it is also possible to get the all the GPS information, such as latitude, longitude and time, and this data are added to the original product information.

All the data captured by the sensors and the GPRS module and send by the transport vehicles, will be stored in a database every few seconds or even in real time. When the user needs that information, it will be processed, to filter the useful in formation, and it will be shown in the corresponding Web Service.

Web Services [18] are technologies that integrate a set of standards and protocols to exchange data between applications developed in different programming languages and they can run on any platform. We can use the Web Services to exchange data in both private computer networks and the Internet. Interoperability is achieved by open standards. Organizations such as OASIS and W3C are responsible for indicating the type of architecture and Web services regulation.

Web Services are loosely coupled software components that offer standardized interfaces based on mainly two languages: the Web Service Definition Language (WSDL) which is used to define the syntax of the interfaces, and SOAP which defines the format of messages that are exchanged when invoking services.

In the Future Internet [19, 20], real-world devices will be able to offer their functionality via SOAP-based Web Services (WS-*) [21, 22] or RESTful APIs [23, 24], enabling other components to interact dynamically. The functionality offered by these devices is often referred to as real-world services because they are provided by embedded systems that are related directly to the physical world.

The Internet of Things (IoT) [25] is a novel paradigm that is rapidly gaining ground among recent wireless telecommunications. The basic idea of this concept is widespread about a variety of things or objects, such as RFID tags, sensors, actuators, mobile phones, etc., which, through unique addressing schemes, are able to interact with each other and cooperating with their neighbors to achieve common goals.

The main force of the idea of IoT is the high impact that it will have on several aspects of everyday life and behavior of potential users. From the point of view of a private user, the most obvious effects of the introduction of IoT are visible in both working and domestic fields. Likewise, from the perspective of business users, the more obvious consequences will be also visible in fields such as, automation and industrial manufacturing, logistics, business/process management, intelligent transportation of people and goods.

Two Web Services have been implemented as applications of the possibilities offered by the proposed architecture and the use of Big Data. The first one is an application to locate the transport on the route map. The location of all the transports is available through a GIS.

OpenStreetMap (OSM) [26, 27] is used as GIS because it is a collaborative project, which aims to create free editable maps all over the world. Besides, it has an Open Database License (ODbL). OSM services are free for non-profit users, who can improve the maps and the information they include. OSM integrates all the data in a unique and editable database with information of all countries in the world.

The OpenStreetMap Foundation [28] is dedicated to encouraging the growth, development and distribution of free geospatial data and to providing geospatial data that anyone can use and share. It is an international non-profitable organization supporting, but not controlling, the OpenStreetMap Project.

To include several layers in our OpenStreetMap, as point of interest or traffic, for example, OpenLayers [29] is used. OpenLayers makes it easy to enclose a dynamic map into any Web Page. It can display map tiles and markers loaded from any source.

OpenLayers has been developed to further use geographic information of any kind. We decided to use it because it is completely free developed under Open Source JavaScript, released under the 2-clause BSD License (also known as the FreeBSD). OpenLayers API last stable version is 2.13.1 [30].

The second Web Service is intended for customers to check their orders in real time. So, customers only should query a web page in order to get this information that is stored in a database. In addition, they can check other features of their orders such

Table 1 Comparative of technology

	Location technology	Data transmission technology	Data storage location	Data processing location
London iBus	GPS	GPRS/LAN	AVL centre and vehicle	AVL centre and vehicle
Sydney PTIPS	GPS	GPRS	AVL centre and vehicle	AVL centre
Goo-Tracking	GPS	GPRS	AVL centre	AVL centre
Fleet/Convoy management	GPS	WLAN/WiMAX	AVL centre	AVL centre
RFID and WSN	RFID	WSN	AVL centre	AVL centre
A.V. Tracking	GPS	GSM/GPRS/RFID	AVL centre	AVL centre
TMSystem	GPS	GPRS/WLAN/WSN	AVL centre and vehicle	AVL centre

as temperature, humidity or any other kind of environmental condition depending on the type of product.

1.4 Related Works

At present, beyond the simple management of vehicles [31], which comprise a fleet, different ways of working are essential, related not only to the management of the vehicle itself, but going a step further, related to the goods carried within them. Therefore a new set of technologies is necessary to meet the needs that allow us to be able to offer services that enable monitoring, tracking and traceability of products contained inside the cargo area of vehicles that come into play to implement this new solution.

AVL Systems are being introduced increasingly in many cities around the world [32]. The objective is to improve the efficiency of the road-based goods and passenger transport systems. Satellite-based location and communication systems, particularly the GPS, have been the infrastructure needed for AVL systems. Also, we can find AVL Systems based on RFID technology.

On one hand, most of the AVL Systems are using GPS as the main location technology. London iBus [33] and Sydney PTIPS [34] are examples of public; Goo-Tracking [35] and VANET [36] for fleet control are systems that use GPS as location technology.

On the other hand, we can find AVL systems based on RFID, as systems that use RFID and WSN [37] or automatic vehicle tracking [38].

In a more visual way, we compared the different systems in a table. Table 1 shows a comparative between TMSystem, and the systems mentioned before. In this table, we compare the location technology, data transmission technology, data storage location and data processing location of each system.

Table 2 Comparative of applications

	Real-time/ Estimation location	App. for mobile devices	Show information on a map	Sensors
London iBus	Real time	No	No	No
Sydney PTIPS	Real time	No	No	No
Goo-Tracking	Real time	No	Yes	No
Fleet/Convoy management	Real time	Yes	Yes	No
RFID and WSN	Estimation	No	No	No
A.V. Tracking	Real time	No	Yes	No
TMSystem	Real time	Yes	Yes	Yes

Table 2 shows the different applications that have been developed in each system, as real time/estimation location, app for mobile devices, if the system shows the information on a map and if the system has sensors.

Apart from the systems previously mentioned, in a more professional environment, transport of goods, as SEUR, MRW, Nacex, FedEx, US Postal, etc. show little information about the products that they carry out. They only indicate date, time and place of departure from the origin and the intermediate points; and the date, time and point of delivery, too.

They do not show neither information from the rest of the transport of the products, nor information from sensors within the transport vehicles. For example, if the products are perishable or special loads that must be continuously monitored to make the process of traceability correctly, they do not show data of temperature and humidity sensors.

2 Proposed Architecture

The proposed architecture must control the location and environmental conditions of any product at any time. To achieve this goal, the system is distributed into four layers, as shown in Fig. 1.

The first layer is made by the products themselves and/or the boxes where they are. The second layer corresponds to the pallets/containers of the product boxes that can be formed by non-homogeneous loads. The third layer is made up of the transportation methods to move the pallets/containers from its origin to destination warehouses. And finally, the fourth layer consists of several Web Services where the final applications are implemented.

Fig. 1 Four-layer model

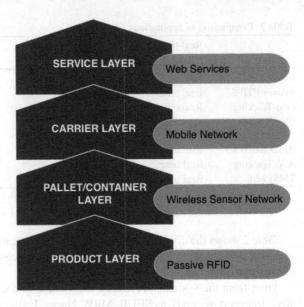

2.1 Product Layer

The passive tags attached to each unit of product, as shown in Fig. 2, constitute the first layer, and can communicate to a closer reader. No additional devices are required for these tags, as their only task is sending their own identification codes to the antenna. Furthermore, reading all the passive tags in a pallet usually require going through a reading arch with several antennas.

In the picking processes, each pallet can be made of different units of product, layers of the same product, or any other option for mixing the products in order to complete specific demands. So, the system collects all the data from the passive tags through the readers implemented in the first layer.

2.2 Pallet/Container Layer

WSN [39] nodes, as the one shown in Fig. 3, made up the second layer and contain the whole information related to the pallet. They communicate to each other creating a wireless network.

WSN nodes have a double mission: on one hand, they store the information regarding RFID tags within the pallet and, on the other hand, they are able to create a radio network that can reach a wider range space, like a warehouse, a ship, a truck, etc. [40].

Fig. 2 WSN node in a pallet

Fig. 3 RFID tags in products boxes

2.3 Carrier/Transport Layer

A Personal Area Network (PAN) coordinator is used to manage the information requested or sent by the nodes within the network. This coordinator is a special WSN node that incorporates a GPS (Global Positioning System) sensor and a GPRS (General Packet Radio Service) module. It is capable of sending the information in real-time, so that the status of the load is available in the logistics database.

Besides, its exact position is continuously monitored, so the tracking of the load on land, at sea, or even by air can be worldwide performed in this way.

2.4 Web Services Layer

The fourth layer is the web service layer. As application examples, two web services have been implemented.

The first Web Service that we implement consists of a service to locate the transport in real time by means of a GPRS module and OpenStreetMap using OpenStreetMap API v0.6 [41], which is responsible for displaying the GPRS module information on the Web Application. Through this Web Application, customers can access to all data, and they can know where their order is at any time in an accurately way. In addition, they can query information about the route, the estimated arrival time to the warehouse and the exact location on the map of the transport route.

On the other hand, by means of the second Web Service the customers can check their orders anytime. Furthermore, several RFID readers are used to collect all the data at the manufacturing warehouse and this information is stored in the WSN node of the second layer, so goods can be real-time monitored by the customers.

Automatically, when customers check their order, the Web Service reads the information from the WSN node and compares it to the ordered list of items, which is stored in the database. This information about the customers order will also be displayed on the Web Application. For example, if the customer orders 10 products and the RFID readers only read 8, the Web Application will detect which products have been received within the shipment and which not.

One of our goals is that the Web Applications can be correctly displayed on any device (computer, tablet or mobile phone) and by any web browser so the W3C standards [42] are fulfilled, as they are compatible with several web browsers. Consequently, customers can check their orders from any place, using any device connected to Internet.

A unique username and a personal password is provided to each customer, so he can only check his own order and access his own information, as this data is stored in the system database.

Fig. 4 Block diagram of nodes

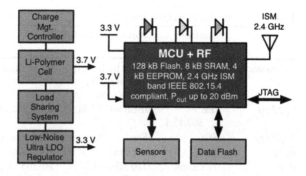

3 System Description

Once the theoretical model of the system has been presented, in this section we describe how the real system has been implemented and what components and protocols have been used in our system.

As previously mentioned, RFID passive tags are a good way for tracking products along the distribution process and they are used in the first layer of our model. RFID passive tags are cheap, can store up to 128 Kbytes of data, and one RFID reader is capable of reading several tags at the same time.

Besides, passive RFID tags do not need maintenance and are more widespread than its active counterparts, so warehouses, generally, possess the required handling equipment. However, it would be possible to use active RFID tags, if needed. WSN nodes are part of the second layer. These nodes contain information regarding all the products within pallet. Initially these data consist on the identification of the products. Additionally, the situation of the different layers of products within the pallets created during the picking can be added to the WSN node.

Other data provided by sensors can be integrated into the WSN node data, such as temperature or humidity, for example. Any problem in the picking process along the distribution chain will be recorded in the WSN node memory, so we can check this information using any device, like a computer, tablet or even a mobile phone.

Basically, a WSN node consists of a microcontroller, a flash memory, a radio transceiver, a power source and several sensors. Figure 4 shows the block diagram of a WSN node. Each WSN node is capable of monitoring several parameters by its enclosed sensors, and it stores this information in its flash memory. So, when it joins a network, the node sends these data to the network coordinator. A critical issue to be considered is the battery life of the WSN nodes.

To reduce the power consumption the WSN node will be asleep (low-power mode) most of the time, and it will only wake up to acquire the sensor data or to talk to another node.

Several sensors have been included in the WSN nodes to get information related to the environmental conditions of the pallet. In our first prototype, only temperature

Table 3 Comparative of wireless protocols

	Bluetooth	Wireless protocol UWB	Wi-Fi	ZigBee
Organization	Bluetooth SIG	UWB forum and and media alliance	Wi-Fi alliance	ZigBee alliance
Standard	IEE 802.15.1	IEEE 802.15.3a	IEEE 802.11 a/b/g/n	IEEE 802.15.4
Topology	Star	Star	Medium-dependent	A11
Range	10 m	4–20 m	10–100 m	10–300 m
Rate	723 Kbps	110 Mbps–1.6 Gbps	10–105 Mbps	250 Kbps
Comsumption	Low	Low	High	Very low
Nodes (max.)	8	128	32	65000

and humidity sensors were implemented. Afterwards, if necessary, more types of sensors may be added depending on the products and according to future needs.

In the third layer, pallets are grouped together and the WSN nodes form a wireless network. Table 3 shows a comparison of the major wireless protocols.

Apart from what is inferred from this table, it is important to point out the following features of the ZigBee protocol [43] from the point of view of the proposed system:

- The use of license-free frequency bands: 2.4 GHz or other of the sub-GHz regional bands.
- Specially designed for implementation of sensor, monitoring and control applications.
- Low complexity (low memory request).
- Low power (devices can be powered by batteries).
- Mesh networks (that feature is not found in the majority of wireless network standards):
  - Self-creation.
  - Self-reorganization.
  - Multi-hop routing protocol (Ad-hoc On-Demand Distance Vector).

Therefore, the ZigBee protocol stands out as the most reasonable choice to implement the required monitoring system. This system will be based on ad hoc, low demand and a low power-consumption network. ZigBee is a constantly evolving standard, with a broad support from the semiconductor industry that is being adopted for standardizations on wireless sensor networks.

The communication range of the WSN nodes depends on the environmental conditions. A ZigBee network reaches distances over 200 m among nodes (this distance is about 6 m in case of passive tags and RFID readers). This range can be greatly increased at the cost of reducing battery life. However, if it is necessary, a multi-hop network by means of routers can be established in order to communicate with the network manager (PAN coordinator).

Sometimes, the ZigBee network cannot be directly connected to a computer. In these cases, the PAN coordinator is connected to a GPRS module that can communicate over big distances in real time. By using this module, it is also possible to indicate the GPS coordinates where the load is located, so the traceability data are added to the products information.

Also, the WSN nodes are used as intermediate storage devices. Wherever there is no coverage, and the GPS/GPRS module is unable to send the location of a transport, the WSN node will send the GPS information when the GPRS network is available again.

The information provided by the GPS connected to the PAN coordinator can be used for locating the transport on an OpenStreetMap application using the OpenStreetMap API v0.6. The location information obtained through the GPRS module will be used in the fourth layer of our model.

All the data (Big Data) provided the GPS/GPRS module with the location information, and all the data obtained by the different sensors included in every WSN node, will be sent in real-time or every few seconds to the Server and to be stored in a MySQL database.

Once this information is stored in the database, when it is requested by the customer through one of the Web Services, these data will be processed and filtered in order to provide the requested information to the customer: location of the transport, real route, estimated time of arrival, estimated distance to the destination warehouse, information collected by the sensors at the time of the query, etc.

With all this data, we can also perform the traceability of any kind of product during its life cycle. Depending on the specific features of each product, for example, if there are perishable goods in the transport, we can query at any time its state of preservation, since it leaves the origin warehouse until it reaches the destination, and in the route between both places. If there is any event or alarm condition detected during their life cycle, it would be stored in the WSN node.

On the other hand, the delivery routes are stored in other databases. The storing of these routes will be used for optimizing the delivery of products in a more efficient way in future distributions, using shorter or faster routes, for example.

To query this information, we have developed a Web Application, with two Web Services, based on regular client-server architecture where the client requests through a web browser for services from the web server, as can be seen on Fig. 5. In our local network we use a database server that provides access, security and storage for all the data of our system. All the information provided by the GPRS module is stored in a database.

The regular client-server works as follows: the client requests through a web browser the services from the Web server. In our local network we have a database server that provides access, security and storage for all the data of our system. The information provided by the GSM/GPRS module and stored in a database, which will be accessed by the OpenStreetMap application, and thus it will be able to display all the information in the Web Application.

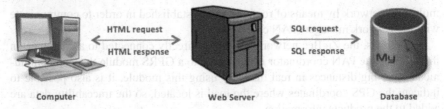

Fig. 5 Client-server architecture

4 Web Services Implemented

In this section, we describe in detail the two Web Services we have proposed. As previously mentioned, we have developed two Web Services in the fourth layer of our model. Afterwards, more Web Services can be added easily.

One of the goals is that customers can only query their own invoices and their own routes on the map, but not the orders or routes of other clients. Therefore, they have a profile in the system with several information, such as: name, surname, address, phone number, fax, Web Page, e-mail, etc.

For this reason and shared by the two Web Services we have implemented, each client has a unique username and a password to log into the Web Services system. With this information, the clients obtain privacy and security in their orders and in the routes in which the orders are shipped.

The username of each customer will be associated with their orders and their order routes in the database, as we can see in the Fig. 6. Additionally, more fields can be defined in each table and more tables can be added, if necessary.

Futhermore, we can denote that the system can be adapted to transport any kind of products: perishable, construction, computers, special loads, medicine, refrigerated, etc.

As mentioned, another of our main goals is the Web Application can be accessed from any compatible device and from any web browser. To achieve this objective, we are using HTML 5 and CSS3 standards and we will try to develop a compatible version with the most commonly used web browsers (Chrome, Firefox, Safari and Internet Explorer), also compatible with smart phones, tablets and computers (independently on their operative systems).

4.1 Location Web Service

The first Web Service is proposed for real-time location of transports and products on a map. In addition, the clients can query the information from the different sensors that are enclosed in the products they ordered.

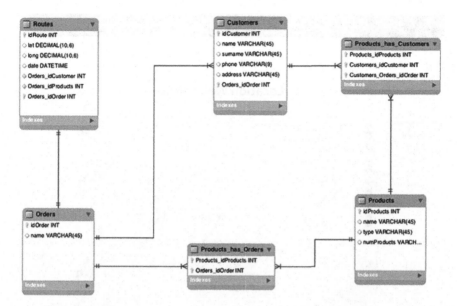

Fig. 6 MySQL diagram

To develop this first Web Service, the GPS data (latitude, altitude, longitude, time, etc.) is necessary, providing the system with the information that we need to locate the transport. The system processes this information to obtain the useful data, and calculate the estimated time left to arrive and the estimated distance to the destination warehouse. These data will be stored in a MySQL database [44]. In addition, the routes used by the products to be carried out, will be stored in another database for future implementation of new Web Services.

MySQL is used as the database for the management system, because it is easy to configure and is a widespread database. Besides, more fields can be added, if needed. In addition, we create tables for customers and orders.

Once the information is stored in the database, we are able to read this data in order to displaying it on the map. Previously, we need to do queries to the database using PHP [45] for this purpose. Thus, we can locate the transport in real time on the OpenStreetMap application. Besides, another type of information can be displayed through the markers or balloons available on OpenStreetMap.

OpenLayers has few layers that can be enabled or disabled by the user. In our OpenStreetMap application and through OpenLayers, the origin and destination warehouses are displayed in the Marker layer. Besides, we can see the Route layer and we can switch the map on the Map Base layer. Also, more layers can be added, if necessary.

An example of a route in the ITAV Web Services Application is shown in Fig. 7. We can see the origin and destination warehouses and the transport route. Therefore, by clicking on the markers or on the route, we can check several data about the order:

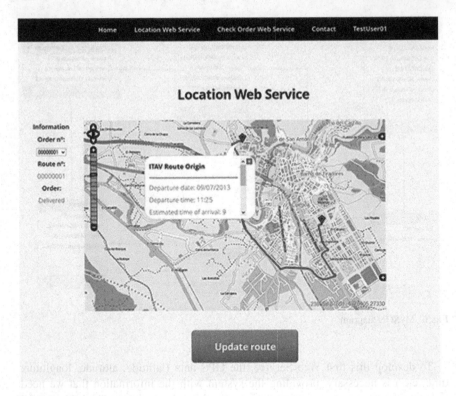

Fig. 7 Location web service

information provided by the sensors, date and time of departure of the order, the time left to arrive, etc.

One of the most important issues is that the complete route information in real time can be shown. So, the application automatically submits the query every few seconds (20 or 30 s, usually) to check for any changes in the database, and, if necessary, the route will be updated. A button has been also implemented in order to update the information when the user needs it.

This Web Service also shows the number of the current and previous orders, besides the number of the route used by the carrier in the different orders. Also we can check the delivery information about a specific order: the Web Service will indicate either if the order has left the origin warehouse or if the order has been delivered to the destination warehouse. The Web Service will not display orders that are still at the store.

Although the Web Service has been developed using the OpenStreetMap API, Google Maps [46, 47] is added as layer of our GIS. More layers, as traffic information for example, can be added if necessary for improving the efficiency of the logistics in the distribution of goods.

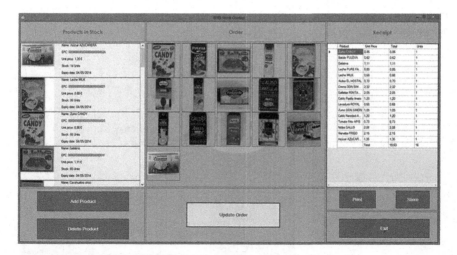

Fig. 8 RFID stock control application

4.2 Check Order Web Service

For the second Web Service, the information related to the customers' order and the data provided by the RFID readers when the order departs from the manufacturing warehouse is required. This information is also stored in the WSN node.

Therefore, this Web Service is splitted into two steps: the application for obtaining the data from the RFID readers and the information of the products ordered by the customer.

4.2.1 RFID Stock Control

First of all, the customer must place an order, which will be stored in the database. Secondly, this information will be compared to the data provided by the RFID readers. The order may be made in several ways: through a Web Page, by telephone, e-mail, etc. Figure 8 represents an example of a warehouse that supplies products to grocery stores and supermarkets.

The next step is the development of an application to collect the RFID readers' information. We have implemented a Visual C# application, RFID Stock Control, to get all the information provided by the RFID readers. By means of this software, the stock control of the warehouse is available, so the invoice can be printed, stored in the database, and also at the WSN node.

As seen on Fig. 8, the RFID stock control application is displayed on a window splitted into three sides: left, middle and right.

Fig. 9 RFID stock control: new product

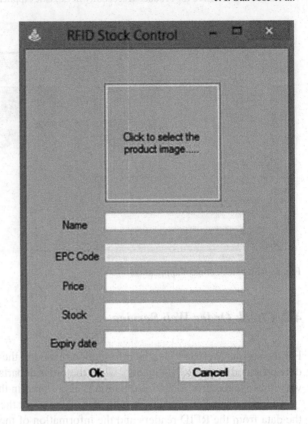

- **RFID Stock Control: Left side**

The left side of the application shows the available stock of all products that is also stored in the warehouse. In Fig. 9 we can see the application section that is used to add products to the stock. Also, the application allows delete products in the database.

The application can display more information. If the stock of any product is near to be finished, the application will show a message with the product and the number of units left in the warehouse. Similarly, if some products are near to their expiration date, the application will show a message with the products and their expiration date.

By default, these messages will be displayed with less than 10 units of any products or when the expiration date of products is less than 15 days. All the parameters are configurable and adaptable to any kind of products.

The products may show large amounts of information, as we can see on the Fig. 10, like: name, EPC code, price, stock and expiration date. Besides, more information can be added, depending on the product.

Name: Flan ROYAL

EPC: 0000000000000000000000000A

Unit price: 0,49€

Stock: 50 units

Expiry date:

Fig. 10 Product example

| | Home | Location Web Service | Check Order Web Service | Contact | TestUser01 |

Check Order Web Service

Customer Order n° 00000001 ▼   RFID Read Order

Product	Unit Price	Total	Units	Product	Unit Price	Total	Units
Zumo CANDY	0,86	0,86	1	Zumo CANDY	0,86	0,86	1
Batido PULEVA	0,62	0,62	1	Batido PULEVA	0,62	0,62	1
Gelatina	1,11	1,11	1	Gelatina	1,11	1,11	1
Leche PURE FAMILIAR	0,85	0,85	1	Leche PURE FAMILIAR	0,85	0,85	1
Leche MILK	0,68	0,68	1	Leche MILK	0,68	0,68	1
Alubia EL HOSTAL	0,70	0,70	1	Alubia EL HOSTAL	0,70	0,70	1
Crema DON SIMON	2,32	2,32	1	Crema DON SIMON	2,32	2,32	1
Galletas FONTANEDA	2,95	2,95	1	Galletas FONTANEDA	2,95	2,95	1
Caldo Paella Aneto	1,20	1,20	1	Caldo Paella Aneto	1,20	1,20	1
Levadura ROYAL	0,68	0,68	1	Levadura ROYAL	0,68	0,68	1
Zumo DON SIMON	1,05	1,05	1	Zumo DON SIMON	1,05	1,05	1
Caldo Navidad Aneto	1,20	1,20	1	Caldo Navidad Aneto	1,20	1,20	1
Tomate Frito APIS	0,73	0,73	1	Tomate Frito APIS	0,73	0,73	1
Nidos GALLO	2,00	2,00	1	Nidos GALLO	2,00	2,00	1
Total	16'95 €		14	Total	16,95 €		14

Order: OK

Update Order

Fig. 11 Check order web Service

- **RFID Stock Control: Middle side**

On the middle side of the window, the application displays the units detected by the RFID readers when all the products go through the antennas located in the RFID arch. The button Update Order can be used to reload the order in case any product was undetected.

- **RFID Stock Control: Right side**

The right side shows all the products in the invoice. Besides, it allows to print and to store the invoice in the MySQL database. When the order has been validated and stored in the database, the application automatically updates the stock of all products detected.

4.2.2 Web Service Description

Once the RFID Stock Control Application is described, we will introduce the Check Order Web Service, as we can see in Fig. 11. In this Web Service, the first and second level information is provided by the system.

The Web Service queries the database about the orders and compares the original (right side) in the warehouse to the real one (left side) ordered by the customer. If the order is correct, the Web Application will display a message on the screen indicating this condition. If it is not correct, the Application will display a list of the missing products. Also, the customer can check all the received orders.

In addition, the query can return more information related to the products, such as, for example: expiration date, order in which they are placed or any type of information that the customer needs about the units of product that he has requested.

5 Conclusions

Most companies need their products to be monitored during production, storage and distribution phases. Sometimes, this monitoring does not only include its location and real-time working, but also the capability to get information about the physical situation of the products.

Customers can also use the system to collect some relevant information about the products they buy, and check if they have correctly received their orders before opening them.

The proposed solution consists of four different layers of identification and wireless communications, with passive RFID systems, WSN nodes, ZigBee networks, GPS real-time location and web services, for improving the performance of extensive tracking for special loads. Besides, an absolute traceability and visibility, which includes real-time location of products, can be obtained during distribution or storage and the ordering processes can be effectively automated.

As we use RFID tags, we also solve the tracing and tracking problem of any kind of products and if necessary we will add different types of sensors to WSN nodes to check temperature, pressure, humidity, shock or any other parameter.

The proposed system, TMSystem, provides a good reliability and quality control, as it is even able to detect errors in any point of the chain. It will be implemented for the location of any type of products by means of passive tags.

Special data collection using WSN nodes and the capacity to monitor the whole supplying chain in a company for manufacturing any kind of products is also provided by the traceability system. By this system, we will try to decrease the number of errors that exists in the orders made by customers.

We will also try to solve two problems by implementing two Web Services that we are proposed: first, the location of the transport in real-time and second, the checking of products at any time.

One of our goals is that our Web Application can be correctly displayed from any device (computer, tablet or mobile phone) and using any web browser. Thus, we use HTML 5 and CSS 3. The Web Applications is correctly working for Mozilla Firefox, Google Chrome, Safari, and Internet Explorer at this time. Actually, we are working in a mobile version.

5.1 Discussion

In this section, we will perform a comparison among all the systems mentioned in this chapter and ours, indicating the advantages and disadvantages of the TMSystem.

The advantage of TMSystem is that you can use existing city infrastructure to transmit information via WLAN or WSN. When the user is not in the city or cannot transmit via these networks, the data can be stored and transmitted later on through GPRS, when possible.

Our system can display the location of the vehicles on a map in real time. This is also performed by Goo-Tracking and Fleet/Convoy Management System. Also, the data collected to the sensors that continuously control the temperature and humidity of the products that are being transported can be queried at any time.

As in the case of Convoy/Fleet Management System, we also have developed an application for mobile devices, both tablets and phones.

5.2 Future Work

As future work, we will develop more Web Services, both for customers and carriers, in order to reduce the delivery time, and, also, we will add other types of sensors for transport of any kind of goods.

Additionally, we will add sensors in the vehicles to get traffic information in real time and this information can be sent this to other vehicles in our fleet through WSN. Also, more layers in our system can be added.

We are now working with OpenStreetMap and OpenLayers for adding new layers to improve the transport, in order to do it in a quicker and more efficient way. For that, a traffic layer is added to the route map. Thus, the carriers will select, for example, the faster route with less traffic for each case.

The development of the app for mobile devices and for any operating system (iOS, Android, and Windows Phone) is also pending to be completed, though the Android app is currently in testing phase.

Our final intention is to develop an architecture and software tools in order to implement tracking and tracing applications in different fields: logistics, people tracking and Wireless Sensor Networks.

References

1. Burlarley-Hyland, A., Toothill, B., Das, A.: Use of automatic vehicle location systems, geographic information systems, and global positioning systems to increase efficiencies and decrease cost in maintenance management. Maintenance Management, pp. 46–54 (2012)
2. Aloquili, O., Elbanna, A., Al-Azizi, A.: Automatic vehicle location tracking system based on GIS environmentt. IET Softw. 3(4), 255–263. ISSN: 1751–8814 (2009)
3. Grewal, M., Weill, L., Andrews, A.: Global Positioning Systems, Inertial Navigation, and Integration. Wiley-Interscience, New York. ISBN: 978-0-470-04190-1 (2007)
4. Chang, K.: Introduction to Geographic Information Systems. McGraw-Hill. New Yotk. ISBN: 978-0-071- 26758-8 (2010)
5. Eberspächer, J., Vöagel, H., Bettstetter, C.: General Packet Radio Service (GPRS). GSM Switching, Services and Protocols, 2nd edn, pp. 241–269. Wiley Online Library, New York. ISSN: 1751–8814 (2001)
6. Glidden, R., Bockorick, C., Cooper, S., Diorio, C., Dressler, D., Gutnik, V., Hagen, C., Hara, D., Hass, T., Humes, T., Hyde, J., Oliver, R., Onen, O., Pesavento, A., Sundstrom, K., Thomas, M.: Design of ultra low cost UHF RFID tags for supply chain applications. IEEE Commun. Mag. 8, 140–151 (2004)
7. Ahson, S., Ilyas, M. (eds.): RFID Handbook. Applications, Technology, Security and Privacy. Ed. CRC Press. Boca Raton (2008)
8. Garcia, A., Chang, Y., Valverde, R.: Impact of new identification and tracking technologies in a distribution center. Comput. Ind. Eng., ScienceDirect 51(3), 542–552 (2006)
9. EPCGlobal, Inc.: EPCGlobal Frequently Asked Questions. http://bit.ly/153I10S (2007)
10. EPCGlobal Inc.: Electronic Product Code (EPC): An Overview. http://bit.ly/1fgdPix (2007)
11. Wong, C.Y., McFarlane, D.: Radio frequency identification data capture and its imact on shelf replenishment international. J. Logistics: Res. Appl. 10(1), 71–93 (2007)
12. Kortuem, G., Kawsar, F., Fitton, D., Sundramoorthy, V.: Smart objects as building blocks for the internet of things. IEEE Internet Comput. 14, ISSN: 1089–7801 (2010)
13. Vasseur, J.-P., Dunkels, A (eds.).: Interconnecting Smart Objects with IP: The Next Internet. Elsevier, Amsterdam. ISBN: 978-0-12-375165-2 (2010)
14. RFID Journal. What is RIFD? http://www.rfidjournal.com/article/view/1339 (2005)
15. Yick, J., Mukherjee, B., Ghosal, D.: Wireless sensor Network survey. Comput. Netw. (2008)
16. Sandell, F., Ghavibazou, M., Bolbat, M., Wei, L.: ZigBee wireless sensor networks: ideas for further development. Network 1(2), 3 (2011)
17. Farahani, S. (ed.): ZigBeeWireless Networks and Transceivers. Newnes. London. ISBN: 978-0-7506-8393-7 (2011)
18. Guinard, D., Trifa, V., Karnouskos, S., Spiess, P., Savio, D.: Interacting with the SOA based internet of things: discovery, query, selection, and on-demand provisioning of web services. IEEE Trans. Serv. Comput. 3(3), 223–235 (2010)
19. Guinard, D., Trifa, V., Thomas, P., Olivier, L: Towards physical mashups in the web of things. In: Proceedings of INSS 2009. IEEE Sixth International Conference on Networked Sensing Systems. Pittsburgh (USA) (2009)
20. Tan, L., Wang, N.: Future internet: the internet of things. In: 3rd International Conference on Advanced Computer Theory and Engieneering (ICACTE), vol. 5. pp. 376–380 (2010)
21. Snell, J., Tidwell, D., Kulchenko, P. (eds.): Programming Web Services with SOAP, 1st edn. O'Reilly & Associates Inc. ISBN: 0-596-00095-2 (2002)
22. AlShahwan, F., Moessner, K.: Providing SOAP web services and RESTful web services from mobile hosts. In: Fifth International Conference on Internet and Web Applications and Services (ICIW), pp. 174–179. ISBN: 978-1-4244-6728-0 (2010)
23. Allamaraju, S. (ed.): RESTful web services cookbook, 1st edn. O'Reilly Media, Inc. ISBN: 978-0-596-80168-7 (2010)
24. Zhao, H., Doshi, P.: Towards automated RESTful web service composition. In: IEEE International Conference on Web Services, pp. 189–196. ISBN: 978-0-7695-3709-2 (2009)

25. Aztori, L., Iera, A., Morabito, G.: The Internet of Things: A survey. Computer Networks. **54**(15), 2787–2805 (2010)
26. OpenStreetMap Project. http://www.openstreetmap.org
27. Bennett, J. (ed.): OpenStreetMap: Be your own cartographer. Packt Publishing, Birmingham. ISBN: 978-1-847197-50-4 (2010)
28. OpenStreetMap Fundation. http://wiki.osmfoundation.org/
29. OpenLayers: Free Maps for the Web. http://openlayers.org/
30. OpenLayers JavaScript Mapping Library. http://dev.openlayers.org/releases/OpenLayers-2. 13.1/doc/apidocs/files/OpenLayers-js.html
31. Castro, M., Jara, A.J., Skarmeta, A.: Architecture for improving terrestrial logistics based on the web of things. Sensors **12**(5), 6538–6575 (2012)
32. Hounsell, N.B., Sherestha, B.P., Wong, A.: Data management and applicaions in a world-leading bus fleet. Transport. Res. Part C: Emerg. Technol. **22**, 76–87 (2012)
33. Clarke, R., Bowen, T., Head, J.: Mass deployment of bus priority using real-time passenger information systems in London. In: Proceedings of European Transport conference, Leeuwenhorst, Netherlands (2007)
34. Mehaffey A., Jarjees, G.: Preliminary trial results of the public transport information and priority system (ptips). In: Proceedings of 8th World Congress on Intelligent Transport Systems, Sydney (2001)
35. Chadil, N., Russameesawang, A., Keeratiwintkorn, P.: Real-time tracking management system using GPS. GPRS and Google Earth. In: Proceedings of ECTI-CON (2008)
36. Rasheed, A., Haleemah, Z., Hashmi, F., Hadi, U., Naim, W., Ajmal, S.: Fleet & convoy management using VANET. J. Comput. Netw. **1**(1), 1–9 (2013)
37. Ammar Hatem, B., Hamman, H.: Bus management system using RFID in WSN. In: European and Mediterranean Conference on Information Systems 2010 (EMCIS2010) (2009)
38. Bajaj, D., Gupta, N.: GPS based automatic vehicle tracking using RFID. Int. J. Eng. Innovative Technol (IJEIT), **1**(1) (2012)
39. Baronti, P., Prashant, P., Chooki, V., Chessa, S., Gotta, A., Fun Hu, Y.: Wireless sensor networks: a survey on the state of the art. Comput. Commun. **30**(7), 1655–1695 (2007)
40. Javed, K.: ZigBee suitability for wireless sensor networks in logistic telemetry applications. Technical report, IDE0612. Halmstad University. Sweden (2006)
41. OpenStreetMap API v0.6. http://wiki.openstreetmap.org/wiki/API
42. World Wide Web Consortium (W3C) Standards. http://www.w3.org/standards
43. ZigBee Alliance.: ZigBee specification 1.0. http://www.zigbee.org (Jan 2008)
44. Oracle. MySQL 5.0 Reference Manual. http://dev.mysql.com (2013)
45. Achour, M., Betz, F., Dovgal, A., Lopez, N., Manusson, H. Richter, G., Seguy, D., Vrana, J., PHP Documentation Group. PHP Manual. http://www.php.net/manual/en/index.php (2013)
46. Svennerberg, G. (ed.): Beginning Google Maps API 3. Apress, New York. ISBN: 978-1-430-22802-8 (2010)
47. Google Maps JavaScript v3 API. https://developers.google.com/maps/documentation/javascript/?hl=en

Disaster Evacuation Guidance Using Opportunistic Communication: The Potential for Opportunity-Based Service

Akihiro Fujihara and Hiroyoshi Miwa

Abstract In recent years, as a practical use of Delay Tolerant Network and Mobile Opportunistic Network, disaster evacuation guidance effective against situations of large-scale urban disasters have been studied. We have proposed a disaster evacuation guidance using opportunistic communication where evacuees gather location information of impassable and congested roads by disaster into their smartphones by themselves, and also share the information with each other by short-range wireless communication between nearby smartphones. Our guidance is designed not only to navigate evacuating crowds to refuges, but also to rapidly aggregate the disaster information. On the other hand, the Great East Japan Earthquake in 2011 taught us a lesson: the only helpful services in disaster situations are services that are daily used by everyone. Inversely services prepared only for disaster situations have never been used in a disaster situation because of lack of maintenance or unawareness by ignorance. To effectively utilise our evacuation guidance, therefore, some service using opportunistic communication should be firstly widespread across the world as an infrastructure and everyone naturally receives much benefit from the service in daily life. In this chapter we consider a possibility of the service, which we call *Opportunity-based Service* (*OBS*). We discuss some practical usages not only for disaster situations, but also for daily life, for example, an autonomous human navigation avoiding congestion by crowds. Through reviewing our past works, we try to foresee a possible next-generation information communication technology regarding Big Data, IoT, and pervasive computing on smart environments.

A. Fujihara (✉) · H. Miwa (✉)
Graduate School of Science and Technology, Kwansei Gakuin University, 2-1 Gakuen,
Sanda, Hyogo 69-1337, Japan
e-mail: cuq54312@kwansei.ac.jp; afujihara@kwansei.ac.jp

H. Miwa
e-mail: miwa@kwansei.ac.jp

N. Bessis and C. Dobre (eds.), *Big Data and Internet of Things:*   425
A Roadmap for Smart Environments, Studies in Computational Intelligence 546,
DOI: 10.1007/978-3-319-05029-4_18, © Springer International Publishing Switzerland 2014

1 Introduction

Delay- and Disruption-Tolerant Networks (DTNs) [6, 37] have been studied in the last 10 years; these systems make the conventional Internet available in challenging or extreme information-communication environments where continuous network connectivity may be unavailable or impossible. Although the concept of the DTN originally comes from the Interplanetary Internet [6], which extends the Internet to wirelessly communicate between planets scattered in space, it is also applied to networks connecting animals, humans, and vehicles with close-range wireless communication, such as ZebraNet [41], Pocket Switched Networks [16], and Data Mule [32]. These networks mark the beginning of Mobile Opportunistic Networks (MONs) [4, 26, 39] as they are known today. This type of network consists of mobile nodes, and these nodes can communicate with each other with close-range wireless communication when they meet. Because their motions are often highly random, every chance encounter is opportunistic. Consequently, communication in MONs is called *opportunistic communication*.

Store-and-forward and store-carry-and-forward are communication methods to route messages from source nodes to target nodes in DTNs. There are many routing protocols using them in DTNs, such as Epidemic routing, Spray and Wait, PRoPHET, and Maxprop [42]. These methods are suitable for routing in the current MONs. Smartphones are increasingly available around the world, from cities in developed countries to rural areas in developing countries, and most modern smartphones enable close-range wireless communications, such as Bluetooth and Wi-Fi. This fact has led to increased interest in MONs from both researchers and the general public.

However, there are currently few services that make good use of MONs and opportunistic communication. In this early stage, one of the few services that successfully uses opportunistic communication is the Nintendo DS's StreetPass. Nintendo DS is a Japanese portable game device developed by Nintendo [35], and StreetPass is an opportunistic communication network that is available on the Nintendo DS. In fact, some Nintendo DS games need to opportunistically communicate with other Nintendo DS devices to progress along their storylines. At first, this experimental challenge was novel enough to change the style of playing games. In Japan, many public places became designated locations where a gamer could wait for opportunistic communications with other Nintendo DSs, and at the time, waiting for this reason became a large social phenomenon. Opportunistic communication by the Nintendo DS is interesting, but it is only available within games, which is the weak point of the device. If the device were able to freely send any message, it would be available in many other daily situations. Opportunistic communication can be used to scan nearby devices and count how many humans are carrying their smartphones. Data gathered by this type of scan could estimate the degree of congestion, although we would have to assume that most humans are carrying their smartphones and keep Bluetooth and Wi-Fi switched on by default. By uploading these data, we could identify congestion hotspots and navigate to avoid them. In disaster situations, this type of navigation could be used to give disaster evacuation guidance and move many crowds of evacuees to the nearest refuges. In addition, disaster information sharing is also available by opportunistic communication.

As mobile devices have advanced, opportunistic communication has become available in smartphones and tablets. As of version 4.0, Google's Android OS supports (insecure) Bluetooth communication without pairing and infinite visible mode. Additionally, Bluetooth version 4.0 supports a low-energy mode that can connect with visible Bluetooth devices by scanning devices in close proximity, which is necessary for opportunistic communication; this process occurs without consuming much of the restricted battery power of portable devices. Qualcomm provides Alljoyn, which is an application development platform to support opportunistic communication with Wi-Fi and Wi-Fi direct. These trends indicate that the world will gradually utilise opportunistic communication as a versatile service.

When long-range mobile communication such as GSM, 3G, or LTE is available from portable devices, we do not need to use opportunistic communication to access the Internet. However, when mobile communication is disabled, a chain of opportunistic communication in neighbouring devices can connect users to the Internet. Although mobile coverage is available in most cities, it can be disabled by a large number of simultaneous accesses from a crowd of people in areas where, for example, a large festival is being held or disaster damage is critical. In these situations, MONs are effective for sending messages outside of affected areas because these areas are dense. Indeed, it is highly probable that many devices will come in sufficiently close proximity for opportunistic communication.

There are known issues preventing the practical use of MONs and opportunistic communication. First, as a fundamental property of opportunistic communication, the smaller the number of users that join in the network is, the weaker the effects of opportunistic communication become. Therefore, to increase the network size, we need to create a mechanism that encourages users to join the network, with increased profits awarded for increased time on the network. It is also important to ensure the security of opportunistic communications with the vast amount of encountered anonymous users, i.e., to keep users free from viruses and wiretapping attacks. Solutions to these problems are necessary before opportunistic networking can be widespread and useful in our society.

Recently, we have published a series of papers on disaster evacuation guidance using opportunistic communication [7, 9, 10]. In these papers, we proposed a guidance system for not only autonomously navigating crowds of evacuees to refuges but also rapidly aggregating real-time disaster information. When following this type of guidance, evacuees naturally collect the locations of impassable and congested roads in their smartphones and share them with nearby smartphones using opportunistic communication. Using the collected information, the guidance proposes an effective shortest-path-based evacuation route to the nearest refuge and avoids known impassable and congested roads in advance. By simulating a simple mathematical model of urban disaster scenarios, we have shown numerically that the guidance reduces the average evacuation time even when the effects of congestion by evacuees are applied. Furthermore, the disaster information can be naturally aggregated into refuge areas using the mobility patterns of evacuees, which are collected by the guidance system.

As we have mentioned above, MONs have great potential for human navigation services, which implicitly have a high affinity for Location-based Services (LBSs). However, as we consider navigation services from the viewpoint of MONs, we can also use the terminology *Opportunity-based Service* (*OBS*) because opportunistic communication plays a strong role in obtaining neighbourhood information. Therefore, in this chapter, by reviewing our previous works on the disaster evacuation guidance and considering the potential of OBSs, we consider some applications of MONs on services that are useful in normal and disaster situations. The goal is to foresee a possible next-generation information communication technology regarding that is related to recent trends such as Big Data, the Internet of Things (IoT), and pervasive computing in smart environments.

The rest of the chapter is organised as follows. In Sect. 2, we propose an OBS by providing some examples of its application both in our daily lives and in disaster situations. In Sect. 3, we review some previous works on emergency evacuation guidance before we explain our disaster evacuation guidance system using opportunistic communication. Then, in Sect. 4, we also mention the direction of some recent studies on human mobility patterns that are closely related to opportunistic communication. In Sect. 5, we explain disaster evacuation guidance as an example of an OBS, and in Sect. 6, we describe a mathematical model to simulate a disaster evacuation guidance system and show some numerical results from our recent works. Finally, in Sect. 7 we summarise disaster evacuation guidance system and our OBS from the viewpoint of certain recent trends, such as Big Data, IoT, and pervasive computing in smart environments.

2 Opportunity-Based Services

Here, we propose an OBS as an extension of a LBS to take advantage of MONs and opportunistic communication. An LBS is a service that utilises location information from GPS or other location-related devices, such as navigation, location detection, and tracking. Today, LBSs are actively used in car navigation, finding lost smartphones, and notifying users when buses and trains are coming; thus, LBSs have become necessary in our life. However, the use of GPS for location detection generally consumes substantial energy, i.e., the batteries of human-carried portable devices, and the detection requires connection time because of the communications with distant satellites and mobile phone towers. Furthermore, the surrounding environmental conditions, such as the weather and buildings, can reduce the accuracy of the location detection. Fundamentally, opportunistic communication with devices in close proximity is a more energy-effective method of detecting location information than the GPS system. Immobile devices (servers) can act as landmarks that provide location information for nearby devices. A good example is PlaceEngine (http://www. placeengine.com), which provides location information using nearby Wi-Fi routers. We also mention that an OBS can be a decentralised method for location detection, whereas PlaceEngine is not fully decentralised. *Local information production for local information consumption* may also be a good methodology for next-generation information communication technologies to avoid location surveillance and enhance the privacy of LBS users.

We consider OBS as a service not only for location detection but also for information dissemination to virally deliver advertisements and other environmental information. An OBS can detect the frequency of encounters with other devices, and this information may help to provide OBSs. For example, we can automatically generate true social networks based on human encounters in daily life and share local information with true local communities. Note that today's SNS does not generate real social networks; the network is biased because the SNS allows users to link even if they never meet and does not automatically link users even if they meet frequently. We can also know the degree of congestion by automatically counting the detected devices in close proximity using opportunistic communication. If we locally share this congestion information, an autonomous human navigation system that avoids congestion in advance may be possible as an OBS. This service would be helpful in reducing the cost of traffic regulation for crowds of people at large events.

Like the Internet, OBSs decentralise the location and human encounter information, whereas most LBSs using GPS centralise the location information. This decentralisation by an OBS also enhances disaster tolerance by supplementing the LBS even if the GPS system is damaged or temporarily disabled. In addition, we mention an application of an OBS in disaster situations: during large-scale disasters, it is very possible that congestion by crowds of evacuees occurs in many places because many evacuees simultaneously rush to refuge areas. Additionally, danger zones due to building collapses or fires hinder their evacuation. However, it is difficult to obtain those disaster updates using mobile phones because mobile communication services are disrupted or temporarily restricted in disaster situations. During the Great East Japan Earthquake, most disaster victims had their mobile phones. Even if mobile communications are disabled, opportunistic communication by Bluetooth and Wi-Fi is available. Thus, it is interesting to consider effective disaster evacuation guidance using opportunistic communication as an OBS.

There are also some existing works that evaluate opportunistic networks and mobile ad-hoc networks in disaster response scenarios where mobile devices wirelessly forward the information generated at the incident location, such as the victim's medical data, to a coordination point for medical care [2, 28, 29].

If everyone shares true social networking information about local communities using OBSs in their daily lives, this information is also helpful for disaster information integration, such as confirming the safety of disaster victims. In this sense, the only helpful service in disaster situations is a service that is used daily by everyone. We will introduce an application of OBSs in disaster situations in detail in Sect. 5.

3 Emergency Evacuation Guidance

In this section, we briefly review studies on emergency evacuation guidance. Emergency evacuation planning is generally formulated as a dynamic network flow problem on an evacuation route network and can be exactly reduced to a linear network flow problem with an induced time-expanded network [14]. However, this reduction enormously enlarges the input of the time graph that corresponds to the problem. If the original evacuation route network G has n nodes, m edges, and an upper bound of the evacuation time T, then the time-expanded network G_T will have at least $(T + 1)n$ nodes and Tm edges. Thus, we need to solve the network flow problem with large inputs from G_T, which is usually difficult in practice.

Therefore, some heuristic algorithms have been proposed to overcome this difficulty. Hamacher and Tjandra [14] wrote a review paper about these algorithms. Lu et al. [23] considered capacity constrained routing algorithms for evacuation planning, and they showed that shortest-path-based algorithms efficiently obtain some heuristic solutions. Hadzic et al. also developed a heuristic technique for solving an extended dynamic network flow problem while considering the shortest or safest evacuation route [13]. In addition, Iizuka et al. proposed a disaster evacuation system that provides information about the appropriate evacuation routes and timings for evacuees to avoid congestion using Ad-Hoc Networks [17]. In fact, they reduced this problem to a distributed constraint optimisation problem and solved it with a heuristic approach.

There are also some simulation-based studies on human behavioural patterns during emergency evacuations. Meguro and Harada [24] simulated human evacuation behaviours in urban space using a potential model that emulates human psychology. The simulation results are applied to the design of urban space to achieve safe and effective evacuations. Wang et al. [38] considered an efficient indoor emergency evacuation using crowd guidance; they treated the situation as an optimisation problem to maximise the sum of egress flows from a building, and they solved it with stochastic dynamic programming.

We also briefly review some implementations of MONs and opportunistic communications that are available in smartphones. Several groups are developing implementations for Android phones. For example, the implementations and their performances are measured with IBR-DTN [40]. Furthermore, PIAX DTN for Android [27] is under development by a Japanese group, and Twimight [15], which is a disaster-tolerant Twitter client, has also been implemented. Whereas in normal situations it behaves like a normal microblogging application, in disaster situations it changes to disaster mode to virally spread and share tweets with opportunistic communication. However, to our knowledge no one has ever developed a smartphone application for disaster evacuation guidance using a DTN and MONs. Consequently, it is necessary to develop navigation software with a disaster mode for future disaster reduction.

4 Human Mobility Patterns and Serendipitous Encounters

An OBS makes use of opportunistic communication in serendipitous human encounters, and human mobility patterns trigger these encounters. Here, we briefly review the recent knowledge about human mobility patterns and serendipitous encounters. With the recent advances in portable devices, wireless communication technologies, and Big Data, it is possible to study the statistical properties of human mobility and encounter patterns. Some mobility models have been proposed to investigate the performance evaluation of wireless sensor and ad-hoc networks whose nodes consist of humans or vehicles [31].

In human mobility patterns, many studies have reported that the average mobility of humans is statistically identical to the Lévy walk [12, 30], where the travel distances between spatial points obey a power-law-type fat-tailed distribution. In addition, the entropy of human mobility patterns is generally small, which means that human mobility patterns are highly biased in small, limited areas [33] because humans move based on their knowledge of the places they have visited in the past. The small entropy of mobility patterns also means that these patterns can easily be predicted using a small number of mobility data samples. One report says that in a dataset where the location of an individual is specified hourly and there is a spatial resolution equal to what is given by the carrier's antennas, just four spatio-temporal points are enough to uniquely identify 95 % of all individuals [25].

Additionally, there are also some studies on human contacts including face-to-face meetings and temporal networks. It was found that the inter-contact times between consecutive human encounters also statistically obey a power-law-type fat-tailed distribution [3, 19]. Using a lifelog experiment dataset from Bluetooth and Wi-Fi, our group also discovered that the human contact frequency, i.e., the number of encounters in the experimental period, statistically obeys a scale-free property [11]. This property implies that there exists a large inequality in human contact frequency. We believe this inequality also results from the biased mobility patterns in human nature. We found that these properties can be explained using a mobility model, which we call the Homesick Lévy walk [11]. These new scale-free properties have attracted increasing attention from the fields of information communication technologies and network sciences in complex systems.

In addition to these results, with our lifelog experiment we also discovered that most serendipitous human encounters in our daily lives inherently have the property of *Ichi-go Ichi-e*, meaning once-in-an-experimental-time. Ichi-go Ichi-e is a famous Japanese proverb that is literally translated as *one chance in a lifetime*, or more idiomatically, as *treasure every encounter, for it will never recur*. Our research group collected data concerning daily human contact using Bluetooth and Wi-Fi wireless communication technologies. Today, billions of electronic devices equipped with Bluetooth and Wi-Fi are used throughout the world. Most of these devices are light and mobile, including mobile PCs and phones, PDAs, tablets, and portable game machines. Therefore, they tend to be carried by humans most of the time. In addition, the communication range of Bluetooth and Wi-Fi is usually on the order of several metres, which is nearly the same as the range that humans can see when observing people in close proximity. Thus, by scanning and logging nearby Bluetooth and Wi-Fi devices, we can collect reasonably well-sampled data concerning human encounters. In our experiment, we used PDAs and smartphones to continuously scan once every 20 s and to record pairs of time stamps and MAC addresses of detected devices, which indicated when a participant of the experiment encounter other people. We began this experiment in 2010, and have continued until now. The experimental periods depended on the participants: they ranged between a minimum of several months and a maximum of more than 4 years. A dozen people (university teachers, students, and company workers) participated in our experiment.

Here, we explain the property of Ichi-go Ichi-e in detail based on a data analysis using the dataset obtained from the lifelog experiment. First, we define $I(t)$ as the ratio of one-time meetings to all meetings that occur until time t. The true ratio of Ichi-go Ichi-e (once-in-a-lifetime meetings) to all encounters can also be denoted as $I(t)$ at $t = T$, where T is the end of life. Because T is too long to perform our experiment, we use $I(t)$ to approximate $I(T)$ and observe the time variance. Some typical time variations in $I(t)$ during the experiment are shown in Fig. 1. Although the initial patterns of $I(t)$ vary strongly from one individual to the next, all of them tend to stabilise as time progresses. For this reason, we roughly assume that each of the time variations in $I(t)$ converges to a fixed point around the ratio $I(T)$. Under this assumption, we may consider the time-averaged ratio over the experimental period, which is denoted as $\langle I \rangle_t$, to be approximately equivalent to the ratio $I(T)$. Averaged over all participants in our experiment, the percentages of Ichi-go Ichi-e meetings using Bluetooth and Wi-Fi are approximately 80–90 %, and hence, the majority of human encounters occur once-in-an-experimental-period. Here, it should be emphasised that an OBS can utilise this vast amount of Ichi-go Ichi-e meetings to calculate the degree of congestion, while the existing routing protocols in DTNs tend to utilise only frequent meetings to efficiently transfer messages.

Next, we also mention the differences in the human mobility patterns during disaster situations. At the time of a disaster, most evacuees go to nearby refuge areas; therefore, the entropy of mobility patterns decreases and the patterns are more easily predictable in general. More specifically, we review human behavioural patterns in disaster evacuation by focusing on a selection of the refuge areas and evacuation routes based on the Great Hanshin-Awaji Earthquake in 1995 [20].

For the selection of a refuge area by evacuees, it was reported that most evacuees made this selection based on their knowledge. They knew an officially recommended refuge area by the local government, which was usually selected from the local elementary and junior high schools. However, if the known refuge area was inaccessible, evacuees were most likely confused about where to evacuate to. In this case, a suggestion of an alternative refuge area is useful for a smooth and prompt evacuation.

In the selection of an evacuation route, it was also reported that most evacuees determined their evacuation route with their knowledge. Some evacuees who had experienced evacuation drills, selected the evacuation route they learned in the drills. Others who had no experience of the drills tended to select a frequently used route from their daily lives. Furthermore, some confused evacuees escaped in a random

Fig. 1 Time variations
in the ratio of $I(t)$ for
individual participants using
a bluetooth and **b** Wi-Fi (*the
line with points*) and their
time-averaged ratios $\langle I \rangle_t$ over
the experimental time (*the
solid line*)

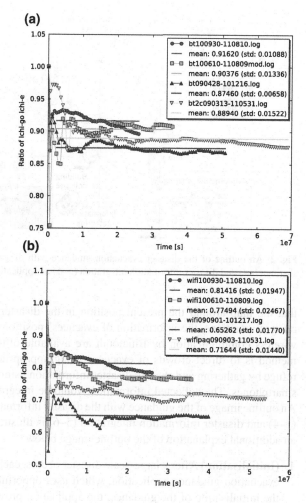

direction and mistakenly travelled to known hazardous areas. Some suggestion of an appropriate evacuation route is also useful to provide evacuees with a smooth and prompt evacuation and reduce their anxiety and stress.

5 Disaster Evacuation Guidance Using Opportunistic Communication

In this section, we propose methods for both disaster evacuation guidance and disaster information integration as applications of an OBS. We assume that mobile cellular communication is completely disabled by an unexpected large-scale disaster, but

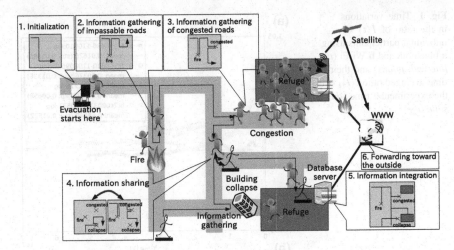

Fig. 2 An outline of the disaster evacuation guidance with disaster information gathering and sharing (*1–4*) and disaster information integration (*5–6*) as applications of an OBS

that the map around the present position in the disaster area, the information of refuges' positions, the information of evacuees' positions from a GPS sensor, and wireless communication by Bluetooth are available offline. The core idea of the method is to direct crowds of evacuees along appropriate evacuation routes to a refuge by gathering and sharing disaster information using those offline functions in smartphones. The collected information would be integrated in some refuge areas. An outline image of the guidance with the disaster information gathering and sharing (1–4) and disaster information integration (5–6) is illustrated in Fig. 2. We provide an additional explanation of the outline image below.

1. **(Initialisation)** After a large-scale disaster occurs, each evacuee runs the disaster evacuation guidance application, which uses opportunistic communication. As the initialisation of the guidance, the application provides the evacuee with the shortest-path-based evacuation route from the present position and the nearest refuge on the map.

2. **(Information gathering about impassable roads)** If an evacuee discovers an impassable road on the way to the refuge due to a fire or building collapse, he or she records its location information as disaster information and saves it to the application. Using the disaster information, the application immediately calculates and displays an alternative evacuation route and avoids any known impassable roads if necessary.

3. **(Information gathering about congested roads)** If the evacuee comes to a road that is congested with other evacuees on the way to the refuge, he or she also records its location information as disaster information and saves it into the application. Using the disaster information, the application immediately calculates and displays an alternative detour evacuation route that avoids any recently congested roads.

4. **(Information sharing)** When the evacuee encounters other evacuees opportunistically, the application virally shares with them all the disaster information that has been gathered in the past via short-range wireless communication. If new, critical disaster information is loaded into the application, it immediately calculates an displays an alternative evacuation route that avoids any known impassable and recently congested roads.

5. **(Information integration)** As soon as the evacuee arrives at the refuge, the application uploads the disaster information that is gathered and shared on the evacuation route to a database server in the refuge for integration purposes.

6. **(Forwarding toward the outside)** For each refuge, the integrated disaster information is reliably forwarded toward the outside of the disaster area by DTN technologies that are stable against disaster, such as satellite-based communications or resilient communication networks between base nodes, e.g., the Virtual Segment [5, 21]. After this forwarding, we can reuse the information in safety confirmations and rescue plans for evacuees.

In the fourth step, we adopt Epidemic routing [36] for the information sharing. Epidemic routing is a routing protocol in DTN that copies any novel information to every encountered node. Observe that this type of data sharing generally causes flooding of forwarding messages. However, in this scenario, the shared data are about the positions of impassable roads, where the number of impassable roads is less than the number of roads around the present position. Because there is this clear upper bound to the data and because the position data are just small text messages, flooding by Epidemic routing is avoidable by preparing a sufficient buffer size in the memory of mobile phones. Another reason to use Epidemic routing is that the fast spread of disaster information is essential for selecting efficient and appropriate evacuation routes for evacuees.

Note that our evacuation guidance becomes invalid if there is no clear path to any refuge site, which is outside of the scope of our research in the simulation. A practical answer to this issue is to provide an alternative route that has the minimum number (or cost) of impassable roads along the way to refuge areas.

Our proposal also includes an efficient integration of disaster information in the early stages of a disaster. It is known that information gathering by opportunistic communication between randomly moving nodes takes a considerable amount of time because the information gathering time generally obeys a fat-tailed distribution [8]. In the information gathering and sharing phase (1–4), evacuees do not move randomly, but their motions become simpler, i.e., they simply gather around some refuge areas. Using these human mobility patterns in the evacuation situations, it is possible to collect the information efficiently by simply preparing some disaster-resistant backbone networks connected to every refuge area. Therefore, in the information integration and forwarding phase (5–6), the integrated disaster information can be reliably transferred. We believe that this *carry, store, and forward strategy* (carry by evacuees, store at refuge areas, and forward outside disaster areas) will be successful for information integration in a disaster scenario.

Fig. 3 An example of an
urban road graph where the
size of the graph $L = 3$, the
number of refuges $a = 2$,
and the number of impassable
roads $b = 4$

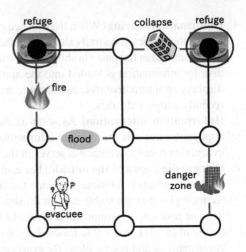

We also mention an exact method to integrate safety confirmation information.
If opportunistic communication is performed by Bluetooth, each device has a MAC
address, which can serve as an identifier for each evacuee. Therefore, if we share
the MAC addresses of people in local communities using an OBS, we can use them
to confirm someone's safety by searching the gathered information of the history of
opportunistic communication in refuge areas.

6 The Effects of Disaster Evacuation Guidance: Simulations

In this section, we introduce a mathematical model of the guidance explained in the
previous section to numerically evaluate the performance of the evacuation time. We
model the disaster scenario using an urban road graph, and we explain the evacuee's
mobility following the directions, the types of disaster information, and information
gathering and sharing. After defining the mathematical model, we provide some
numerical results of the effects of our disaster evacuation guidance system.

6.1 Modelling Disaster and Evacuation Scenarios

First, we consider a model of urban road networks. Specifically, we consider an urban
road network as a square grid graph with size L. The edges of the graph represent
roads, and the vertices are the intersections and corners of roads. This graph consists
of $n = L^2$ nodes and $m = 2L(L - 1)$ edges. On this graph, we give a refuge nodes
as the destination nodes of evacuees, and we remove b edges as impassable roads
from the graph. An example of the type of urban road graph that we consider is
illustrated in Fig. 3. Here, we implicitly assume there exists at least one evacuation

Fig. 4 An urban road graph of the Sendai city centre. The *four small circles* indicate refuge areas

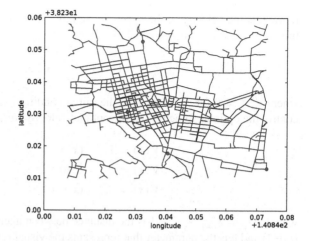

route to an evacuation site. To guarantee this assumption, b edges are removed while maintaining the connectivity of the graph. Because the connected graph that has the least number of edges is a tree graph [and without loss of generality, a tree graph consists of n nodes and $(n - 1)$ edges], the number of removable edges can vary within the range $0 \leq b \leq (L - 1)^2$.

We also introduce the graph of the Sendai city centre, which was generated from datasets of real road networks in Sendai, Japan using OpenStreetMap (http://www. openstreetmap.org), as shown in Fig. 4. This graph consists of $n = 1,561$ vertices and $m = 1,895$ edges, and while maintaining its connectivity, a maximum of $b = 335$ edges are removable. In the graph, four refuge areas $a = 4$ are placed based on the position information of real refuge areas during the Great East Japan Earthquake.

Next, we model an evacuee's mobility when following the guidance. To incorporate a large number of people into the model, we assume that as an initial condition, there is one agent that represents an evacuee or a group of evacuees for each node on the graph. Therefore, there are n evacuees in total on the graph. Additionally, the initial evacuation route is determined by a shortest path from the present node to the nearest refuge node. No agent knows which edges are impassable or congested at the beginning of the evacuation.

To compare the algorithms for managing congestion information, we need to introduce evacuee movement such that congestion naturally emerges. If the density of evacuees in an evacuation route exceeds a certain threshold value, congestion emerges. The evacuees in the congested region either cannot move or move more slowly than usual because of the excluded volume effect between the bodies of evacuees. To consider this effect, a congestion model that incorporates a cellular automaton should be used. Here, we explain the Stochastic Optimal Velocity (SOV) model [18], which is a congestion model that includes the effect of acceleration. The SOV model is a stochastic cellular automaton on a finite one-dimensional lane where each agent i steps forward through each box one by one with a probability v_i, which represents the velocity of the agent. To be more specific, the velocity of agent i at time $(t + 1)$, or v_i^{t+1}, is described as follows:

$$v_i^{t+1} = (1 - r)v_i^t + rV(\Delta x_i^t), \tag{1}$$

where $v_i^{t=0} = 0$ for $1 \leq i \leq n$ as the initial condition, r is a weight parameter to control the optimal velocity, and the optimal velocity function V is given by

$$V(x) = 1, \quad (x > c) \tag{2}$$
$$V(x) = x/c, \quad (0 \leq x \leq c) \tag{3}$$
$$V(x) = 0, \quad (x < 0) \tag{4}$$

where $\Delta x_i^t = x_{i+1}^t - x_i^t - 1$ is the distance between agent i and the forward agent $(i + 1)$ and c is the parameter that represents the visual range of evacuees.

However, in this model, it is rare for multiple evacuees to arrive at the same node of the graph. Our model assumes that evacuees on the same node can share disaster information with each other. Thus, we consider a slightly modified version of the SOV model to share disaster information between evacuees at the same node of the graph. In the modified model, the velocity is described by

$$v_i^{t+1} = \max\{(1 - r)v_i^t + rV(\Delta x_i^t), 0\}, \tag{5}$$

where

$$\Delta x_i^t = x_{i+1}^t - x_i^t, \tag{6}$$
$$V(x) = 1, \quad (x > c) \tag{7}$$
$$V(x) = x/c, \quad (0 \leq x \leq c); \tag{8}$$
$$V(x) = -2/r. \quad (x < 0) \tag{9}$$

The characteristic feature of this model is that overtaking evacuees in the forward nodes is prohibited, whereas in the SOV model the evacuee may pass other evacuees while moving forward. The main reason for this modification is account for congestion; overtaking is impossible because a large number of crowded evacuees creates a wall that blocks the progress of other evacuees. In the modified SOV (mSOV) model, the evacuee moves forward step by step on his or her shortest-path-based evacuation route on the graph. The image of the mSOV model is summarised in Fig. 5.

Here, we discuss the type of disaster information and route modification in the model. In the disaster scenario, two types of disaster information are considered: the locations of impassable edges and congested edges. We assume that the impassable edges are provided beforehand and their location information does not change over time, which is the scope of our research. Note that no congested edge exists at the initial time, but they appear depending on the density of the evacuees on the evacuation route. An evacuee node considers an edge congested when there are more

Fig. 5 The modified stochastic optimal velocity (mSOV) model on a one-dimensional lane (or a graph of a one-dimensional path)

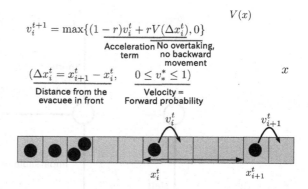

$$v_i^{t+1} = \max\{(1-r)v_i^t + rV(\Delta x_i^t), 0\}$$

$\underbrace{}_{\substack{\text{Acceleration} \\ \text{term}}}$ $\underbrace{}_{\substack{\text{No overtaking,} \\ \text{no backward} \\ \text{movement}}}$

$$(\Delta x_i^t = x_{i+1}^t - x_i^t, \quad 0 \leq v_*^* \leq 1)$$

$\underbrace{}_{\substack{\text{Distance from the} \\ \text{evacuee in front}}}$ $\underbrace{}_{\substack{\text{Velocity} = \\ \text{Forward probability}}}$

than θ_c evacuees in total on the next c nodes of the evacuation route. As time passes and the density of the evacuees on the evacuation route decreases, the congestion will naturally disappear. However, the evacuee does not know the time that congestion will resolve based on only local information gathering and sharing. For convenience, we introduce the congestion effective time t_{eff} to avoid congested evacuation routes. When the effective time passes after the evacuee encounters congestion, the evacuee neglects the information about the congested edge as if the congestion on the edge has been resolved.

The information sharing using opportunistic communication is carried out between evacuees on the same nodes, and they share all the known disaster information with each other.

When the evacuee obtains new disaster information that blocks safe passage of the evacuee on his or her evacuation route, an alternative shortest path to a refuge is calculated; this path avoids all the impassable and congested edges. This calculation will be repeated whenever new information is obtained, and if an alternative route exists, it becomes the new evacuation route. The complexity of the rerouting algorithm is the same as the complexity of the shortest-path algorithm, which is also true for Dijkstra's algorithm. An example of the route modification is illustrated in Fig. 6.

6.2 Performance Evaluation

We investigate the performance of the proposed guidance. As the performance measure, we focus on the evacuation time, which is defined by the elapsed time from the start of the evacuation at the initial position to the termination time of the evacuation at a refuge site. To simulate the performance, we programmed our original simulator, which includes all the functions explained in the previous section, from scratch using the Python and C programming languages. The numerical simulations were performed on our stand-alone PCs. In our simulations, the unit time is defined

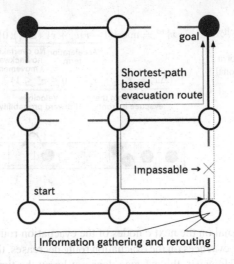

Fig. 6 An example of the shortest-path-based evacuation route on the road graph in Fig. 3. The first evacuation route, which needs a modification because of the impassable edges in the middle of the route, is illustrated with the *dotted line*. The new evacuation route is selected by avoiding all the impassable and congested edges to arrive at a refuge via the shortest path. The actual movement of the evacuee is indicated by the *solid line*

by timesteps such that at each step, the evacuees move to the next node in the road graph or stay at the present node due to congestion. We compare the evacuation time in three cases: (1) no opportunistic communication is allowed (no opp. comm.) T_1, (2) opportunistic communication is allowed (opp. comm.) T_2, and (3) all nodes *a priori* know all the impassable roads in the graph (fastest) T_3. In all the simulations, we repeat the evacuation scenario N times while changing the positions of removed edges to average the position effect of removed edges on the performance.

We also define a reduction ratio of the average excess evacuation time R as

$$R = (\langle T_2 \rangle - \langle T_3 \rangle)/(\langle T_1 \rangle - \langle T_3 \rangle), \tag{10}$$

where $\langle T_1 \rangle$ is the average evacuation time with no opportunistic communication, $\langle T_2 \rangle$ is the corresponding time with opportunistic communication, and $\langle T_3 \rangle$ is the evacuation time in the case that evacuees know *a priori* the positions of all the impassable roads. The average evacuation time is calculated by averaging the evacuation time over all the evacuees in the model.

We show the relations between the number of removed edges and the evacuation times in the grid graph in Fig. 7. We also show numerical results for the Sendai city road graph in Fig. 8.

We find that in both graphs, the effect of opportunistic communications considerably reduces the average evacuation time. As shown in Fig. 7b, the reduction ratio R reaches 50–70 %, and it tends to increase as the number of impassable roads increases. Because the fluctuation of the maximum evacuation time is very large,

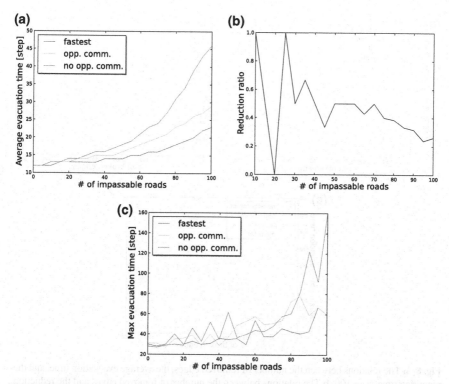

Fig. 7 Numerical results with the effect of traffic volume in an 11 × 11 square grid graph and $N = 1,000$ repetitions. **a** The relationship between the number of removed edges and the average evacuation time. **b** The relationship between the number of removed edges and the reduction ratio in Eq. 10. **c** The relationship between the number of removed edges and the maximum evacuation time. Three cases are compared: (1) allowing no opportunistic communication (no opp. comm.), (2) allowing opportunistic communication (opp. comm.), and (3) all impassable roads are known *a priori* (the fastest case)

averaging the maximum evacuation time over 1,000 repetitions is not enough to smooth the change when increasing the number of impassable roads. However, the maximum evacuation times in opportunistic communication are almost comparable to or less than those times that occur in the case of no opportunistic communication.

Next, we investigate how the effects of both information gathering and sharing congestion-location information reduce the evacuation time more substantially by switching to another detour evacuation route before or when the evacuees encounter congested areas. To investigate these basic differences, we consider the following three simple detour-route switching algorithms to compare the performance.

1. No congestion-location information sharing is performed, and the route is switched to a modified shortest-path route that avoids the most congested roads whenever evacuees encounter congestion.

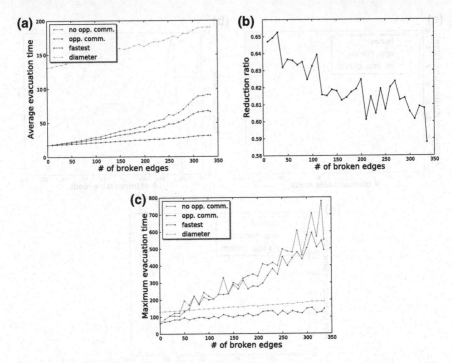

Fig. 8 **a** The relations between the number of removed edges, the average evacuation time, and the repetition time $N = 100$. **b** The relations between the number of removed edges and the reduction ratio in Eq. 10. **c** The relations between the number of removed edges and the maximum evacuation time. Three cases, namely, (1) allowing no opportunistic communication (no opp. comm.), (2) allowing opportunistic communication (opp. comm.), and (3) all impassable roads are known *a priori* (the fastest case), are compared. The diameter of the graph is also shown for reference (diameter)

2. Congestion-location information sharing is performed, and the route is switched to a modified shortest-path route that avoids the most congested roads whenever evacuees encounter congestion.
3. Congestion-location information sharing is performed, and the route is switched to a modified shortest-path route that avoids the most congested roads whenever new congestion-location information is gathered.

We numerically compare the performance of these three algorithms. We investigate how the measure is changed by increasing the number of impassable edges b. Here, the other parameters in the model are fixed as follows: the size of the graph $L = 10$, the number of refuges $a = 3$, the weight parameter of mSOV $r = 0.5$, the visual range $c = 2$, the congestion threshold θ_c, and the effective time of congestion $t_{eff} = 3$.

We show the numerical results for three cases in Fig. 9: (1) the shared information about congested edges is unused; (2) the shared information about congested edges

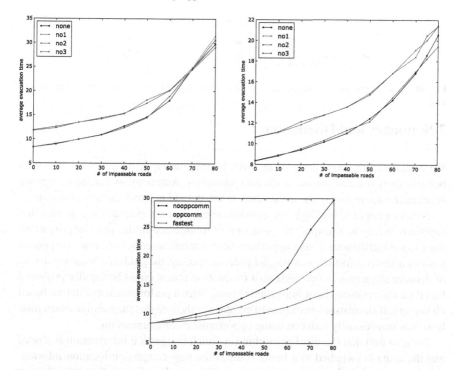

Fig. 9 The relation between the number of impassable roads and the average evacuation time with $N = 100$ repetitions. The first figure (**a**) shows the case where information about impassable edges is not shared, and the second figure (**b**) shows the case where information about impassable edges is shared. Lastly, the third figure (**c**) shows the case where the gathered and shared information about congested edges is unused. In the legend, 'none' means that no detour route is selected and the evacuee continues on the current evacuation route. Furthermore, 'no1', 'no2', and 'no3' mean the respective numbered strategies, 'nooppcomm' means that no information sharing regarding impassable edges is performed, and 'oppcomm' means this information sharing is performed. In addition, 'fastest' means the case when all the evacuees initially know all the impassable edges in the graph, which indicates the lowest limit of the average evacuation time

is used; (3) neither gathered information nor shared information are used. As shown in the figures in the top and middle of Fig. 9, we can see that the third algorithm is better than the others, and furthermore, the first and second ones are competitive, which is a common trend independent of the effect of the information sharing about congested edges. Comparing the third algorithm to the case where information about impassable roads is used but congestion information is not, we see no clear difference in the average evacuation time, which is also a common trend. As shown in the figure at the bottom of Fig. 9, we observe that sharing information about impassable edges notably shortens the average evacuation time, while sharing information about congested edges apparently has no clear effect on the evacuation time. We find similar trends in the maximum evacuation time.

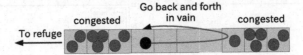

Fig. 10 An example where evacuees are forced to go back and forth between congested areas in vain

7 Summary and Discussions

We considered the possibility of an OBS by introducing some practical examples both for daily use and for use in disaster situations. Autonomous human navigation is necessary for reducing the congestion in urban road networks in any situation.

As an example of an OBS, we considered the disaster evacuation guidance that functions during an unexpected disruption of mobile communication infrastructure due to a natural disaster. Using opportunistic communication between mobile phones to share disaster information through Epidemic routing, the guidance became tolerant of disaster situations in the sense that evacuation routes could be rapidly proposed based on information about impassable roads. With a performance evaluation based on numerical simulations in road map graphs, we showed that the average evacuation time was considerably reduced using opportunistic communication.

We also find that the third algorithm (in which congestion information is shared and the route is switched to a better route when new congestion-location information is received) is better than the others. This result indicates that later detour-route switching decisions around congested areas degrade the performance because repeated evacuation route changes force evacuees to go back and forth between some congested areas in vain, as shown in Fig. 10. We compare the third algorithm and the case where only information about impassable roads is used. We find no clear difference in the observed evacuation time. This result demonstrates that the three algorithms for congestion control have less impact on the evacuation time than the impassable roads information.

For future work, the reliability of our disaster evacuation guidance must be improved. For example, it is vulnerable to the spread of false information to misguide evacuees to dangerous areas. This type of attack can be resisted if we introduce an information-checking mechanism such as polling. It is also important to compare some other dissemination schemes for disaster information, such as Spray and Wait [34], PRoPHET [22], and MaxProp [1], with Epidemic routing. In this study, we assumed that each node has a map of the disaster area and a GPS, which might be restrictive for some evacuees who have basic or feature phones instead of smart-phones. We would like to improve the guidance to include an alternative mechanism for obtaining disaster information when a user has no map and/or no GPS.

We believe that OBSs have strong impacts on Big Data, IoT, and pervasive computing in smart environments. If an OBS is widespread, it is possible to gather 'big data' about opportunistic communication, which possibly includes information about locations, social networking, and serendipitous encounters between animals, humans, and things. In this sense, an OBS is a possible candidate to connect all the things in the world, as is the concept of IoT. Moreover, people can automatically record a lifelog using an OBS. From the viewpoint of pervasive computing in smart environments, the function of crowd navigation and disaster evacuation guidance using an OBS should be incorporated to move comfortably within cities.

Acknowledgments This work is partially supported by Japan Society for the Promotion of Science through Grant-in-Aid for Scientific Research (C) (23500105) and Grant-in-Aid for Young Scientists (B) (25870958).

References

1. Burgess, J., et al.: MaxProp: routing for vehicle-based disruption-tolerant networks. In: Proceedings of IEEE INFOCOM, pp. 398–408 (2006)
2. Campillo, A.M., Crowcroft, J., Yoneki, E., Marti, R.: Evaluating opportunistic networks in disaster scenarios. J. Netw. Comput. Appl. **36**, 870–880 (2013)
3. Chaintreau, A., et al.: Impact of human mobility on opportunistic forwarding algorithms. IEEE Trans. Mob. Comput. **6**(6), 606–620 (2007)
4. Denko, M.K. (ed.): Mobile Opportunistic Networks: Architectures, Protocols and Applications. CRC Press, Boca Raton (2011)
5. Dou, R., Fujihara, A., Miwa, H.: Algorithms for the base node location problem in the virtual segment method in store-carry-forward routing schemes. In: 2nd International Conference on Intelligent Networking and Collaborative Systems (2010)
6. Farrell, S., Cahill, V.: Delay- and Disruption-Tolerant Networking. Artech House, Norwood (2006)
7. Fujihara, A., Miwa, H.: Effect of traffic volume in real-time disaster evacuation guidance using opportunistic communications. In: 2012 Third International Conference on Intelligent Networking and Collaborative Systems (INCoS2012)
8. Fujihara, A., Miwa, H.: Efficiency analysis on an information sharing process with randomly moving mobile sensors. In: SAINT2008, IEEE Computer Society, pp. 241–244 (2008)
9. Fujihara, A., Miwa, H.: On the use of congestion information for rerouting in the disaster evacuation guidance using opportunistic communication. In: ADMNET2013, IEEE-COMPSAC2013 Workshop (2013)
10. Fujihara, A., Miwa, H.: Real-time disaster evacuation guidance using opportunistic communications. In: The 2012 International Symposium on Applications and the Internet (SAINT2012)
11. Fujihara, A., Ono, S., Miwa, H.: Optimal forwarding criterion of utility-based routing under sequential encounters for delay tolerant networks. In: Third International Conference on Intelligent Networking and Collaborative Systems (INCoS) 2011, pp. 279–286 (2011)
12. González, M.C., et al.: Understanding individual human mobility patterns. Nature **453**, 779–782 (2008)
13. Hadzic, T., Brown, K.N., Sreenan, C.J.: Real-time pedestrian evacuation planning during emergency. In: IEEE International Conference on Tools with, Artificial Intelligence, pp. 597–604 (2011)
14. Hamacher, H.W., Tjandra, S.A.: Mathematical modeling of evacuation problems: a state of the art. Pedestrian Evacuation Dyn. pp. 227–266 (2002)
15. Hossmann, T., Legendre, F., Carta, P., Gunningberg, P., Rohner, C.: Twitter in disaster mode. In: ExtremeCom2011 (2011)
16. Hui, P., Chaintreau, A., Scott, J., Gass, R., Crowcroft, J., Diot, C.: Pocket switched networks and human mobility in conference environments. In: WDTN '05 Proceedings of the 2005 ACM SIGCOMM Workshop on Delay-Tolerant Networking, pp. 244–251 (2005)
17. Iizuka, Y., Yoshida, K., Iizuka, K.: An effective disaster evacuation assist system utilized by an Ad-Hoc network. In: HCII 2011. CCIS 174, pp. 31–35 (2011)

18. Kanai, M., Nishinari, K., Tokihiro, T.: Phys. Rev. E **72**, 035102(R) (2005)
19. Karagiannis, T., et al.: Power law and exponential decay of intercontact times between mobile devices. IEEE Trans. Mob. Comput. **9**(10), 1377–1390 (2010)
20. Kashiwabara, S., Ueno, J., Morita, T.: *Studies on refuges in the Great Hanshin-Awaji Earthquake* (Japanese), Osaka University Press, http://www.osaka-up.or.jp/kokai.html (1998)
21. Kimura, Y., Tsuji, H., Miwa, H.: Performance evaluation of virtual segment method based on actual road network topology at the time of the Great East Japan Earthquake. In: 4th International Conference on Intelligent Networking and Collaborative Systems, pp. 451–456 (2012)
22. Lindgren, A., et al.: Probabilistic routing in intermittently connected networks. In: Proceedings SAPIR Workshop, pp. 239–254 (2004)
23. Lu, Q., George, B., Shekhar, S.: Capacity constrained routing algorithms for evacuation planning: a summary of results. In: SSTD 2005. LNCS, vol. 3633, pp. 291–307 (2005)
24. Meguro, K., Harada, M.: Urban space design and safety evaluation from the viewpoint of evacuation behavior of users. Bull. Earthquake Resistant Struct. Res. Center (32), 137–144 (1999)
25. de Montjoye, Y.-A., Hidalgo, C.A., Verleysen, M., Blondel, V.D.: Unique in the crowd: the privacy bounds of human mobility. Sci. Rep. **3**, 1376 (2013)
26. Pelusi, L., Passarella, A., Conti, M.: Opportunistic networking: data forwarding in disconnected mobile ad hoc networks. Commun Mag, IEEE **44**(11), 134–141 (2006)
27. Piax, DTN for Android, http://www.piax.org/?PIAX+DTN+for+Android
28. Reina, D.G., Toral, S.L., Barrero, F., Bassis, N., Asimakopoulou, E.: Modelling and assessing ad hoc networks in disaster scenarios. J. Ambient Intell. Humanized Comput. **4**, 571–579 (2013)
29. Reina, D.G., Toral, S.L., Coca, J.M.L., Barrero, F., Bessis, N., Asimakopoulou, E.: An evolutionary computational approach for optimizing broadcasting in disaster response scenarios. Appl. Soft Comput. **13**, 833–845 (2013)
30. Rhee, I., et al.: On the levy-walk nature of human mobility: do humans walk like monkeys? In: Proceedings of IEEE INFOCOM, pp. 924–932 (2008)
31. Roy, R.R.: Handbook of Mobile Ad Hoc Networks for Mobility Models. Springer, Berlin (2011)
32. Shah, R.C., Roy, S., Jain, S., Brunette, W.: Data MULEs: modeling a three-tier architecture for sparse sensor networks. Ad Hoc Netw. **1**, 215–233 (2003)
33. Song, C., et al.: Limits of predictability in human mobility. Science **327**(5968), 1018 (2010)
34. Spyropoulos, T., et al.: Spray and wait: an efficient routing scheme for intermittently connected mobile networks. In: Proceedings of ACM SIGCOMM Workshop on Delay-Tolerant Networking, pp. 252–259 (2005)
35. StreetPass, http://en-americas-support.nintendo.com/app/answers/landing/p/48/c/231 or http://www.nintendo.co.uk/Nintendo-3DS/Features/StreetPass/StreetPass-114943.html
36. Vahdat, A., Becker, D.: Epidemic routing for partially connected ad hoc networks. Technical Report CS-2000-06. Department of Computer Science, Duke University (2000)
37. Vasilakos, A., Zhang, Y., Spyropoulos, T.V.: Delay Tolerant Networks: Protocols and Applications. Wireless Networks and Mobile Communications Series. CRC Press, Boca Raton (2012)
38. Wang, P., Luh, P.B., Chang, S.-C., Marsh, K.L.: Efficient optimization of building emergency evacuation considering social bond of evacuees. In: IEEE International Conference on Automation Science and Engineering (2009)
39. Woungang, I., Dhurandher, S.K., Anpalagan, A., Vasilakos, A.V. (eds.): Routing in Opportunistic Networks. Springer, Berlin (2013)
40. Yanggratoke, R., Azfar, A., Marval, M.J.P., Ahmed, S.: Delay tolerant network on android phones: implementation issues and performance measurements. J. Commun. **6**(6), 477–484 (2011)
41. Zhang, P., Sadler, C.M., Lyon, S.A., Martonosi, M.: Hardware design experiences in ZebraNet. In: SenSys '04 Proceedings of the 2nd International Conference on Embedded Networked Sensor Systems, pp. 227–238 (2004)
42. Zhang, Z.: Routing in intermittently connected mobile ad hoc networks and delay tolerant networks: overview and challenges. IEEE Commun. Surv. Tutorials **8**(1), 24–37 (2006)

iPromotion: A Cloud-Based Platform for Virtual Reality Internet Advertising

Markos Zampoglou, Athanasios G. Malamos, Kostas Kapetanakis,
Konstantinos Kontakis, Emmanuel Sardis, George Vafiadis,
Vrettos Moulos and Anastasios Doulamis

Abstract We present a large-scale platform for distributing Virtual Reality advertisements over the World Wide Web. The platform aims at receiving and transmitting large amounts of data over mobile and desktop devices in Smart City contexts, is based on a modular and distributed architecture to allow for scalability, and incorporates content-based search capabilities for Virtual Reality (VR) scenes to allow for content management. Data is stored on a cloud repository, to allow for a large amount of VR material to be kept and distributed, and follows a service-based approach of independent subsystems for the management, conversion and streaming of information. In order to function over a wide range of used end-devices, from mobile phones to high-end desktop PCs, the system is based on HTML5 technologies, and implements a remote rendering server to alleviate the computational burden on the end device. Furthermore, an extension of the MPEG-7 standard is used for the description and retrieval of 3D scenes from the cloud, and we have further ensured compliance of our system with a number of other structure and communication standards, to ensure

M. Zampoglou (✉) · A. G. Malamos · K. Kapetanakis (✉) · K. Kontakis
Multimedia Content Laboratory, School of Applied Informatics and Multimedia,
TEI of Crete, Heraklion, Greece
e-mail: mzampog@epp.teicrete.gr
http://www.medialab.teicrete.gr

K. Kapetanakis
e-mail: kapekost@gmail.com

A. G. Malamos
e-mail: amalamos@epp.teicrete.gr

E. Sardis · V. Moulos · G. Vafiadis
Institute of Communication and Computer Systems, ICCS, 9, Iroon Polytechniou Str.,
Athens, Greece
e-mail: sardism@mail.ntua.gr
http://www.iccs.gr/eng/

A. Doulamis
Department of Production Engineering and Management, Technical University of Crete,
Chania, Greece

N. Bessis and C. Dobre (eds.), *Big Data and Internet of Things:*
A Roadmap for Smart Environments, Studies in Computational Intelligence 546,
DOI: 10.1007/978-3-319-05029-4_19, © Springer International Publishing Switzerland 2014

extensibility and reusability of the sub-modules. The platform is a research work in progress: we present the subsystems already implemented, plan our next steps and describe our contributions to research.

1 Introduction

Web advertising has been an important commercial activity since the World Wide Web first rose in popularity, and still remains a fast growing field. It is indicative that in 2012, internet advertising revenues in the USA totaled a sum of $36.57 billion, up 15 % from the $31.74 billion reported in 2011 [1]. Out of that sum, 33 % was brought by display-related advertising, i.e. web-based promotion, such as banners or digital video, while mobile advertising—which is measured separately, although it is still based on visual media—amounted to another 9 % of total revenues. Mobile-based advertising is also currently the fastest growing market, having almost doubled its share since 2011, starting from a virtually zero share in 2009.

In display-based web advertising, products are typically promoted through the traditional mediums of text, image, video and sound. Using multimedia objects, promoters aim to appease potential customers, and at the same time inform them on the features of the products promoted. There exists one type of multimedia information, however, that has only recently begun to see application in advertising: Virtual Reality (VR). The term refers to synthetic, interactive graphical environments, which allow the user to experience and interact with 3D models of objects, scenes, or even entire worlds.

In the past, offering VR technologies over the web for something as widespread and commonplace as advertising was almost unthinkable, due to the absence of the necessary technologies, and the high computational and bandwidth requirements involved. However, VR technologies have taken leaps in the recent years, allowing for rich, immersive experiences to be provided at relatively modest computational costs. In parallel, a number of different technologies for Web-based 3D graphics have appeared, allowing VR worlds to be displayed within web browsers. These include the VRML and X3D languages, the X3DOM and XML3D extensions to HTML, and the increasingly popular WebGL JavaScript API. The research focus on Web 3D and the ensuing advent of a number of powerful technologies has finally opened the prospect of integrating of 3D scenes in web pages, and, as a result, applying them for the needs of web advertising.

Although VR advertising has only recently began to become realistically feasible, it has been the object of speculation and experimentation between marketing researchers for more than a decade. The potential benefits of VR advertising have been extensively studied, within constrained environments, to explore whether such a shift in the technological paradigm would benefit marketing campaigns. We can now claim that the advantages of VR advertising have been quantitatively confirmed: using VR environments for advertising can increase not only the customer's sense of informativeness, but also their actual intention to purchase the product (i.e. the effectiveness of the campaign) [2]. Furthermore, it can encourage customers to invest more time in browsing and exploring options, and inspire increased loyalty [3].

While the positive effects of 3D advertising do not extend to all types of marketable products, as some product types, such as food, cannot be experienced virtually in a meaningful way [4], the overall advantages of 3D browsing have been repeatedly demonstrated [5]. Furthermore, while a part of the positive effect of 3D advertisements can be attributed to the novelty of the medium, it has been demonstrated that, even when the sense of novelty wanes, 3D advertising can retain its effectiveness, provided that we take advantage of a VR environment's ability to offer not only a more enjoyable experience but also more information about the product [6]. For the interested reader, a recent paper [5] includes an extensive bibliographic review of marketing research on the effectiveness of VR advertising.

Besides the novelty of the experience, and the increased informativeness of interactive advertising, internet-based VR advertising offers one additional significant advantage: content personalization. By collecting data from potential customers, a VR advertising platform can adapt the content, both in terms of products/services and the form of experience, to a specific user's profile. The ubiquity of networked mobile devices, combined with the anticipated increased presence of networked sensors and display devices throughout todays "smart city" modern environments can give a system the ability to collect mass-scale data on customer behavior, and provide adapted marketing experiences to them through various media, in a variety of locations and hardware interfaces. The advances in content description research can help organize large-scale collections of VR advertisements, and manage them to serve the appropriate content to each user, provided a sufficiently platform-independent approach is used, to bypass the device variability inherent in such contexts.

In our work, we are building a platform that allows marketers to offer VR advertisements via the Web, which is today a common approach for device-independent applications. For the rest of this chapter, while examining advertising practices, we have to make a distinction between two basic types of web advertisements: In the first case, product promotion is achieved by *actively* inserting banners or thumbnails within—often unrelated but preferably partly relevant—web pages, which promote goods and services to the end users.

Normally, such a banner gives a rough idea of the advertised product, and provides a hyperlink to take interested viewers to a separate web page, dedicated to promoting the product. This takes us to the second typical mode of web advertising: since potential customers often actively search for products they desire, product providers commonly offer dedicated pages presenting and describing these goods and services in detail. Such pages offer *passive* promotion, in the sense that a user has to intentionally seek them to experience their content.

We are interested in building a framework for web-based VR marketing that will support both passive and active advertising. However, an important aspect of any web-related application, especially market-oriented ones, is flexibility and platform independence. Users nowadays can access the web using a wide variety of different end devices, ranging from high-end desktop computers to lightweight, low-energy mobile devices. A successful platform must be able to handle this inherent variability

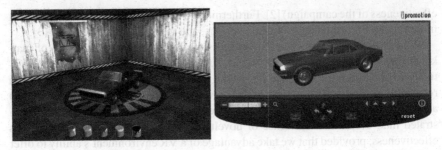

Fig. 1 Examples for the iPromotion large scale scenario interface (*left*), and the Web/Mobile small scale scenario interface (*right*)

in order to satisfy (ideally) all user needs. As indicated by the rise of the mobile advertising income share, a marketing company today expects its campaigns to reach users regardless of end-device type and limitations, and offer them the fullest experience possible in order to maximize reach and, as a result, profits.

In this chapter, we present our progress towards the building of a VR marketing system, which we have named iPromotion. Our system aspires to play the role of a large-scale platform offering virtual reality e-marketing services, supporting high application variability and also content personalization. We want to provide an encompassing framework that can support the VR web advertising process throughout the life-cycle of a promotional interactive Web 3D scene, from its uploading to a server, to editing and preparation for publication, to storage, search and retrieval, and finally to distribution and display to the end users.

To further clarify the extent of the platform's scope and its extreme adaptability, we will proceed to conclude this introduction with two indicative use-case scenarios. These scenarios cover two opposing sides of the "computational load" versus "richness of experience" axis: one is a mobile-based scenario, and the other is a large-scale scenario.

In the first scenario, we have a user handling a typical mobile touch-screen device, such as a mobile phone. A small banner is shown on the phone's display, containing a 3D model of a product. The banner allows the user to manipulate the model and interact with it as intended by the model designer, using an interface appropriate for the input device (Fig. 1). The experience is real-time and appears seamless, without exposing the user to lag or high loading times -on the other hand, the small size of the display screen and the small number of input options offered by the device make for a relatively simple experience, focused on high responsiveness and intuitive control.

In the second, large-scale scenario, a user -or group of users- apply a content-based search algorithm, and browse a large 3D scene database to select a VR world that matches their desires. They then proceed to display it on an interactive touch table, or a wall projection. The 3D world allows for a large number of complex interactions, controlled by a wide range of input devices, from the typical mouse-and-keyboard setup, to motion detection cameras. The large resolution of the display device and the high interaction capabilities offered by the input devices mean that focus is placed on immersiveness and richness of experience, while the presence of significant local computational resources can be assumed.

From these two scenarios, it can be seen that the iPromotion platform is expected to handle a broad range of different settings, varying in computational resources, user expectations and hardware specifications. The platform architecture is designed with a three major factors in mind: extreme device and environment adaptability, scalability, and content-awareness. The first is achieved through a combination of remote model storage and scene rendering, MPEG-21 multimedia adaptation, and a service-oriented platform-independent philosophy. The second is achieved through cloud-based storage and cloud-based computation, while for the third we adapt the existing MPEG-7 multimedia content description technology, in order to index semantic and content metadata for Virtual Reality scenes. This allows us to give both end-users and content providers the power to personalize their experience according to their particular profile, preferences, and requirements.

By incorporating scalable cloud-based storage, content-based description and delivery, multimedia adaptability and device independence, the framework can serve to coordinate a constant process of personalizing and distributing VR advertisements on user's mobile and fixed devices. This can be combined with the collection of user behavioral and preference data on a massive scale, in order to build personalization profiles, upon which the appropriate content can be provided to the corresponding users.

The rest of the chapter is organized as follows: Sect. 2 explores previous research in the fields we are interested in Sect. 3 presents the overall platform architecture. Section 4 offers an in-depth description of the most important platform components and the technologies we have developed. Finally, Sect. 5 explains the current state of our project, and gives our future directions.

2 Background and Previous Work

iPromotion is a large-scale effort, integrating the fruits of multiple research directions in one, broad-scope platform. As a result, we are simultaneously focusing on multiple research fields in order to develop the necessary technologies for our aims. In the context of the work presented in this chapter, the background upon which our research is based concerns in three distinct areas: remote rendering, content-based scene indexing and cloud storage.

2.1 Remote Rendering

An important aspect of the iPromotion platform is its ability to function over radically different client devices. We expect the platform to be able to distribute content to a wide range of systems, ranging from mobile devices, to laptop and desktop PCs, to high resolution touch tables. The first decision we have taken in order to achieve this, is to base the service client entirely on HTML5 technologies.

Currently, there exist a number of Web3D technologies being developed (such as X3DOM and XML3D, but also the directly accessible WebGL framework), which aim to allow any of today's popular browsers to display VR scenes without the need for add-ons or other extra software. As a result, the technological environment is becoming ripe for a platform that will allow us to integrate our VR ads in web pages in a non-obtrusive way, where the inherently multi-platform nature of HTML can solve the majority of today's compatibility issues.

In the case of high-end devices, such as PCs, this approach is often adequate for our needs: such devices usually possess sufficient computational power to display a VR scene written in, say, X3DOM, within a web page, innately and seamlessly. The case of mobile devices, however, raises multiple considerations. Mobile devices carry a number of limitations [7] with respect to computational power, physical display size, power supply and input modalities. It is often impossible to render a 3D scene into a sequence of images on a mobile device, while on the other hand, as we explained in the case of the mobile-based scenario, the quality of any image destined for a mobile device screen does not have to be particularly high, making them relatively easily transferable via network. These characteristics have given rise to the research field of remote rendering.

The concept of remote rendering is relatively simple: the computational cost of rendering is deferred to a remote server, which produces the visual output of the rendering process, and transfers it to the client device. Of course, when the user desires to display a single, non-interactive view of a scene, or a predetermined animation, the solution is simple: a pre-rendered version can be stored and displayed at any time. It is when real-time interactivity is required that remote rendering becomes a valuable option. In this scenario, the user interacts with the scene on the client device, which then sends the interaction commands to the remote server. The remote server adjusts the scene accordingly, re-renders it, and forwards the new output to the end-user. This process, however, in order to work in real-time, has to occur multiple times per second without significant delays, so as to provide a flowing experience. This can, on occasion, raise issues of network delays and remote server load.

Multiple approaches for real-time remote rendering for mobile devices have been presented in the past. In one of the oldest cases [8], a system is presented that performs remote rendering and sends streams of images to the client for display. In order to reduce bandwidth usage and increase display speed, image resolution is halved during interactions, and returns to its original value when static. More recent attempts [9, 10] use video streaming instead, which, when utilizing certain formats (such as h.264), can significantly reduce the data size to be transmitted, and give the end user a smoother result. An alternative [11] is to fuse the two approaches, by alternating between images and video.

Besides reducing the computational cost, another major consideration is to keep network traffic to a minimum, and certainly within the available limits. A number of systems have been proposed in the past [10, 12, 13] which adjust the data flow in accordance with the network latency, to provide the best possible user experiences.

However, simply adjusting the frame resolution and rate are not the only ways to deal with bandwidth limitations. In essence, we always have to deal with a tradeoff between the limited bandwidth and the limited computational resources of the client device. Optimally, we would like to take advantage of both resources to the maximum of their capabilities, and a way to achieve that is to leave some parts of the rendering computation to the client device—as much as it can handle—and remotely render the rest of the output.

A set of approaches have been proposed to that effect. In [14] certain line primitives are transferred via the network, 2D vectors are calculated based on them, and are finally sent to the client for rendering. In Sect. 3 we present our own approach, which is also based on an adaptive framework, in order to take maximum advantage of any local resources, while still relying on remote rendering.

2.2 Content-Based Scene Indexing

An increasingly common problem with large multimedia databases is that of indexing by content: search within multimedia databases is still today in most cases text-based, and performed on the multimedia file names or, commonly, on manual metadata annotations. The descriptive text in these cases may or may not coincide semantically with the actual content of the multimedia files, depending on the quality of the manual annotation effort. It would be desirable if multimedia files could be automatically indexed according to their actual content. For example, images could be sorted according to their content's color distribution, or even high-level semantic concepts (e.g. "Airplane", "Crowd", "Fight"), which will be detected automatically by the system and offered to the user for search.

During the last decades, a significant research effort has been dedicated towards content-based (or even semantic) multimedia indexing and retrieval [15–17]. Significant progress has been made towards intelligently organizing image, video and audio collections according to the actual multimedia content, to enable intuitive handling of the database by the users. However, very little effort has been devoted towards achieving the same results on 3D graphics scenes.

This can partly be attributed to the complexity inherent in this multimedia type: 3D scenes contain a wealth of information, in the form of colors, shapes, textures, overall structure, animations, lighting, sound, and on top of that, a list of permitted user interactions and their programmed impact on the scene. Any of these modalities can be the focus of a user query, and an automatic description scheme would have to incorporate all this information, and preferably in the form of meaningful descriptions that can be understood in human language terms.

In the past, a significant amount of research has been devoted specifically to describing the shape characteristics of an isolated 3D model. Part 3 (Visual Descriptors) of the MPEG-7 standard [18], whose introduction has been a significant landmark in the field of multimedia content description, essentially contains two visual descriptors for 3D data, named *3D Shape Spectrum* and *Multiple View*. The former is aimed at expressing a single 3D model's shape. The latter is based on taking multiple 2D rendered snapshots of one (again, isolated) 3D model and describing each resulting image using simple 2D features. This makes it essentially a 2D descriptor of limited relevance to our aims. Besides the descriptors standardized in MPEG-7, a large amount of further research has been devoted to 3D shape description [19, 20], and the field remains active today.

However, as we mentioned above, a 3D scene is far more than the shape of a single model. Certain attempts have been made in the past to describe other aspects of 3D scenes, the most notable among them being 3D SEmantic Annotation Model (3DSEAM) [21] and the Multimedia Interaction Model [22]. The latter is based on a framework which the authors call the Interaction Interface, which aims to provide a unified framework for the description of all available interactions in a 3D scene, and a common dictionary of interaction parameters. 3DSEAM, on the other hand, extends the MPEG-7 standard to describe 3D scene structure and facilitate model reuse.

3DSEAM is designed for scenes written in X3D, a language which inherently encourages heavy content reuse. Since, in X3D, it is common to draw resources from multiple different files (local or online), normally it would take multiple MPEG-7 *MediaLocator* descriptors to reference all the locations of the content resources. With 3DSEAM, a standardized structure description is provided, which allows for controlled referencing of the various resources available. Furthermore, the model incorporates semantic information on the scene, which further allows content-based indexing and retrieval based on object structure and location.

Recently, a framework for the combination of these methods has been proposed [23]. The resulting integrated framework provides an MPEG-7 compliant description scheme for scene structure and object interaction, However, even in this case, a large part of 3D scene content (such as color, motion or lighting) is still left out.

Other approaches at 3D scene description include a method for describing VRML 3D scenes by storing semantic graphs through MPEG-7 [24], a system for semantic annotation and search in 3D scenes [25], which however puts more focus on textual rather than content data, and is designed specifically for 3D architectural reconstructions, and an approach for the automatic extraction of MPEG-7 data from MPEG-4 3D scenes [26] which are syntactically very similar to X3D scenes. While the latter approach is very similar to our aims, it is limited to a small number of scene characteristics, most of which are textual in nature. What we aim for is an integrated description scheme for scene content, including color, shape, animation and structure.

Finally, a different approach worth mentioning is that followed by [27, 28], in which semantic web technologies are used for the description of X3D scenes using an RDF schema ontology. However, these methods basically aim at drawing from already existing textual metadata within X3D scenes, and not describing the actual raw scene content. Furthermore, while RDF Schema-based ontologies are indeed used widely in web contexts, we consider the MPEG-7 description approach more relevant to low-level multimedia content. While MPEG-7 has been criticized in the past for ambiguity in semantic identification, and lack of interoperability with other ontology frameworks [29], MPEG-7 in its original form remains the definitive standard for low-level multimedia content description.

2.3 Cloud Storage

The main focus—and at the same time challenge—of Information and Communication Technology (ICT) is to provide and serve, continually, simply, and without restrictions, services to the users. Cloud computing is a computing paradigm for providing these qualities, whose popularity during the recent years is constantly rising. Characteristic examples—inspired from recent successful startups—include Dropbox, which makes use of a cloud service, Amazon's Simple Storage Service (S3), Aryaka, which specializes on solving application and network performance issues faced by the distributed enterprise, NetFlix which streams video on demand and also stores the data in Amazon Storage Service, OpenStack, offering scalable cloud operating system, and OpenNebula which provides OpenStack-based applications that let businesses deploy and manage private clouds. These successful examples prove that the SPI model [30], Software (SaaS), Platform (PaaS) and Infrastructure (IaaS) provide all the appropriate tools to move an application to the new wave of internet application philosophy.

Following this philosophy, more and more services are reengineered to be cloud-aware. The root of that trend is the simplicity of billing, monitoring and scaling such services. Although it is also possible to have the same features in the traditional services, the advantage of writing/using cloud aware services is that the above features are supported internally by the architecture, which means that is not necessary to implement each one from scratch. Such an approach also allows us to easily handle and use those features without the necessity to understand in detail the specifications of the API that exist in the legacy services.

In this sense, a cloud multimedia framework should address several challenges in order to provide efficient searches and real time responses between the various stake-holders involved. In the data storage phase, a cloud multimedia storage framework should be capable to analyze and consider the relations of the data (MPEG7, X3D and any additional auxiliary files) and also store them in a manner that allows easy retrieval and grouping. Current cloud based data storage algorithms handle files based on usage or using business criteria specified by the owner (Geo-location, Security, and Ownership) [31]. Also there are few cases that metrics from Map-Reduce are considered in order to balance the I/O usage and minimize the network utilization [32]. In the search phase, a well-behaved cloud framework should handle multimedia metatags and content attributes (retrieved by the XML files) for selection. Benchmarks prove that the most time-consuming task is the parsing of the XML files [33]. That step is very important when using XML documents because the only way to extract information from the document and retrieve the values of the attributes is through parsing. There exist a few studies run on key-value data stores, which overcome that issue, but the migration of a legacy dataset to the new

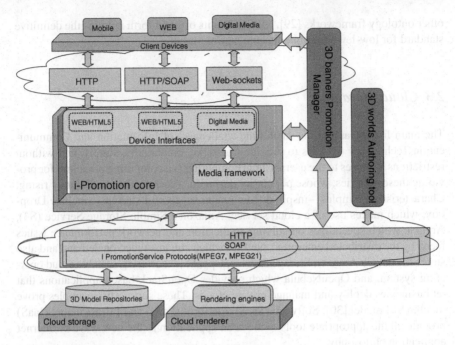

Fig. 2 Overview of the iPromotion platform architecture

philosophy of a key-value storage is a challenging task [34]. To overcome that limitation, in Sect. 4.3 we propose an implementation of appropriate functions using the Map/Reduce technique that could bridge the gap between the two worlds.

3 Platform Architecture

As we described in Sect. 1, we aim at a scalable, multi-platform system, which can store extremely large 3D scene databases and distribute content to end-users in an efficient, intuitive and content-aware manner. To this end, the iPromotion platform adopts a web service-oriented, distributed architecture (Fig. 2) comprising of multiple independent components that communicate with each other and the user. The modularity of the system offers many development and support advantages, but also allows for more efficient distribution of the storage and computational cost, which helps manage the potentially high load of the system.

To allow smooth integration with a range of different devices, we have based the *User Interface* on HTML5 technologies. All client-side applications take the form of HTML pages to account for the client device variability, and different pages can be provided depending on the end device type (e.g. PC or mobile). The 3D scenes are displayed inside a Canvas element within the HTML web page, which interacts with a Web service via SOAP messaging, through Asynchronous Javascript And XML

(AJAX). When a user receives an advertisement, or intentionally opens a 3D world in any device, an HTML page is opened, which begins communicating with the system core via SOAP messages. Depending on the computational capabilities of the end device, the Canvas element can contain a sequence of images, originating from our remote renderer, or a 3D scene rendered locally using X3DOM code.

The HTML5 Canvas element is in this sense an invaluable tool, as it can serve both as a container for the remotely rendered images, or for the X3DOM code running locally. Furthermore, a Canvas can detect user actions (e.g. mouse moves and clicks) and can thus at the same time serve as the user input sensor. The user interacts using a device-appropriate input device (which can be a mouse, touchscreen, or a more specialized tool) through the Canvas, which detects user actions, and displays the renderer output. Using AJAX, the client transmits the user interactions and receives the resulting response commands—or images—which affect the content to be displayed back on the Canvas.

The subsystem responsible for communicating with the client and delivering the final content to it is called *Device Interfaces*. It is responsible for handling traffic from and to the system, and also coordinating any possible format conversions prior to final delivery to the user. When the images or videos from the remote renderer arrive at the Device Interfaces subsystem to be forwarded to the user, their format and encoding parameters might be judged inappropriate for the end device. In this case, when content conversion or adaptation is required, the *Media Framework* receives the files and any necessary instructions, and takes care of the operation. This aspect of the system is particularly important, as it ensures not only format compatibility with respect to the user's device, but also performs multimedia adaptation according to any network restrictions or client-side display and computational limitations, for example in image resolution. We thus ensure that, regardless of the original source material, the output finally delivered to the user is adapted to the particularities of their circumstances.

The Device Interfaces along with the Media Framework form the core of the iPromotion system. The system core manages the flow of information from the storage and rendering units to the user interface. It receives the request for a new session (i.e. the loading of a new scene) from the User Interface, as well as the user interactions with the scene, and forwards them to the interface of the rendering system. In the opposite direction, it receives the rendered images, adapts them accordingly, and forwards them to the User Interface.

The *Rendering Cloud* is based on a modified version of the Xj3D [35] player for X3D scenes. We have repurposed certain parts of the application, so as to override its innate menus and clickable interface with a SOAP message service, which receives all operations externally, over HTTP. The system receives the input commands, modifies the scene accordingly, and sends back a reply which usually contains a confirmation on the operation. Furthermore, the renderer can take the currently rendered frame and send it, over HTTP, converted in text format using Base64 encoding. Using a video format is also an option, leading to smoother motions and better compression.

In designing the platform architecture, besides the actual modules to be implemented, another important aspect we have taken into account is standards compliance.

We have already mentioned the MPEG-7 standard, which we use to describe the content of 3D scenes and perform content-based search in the cloud. We have also incorporated rights management for the repository, to allow multiple content providers to retain their work there, without fear of interference or infringement. Rights management is achieved through the MPEG-21 standard. MPEG-21 is also used for multimedia adaptation, so as to ensure that the content that reaches the end-users is suitable for their device specifications. Finally, MPEG-4 is used to merge this information into unified 3D multimedia objects, incorporating the rights information with the actual scene description. More details on the role these standards play in our platform are given in Sect. 4.1 for MPEG-7 and Sect. 4.2 for MPEG-4 and MPEG-21. Another significant choice we had to make was the standard that would be used for the 3D scenes. We concluded that the X3D standard [36], an XML-based language for forming complex interactive 3D scenes, is the most appropriate for our needs, for a number of reasons:

• X3D and MPEG-4 practically use the same set of commands for forming 3D scenes so interoperability is easy to achieve.
• X3D is a standard designed for web 3D graphics, and as such, is highly appropriate for our needs.
• Finally, X3DOM [37] is a recently developed open-source framework that allows for the direct execution of X3D code within HTML. This means that almost any scene written in X3D can be directly displayed in HTML environments, if the local computational resources allow it. This flexibility makes X3D ideal for our purposes.

Data storage takes place in a STorage-as-a-Service (STaaS) cloud platform. The structure of the 3D scene repositories, which is based on heterogeneous file sizes and also standard relationships between them (MPEG-4, MPEG-7 and X3D files) made clear that there is a gap in current cloud data store systems that cannot be filled with off-the-shelf solutions. The custom framework that we have built and propose to accommodate our needs is based on four layers. The first layer is the storage system, the HDFS by Hadoop, which is designed to provide efficient, reliable access to data. The second layer handles the iPromotion data in a more efficient way, by taking into consideration their uniqueness, as HDFS is not geared to use small files. This is achieved with the help of the object database MongoDB, which receives the data and transforms them into buckets suitable for HDFS. The third layer deals with the assigning problem, where an efficient algorithm [38] was introduced for better utilizing the tasks into the cloud storage system. Finally the third layer is responsible for providing an interface between the platform and all the services (search attribute, calculate values, add/remove 3D scenes, stream data).

Finally, the system also includes two semi-independent units: The first one, called the *Promotion Manager*, is responsible for automatically choosing the appropriate content to display when a banner space is offered for advertising. The choice of which stored scene to display within a page depends on the page content, and any available information we have on the user. The scene choice can be enhanced by the MPEG-7 descriptions we have, so as to efficiently search the entire repository, in order to choose a scene that matches the desired profile. The second, called the *3D Worlds Authoring Tool*, enables content providers to compose scenes using existing models and textures found in the repository. The Authoring Tool is another modified version of the Xj3D player, which offers an intuitive drag-and-drop interface for merging X3D scenes while modifying their size, position and orientation. It also offers powerful content-based search capabilities based on our MPEG-7 descriptors, to allow scene authors to find the desired content within the repository. This encourages content reuse, provided the appropriate access and use rights are available. Both units need to access and search the database, and in this sense, our effort in extending the MPEG-7 for 3D scenes pays off, as it allows us to access the database in a targeted way and retrieve relevant scenes with ease. This can refer to the retrieval of scenes based on high-level concepts (e.g. "a dog"), drawn from Metadata descriptors, or content-based characteristics (e.g. a predominantly blue scene without any animations) drawn from the various content-related MPEG-7 descriptors.

4 Detailed Description of Platform Components

4.1 MPEG-7 Extension for 3D Scenes

As we mentioned in Sect. 3, we have developed a multimedia content description interface for 3D scenes, in order to organize our database and allow intelligent multimedia management through content-based and semantic search and retrieval. This interface is an extension of the MPEG-7 standard, and comprises of a number of repurposed or modified versions of MPEG-7 visual descriptors, with the addition of a number of novel ones. The full extent of our proposed standard is described in [39]. In this section, we will give an overview of the aspects that are relevant to the work presented here.

Our aim was to build a description framework that can cover all significant aspects of a 3D scene. These include geometry, scene structure, texture, lighting, viewpoints, animation, interactivity and metadata. Furthermore, we have extended the *DescriptorMetadata* descriptor from MPEG-7 Part 5, to contain information concerning the particular X3D profile used by the scene, and any scripting code (in Java or ECMAScript) that may be contained in the scene. X3D profiles define subsets of the full list of X3D commands. X3D files that are based on the simplest profiles can run on lightweight players that offer minimal functionalities, while the higher-level profiles require more powerful players that offer the full extent of X3D commands. Similarly, the presence of scripting code in an X3D file requires on the part of the player the ability to compile and execute the corresponding language, which allows for more complex scenes, but increases requirements. Information on either can help us retrieve scenes that match our requirements, not only in terms of content but also complexity.

Overall, we have extended both Part 3 (Visual) and Part 5 (Descriptor Schemes) of MPEG-7. As can be seen in Table 1, our extended MPEG-7 descriptors cover the

Table 1 A list of the MPEG-7 descriptors used in our extended standard

	Name	Usage
MPEG-7 Part 3-Visual	BoundingBox3D	Specifies the position and size of a complex 3D object in a scene, by providing the volumetric coordinates of the group of shape nodes that composite the complex model
	Geometry3D	Describes the types of primitive or complex geometries contained in the X3D scene along with curvature details
	Shape3D	Describes the shape characteristics, in terms of curvature, of a geometry, as a histogram of the values of the Shape Index parameter
	DominantColor	Describes the color of a given geometry or texture image, in (R, G, B) values
	ContentCollection	Describes the scene hierarchy, through the nesting structure of X3D grouping nodes
	Metadata3D	Specifies any additional metadata information provided with an X3D node (such as MetadataFloat, MetadataInteger, MetadataSet)
	Interactivity3D	Describes how an X3D object interacts with other objects in the scene or with the end user, by extracting interaction details from the corresponding X3D ROUTE elements
	MotionTrajectory	Describes the animation characteristics of a 3D moving object within an X3D Scene by determining the type of animation it contains through the X3D Interpolator or Follower nodes bound to that object, along with the motion path keypoint values
	Viewpoint3D	Describes each viewpoint nodes' position, orientation, animation and coordinate properties contained in an X3D file
	Lighting3D	Specifies the type of X3D Light nodes used within an X3D Scene in addition to their distinct characteristics, illumination specifications and whether they are used globally in the scene or scoped to a targeted area of objects in the X3D world
MPEG-7 Part 5-MDS	Profile3D	Describes the X3D profile in use for the file described in order to gain all additional functionality and specify browser and rendering requirements
	Script3D	If the X3D file described contains scripting nodes controlling interactions and animations, this datatype specifies the script class location (internally in the content file or in an external file location) and the scripting language used, in order to define all necessary rendering and control requirements

entire range of possible scene information a 3D scene might contain: geometry position, size, color and shape, texture, scene structure, animation, interactivity, lighting, viewpoints, textual metadata and programming/script complexity. Thus, it is, to our knowledge, the fullest 3D scene description interface to date. Figure 3 shows a sample of MPEG-7 description for an X3D scene, containing the color, size and position description of a single geometric shape, and Fig. 4 shows the 3D Shape Spectrum (Shape3D) descriptor of a complex geometry.

Having a set of content descriptors accompanying each 3D scene allows search and retrieval based either on textual metadata (which should contain meaningful semantic information about the scene content) or any content characteristic of the scene. This means that end-users and content providers can seek scenes based on their color or shape content, but also on their animation characteristics and interactivity potential.

In terms of content management, this is an important aspect for any scalable multimedia repository, since it allows us to overcome the usual limitations of simply relying on manual annotations or—even worse—only on the multimedia file names and directory structures. The cloud repository is aimed at storing a vast amount of 3D scenes, and it would be virtually impossible to access and manage this volume of information simply by serially browsing thumbnails and filenames.

However, designing a theoretical standard that allows us to describe 3D scene content is only a first step towards allowing fast and efficient management of the contents of our repository. It was deemed essential that we also include a method for the automatic extraction of our descriptors from all the scenes in the repository. This means that, whenever a new scene is uploaded in the repository, the content descriptors are automatically and intelligently extracted without any manual intervention on behalf of the scene author. Such an intervention would make the process time-intensive, and could also increase unreliability.

We have thus developed a technique for the automatic extraction of MPEG-7 descriptors from X3D scenes [40]. The principle upon which we base our extraction method is that both X3D and MPEG-7 are XML-based standards. Thus, the descriptor extraction can take the form of a conversion between two XML schemata.

We use eXtensible Stylesheet Language Transformations (XSLT) for the conversion. XSLT is a language that allows us to convert XML files from one schema into another. In our case, we use it to extract the relevant information from the X3D file, and form an MPEG-7 description as per our extended standard. We then proceed to keep both files in the repository, each serving a different purpose: the X3D file is used for display, while the MPEG-7 file is used for description, search and retrieval of the X3D original. Further details on the automatic descriptor extraction algorithm can be found in [40].

```
<Mpeg7 xmlns="urn:mpeg:mpeg7:schema:2004">
  <Description xsi:type="ContentEntityType"
    xmlns:xsi="http://www.w3.org/2001/XMLSchema-instance">
    <DescriptionMetadata>
      <Profile3D>Immersive</Profile3D>
    </DescriptionMetadata>
    <MultimediaContent xsi:type="MultimediaType" id="Demo.x3d">
      <Multimedia>
        <MediaLocator>
          <MediaUri>/webapps/X3DtoMPEG7Tool/Demo.x3d</MediaUri>
        </MediaLocator>
      </Multimedia>
    </MultimediaContent>
  </Description>
  <Description xsi:type="ContentEntityType"
    xmlns:xsi="http://www.w3.org/2001/XMLSchema-instance">
    <MultimediaContent xsi:type="MultimediaCollectionType">
      <StructuredCollection>
        <Collection xsi:type="DescriptorCollectionType" id="Geometries">
          <DescriptorCollection id="Geometry_d1e10" name="Transform_1">
            <Descriptor xsi:type="BoundingBox3DType">
              <BoundingBox3DSize BoxWidth="1"
                BoxHeight="2" BoxDepth="1"/>
              <BoundingBox3DCenter BoxCenterW="0"
                BoxCenterH="0" BoxCenterD="0"/>
            </Descriptor>
            <Descriptor xsi:type="Geometry3DType">
              <Geometry3D ObjectType="Cylinder">
                <DominantColor3D xsi:type="DominantColorType">
                  <SpatialCoherency>0</SpatialCoherency>
                  <Value>
                    <Percentage>1</Percentage>
                    <Index>230 51 230</Index>
                  </Value>
                </DominantColor3D>
              </Geometry3D>
            </Descriptor>
          </DescriptorCollection>
          <DescriptorCollectionRef href="#Transform_1"/>
        </Collection>
        <Relationships/>
      </StructuredCollection>
    </MultimediaContent>
  </Description>
</Mpeg7>
```

Fig. 3 Part of the MPEG-7 description of a simple X3D scene

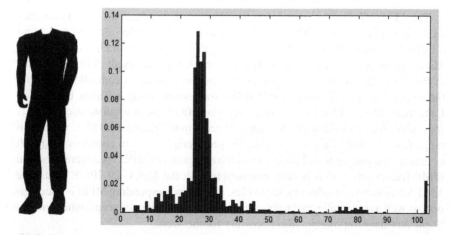

Fig. 4 A humanoid body model and its 3D shape spectrum descriptor (Shape3D), with the bin values expressed through a bar graph (Model from the savage X3D examples archive, https://savage.nps.edu/Savage/)

4.2 MPEG-4 and MPEG-21 Compliance

In order to achieve intelligent and efficient management of the repository, while retaining scalability, we have taken multiple steps to ensure that the X3D scenes contained in the repository are more than an unstructured set of files. This is not limited to the MPEG-7 content descriptions, but also to rights information, as well as integrated texture files, thumbnails and a set of metadata aimed at facilitating adaptation to any end-user requirements. Thus, each file amounts to a complete multimedia object consisting of multiple multimedia resources, in the form of 3D scenes, images and XML metadata. Another important consideration is that the final output to be rendered and sent to the users might not necessarily be the original X3D scenes stored in the cloud. As we have described in Sect. 3 authors are given the option to compose scenes by combining multiple X3D files, using an Authoring Tool. These composite scenes are stored back inside the cloud repository and can be accessed alongside the original X3D scenes. We are thus faced with inherently composite multimedia objects, and our chosen storage standards should reflect this.

The format we have selected in order to incorporate composite multimedia objects based on 3D graphics is the MPEG-4 XMT (eXtensible MPEG-4 Textual format). XMT is an XML-based format described in MPEG-4 part 11: Scene description and application engine [41]. It allows us to describe the structure of multimedia scenes, by incorporating large parts of SMIL, SVG and X3D. Especially with respect to X3D, XMT uses an almost identical set of commands, making the two formats highly compatible. Besides representing simple or composite scenes, however, XMT also allows for the incorporation of more complex multimedia features, such as MPEG-7 descriptions and MPEG-21 rights management, all in the same structure (Fig. 5).

The MPEG-7 description scheme for 3D files, as has been presented, can easily be incorporated in an XMT file as is. Besides that, we are also implementing a rights management system based on the MPEG-21 part 4: Intellectual Property Management and Protection (IPMP) [42]. As content providers all store their scenes in the cloud repository, it is important that access to them is controlled based on the authors' desires. The MPEG-21 IPMP framework, using a Rights Expression Language (REL) and a Rights Data Dictionary (RDD) can regulate access rights for each scene. This allows us to regulate access to the multimedia files, which can range from limited to a single author (for proprietary work) to common use (such as generic backgrounds and widely shared basic models). MPEG-4 contains its own IPMP framework, which is fully compatible with the MPEG-21 IPMP. Thus, the XMT file contains, besides any scene files, also the corresponding MPEG-7 description, as well as the intellectual rights information concerning the scene, while linking to any auxiliary multimedia files necessary for the scene.

In the end, in terms of file storage, the cloud storage repository contains the X3D files, as well as all necessary auxiliary files (e.g. textures) and the corresponding MPEG-4 files linking the X3D scenes with information and supplementary files. It is up to each scene's MPEG-4 structure description to link all the available resources in a complete scene, which can be searched, accessed and rendered (locally or remotely) for display.

A final aspect of MPEG-21 compliance is multimedia adaptation. Part 7: Digital Item Adaptation deals with a structured and standardized framework for communication between a server and a client, in order to adapt in real time—the format and parameters of a multimedia resource. In our platform architecture, the Media Framework module is designed to perform adaptation according to MPEG-21 specifications, so as to achieve maximum flexibility and interoperability by complying with the dominant standard in the area.

4.3 Cloud Storage

A Cloud-based storage system provides appropriate mechanisms to respond to search requests triggered by the application. Our cloud based storage system enhances these mechanisms providing additional features used to orchestrate the scene construction. These mechanisms provide the necessary scene files and retrieve all the auxiliary files referenced by the scene. The output of this request has the form of a package with all required information to reconstruct the scene. In particular, the files that are stored on the cloud include the models for the X3D scenes (.x3d) files, the MPEG-7 scene descriptions, and any auxiliary files necessary for rendering the scenes, such as texture images. Each of these data are filtered and preprocessed by the 3D Model Repository Framework before being saved into the cloud storage. The reason for that is twofold, firstly to provide them with the necessary metatags for indexing and relationship management for structure, and secondly to separate and store them in a manner that is optimized for the size of the files and the usage requirements. Metatags and

```xml
<?xml version="1.0" encoding="UTF-8"?>
<XMT-A xmlns="urn:mpeg:mpeg4:xmta:schema:2002"
 xmlns:xsi="http://www.w3.org/2001/XMLSchema-instance"
 xsi:schemaLocation="urn:mpeg:mpeg4:xmta:schema:2002
 xmt-a.xsd">
<Header>
<InitialObjectDescriptor objectDescriptorID="od1" binaryID="1" >
   <Profiles audioProfileLevelIndication="255"
    visualProfileLevelIndication="254"
    sceneProfileLevelIndication="1" graphicsProfileLevelIndication="1"
    ODProfileLevelIndication="1" />
   <Descr> <esDescr> <ES_Descriptor ES_ID="es1" binaryID="1" >
             <decConfigDescr>
                 <DecoderConfigDescriptor objectTypeIndication="0"
                   streamType="3" >
                     <decSpecificInfo> <BIFSConfig>
                         <commandStream pixelMetric="true">
                             <size pixelWidth="300" pixelHeight="200"/>
                         </commandStream>
                     </BIFSConfig> </decSpecificInfo>
                 </DecoderConfigDescriptor>
             </decConfigDescr>
             <slConfigDescr>
                 <SLConfigDescriptor/>
             </slConfigDescr>
       </ES_Descriptor> </esDescr> </Descr>
</InitialObjectDescriptor>
</Header>
<Body> <Replace>
       <Group> <children>
             <Transform> <children>
                 <Shape>
                     <appearance> <Appearance>
                         <material>
                             <Material diffuseColor="0.2 0.9 0.9"/>
                         </material>
                     </Appearance> </appearance>
                     <geometry>
                         <Box/>
                     </geometry>
                 </Shape>
             </children> </Transform>
       </children> </Group>
</Replace> </Body>
</XMT-A>
```

Fig. 5 MPEG-4 XMT-A representation of a simple X3D scene

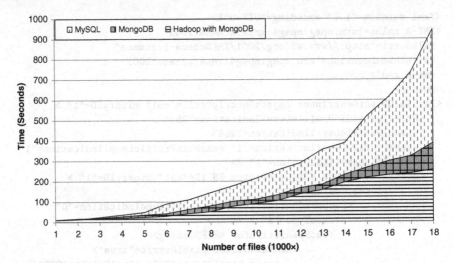

Fig. 6 Comparison of the response time behaviour between MySQL, MongoDB and Hadoop with MongoDB

relationship management are important for the exposed search API that the platform is providing to the higher-level services. It is also crucial in providing compatibility with legacy systems, as it can support extended driver systems, able to use and handle multimedia data stored in traditional ways (files, Relational Databases).

A key part of a responsive and accurate multimedia framework is data handling. Big Data multimedia warehouses like video producers, broadcast providers and 3D model makers update their content quite frequently. Also they enhance and update new types of data into the storing facilities. This data type variance could not be handled by a traditional file system or by a strict database schema in an optimum way. In order to overcome this issue a solution could be a document-oriented database (like MongoDB or CouchDB) which is does not impose strict schema restrictions and has deep query-ability [43]. iPromotion introduces an extra layer over the file system that brings the files, data and images to conform into a semi-structured data object which is more appropriate for handling by a filesystem. That is a result of data integration that MongoDB is doing during the concatenation of all the appropriate files for the 3D model or multimedia. Moreover, the Hadoop Distributed File System (HDFS), which is one of the most common file system for Big Data warehouses can, in this manner, handle and arrange objects much faster—one example is in the case of fragmented files. The results of our experimental runs evaluating system responsiveness are presented in Fig. 6, in which we compare a traditional structured SQL warehouse system against a NoSQL MongoDB and the solution that iPromotion suggests, resulting from a combination of Hadoop and MongoDB.

This integration of technologies, besides boosting performance, also provides the opportunity to combine different datasets. Pig extractions, Hive jobs, Mahout analysis [44] are only few of the types of data that can be used to puzzle a query originating from different processes or even different platforms. By working with the same middleware (Hadoop), cooperation between different types of data becomes significantly easier. Even in cases where the query is so complex that it cannot be deconstructed, Hadoop can provide a number of solutions, such as Cascading, which allow us to work in a higher level thought stream, or Hive, which gives us the advantage of working in an SQL-like language.

Moving to the data processing part of the system, the Hadoop scheduling algorithm was not judged efficient enough for small tasks with respect to our needs. The problem is located in Hadoop's scheduling algorithm, which was made for tasks that are complicated and involve a lot of data. In our case the data, as well as the queries, are small, so the system could support more threads in the cluster. Using the job-scheduling algorithm of Facebook Corona and taking into advantage the manager that manages the cluster, the resources can be controlled and handled more efficiently. Finally, by uncoupling the job tracker from the Hadoop cluster, smaller tasks can finish sooner.

5 Conclusions, Current Progress, and Further Steps

We have presented our work towards an integrated, scalable, cloud-based platform for the distribution of Virtual Reality advertisements in a wide range of environments and devices. A combination of web 3D technologies and remote rendering is used to bring end-users the best VR experience possible, in accordance to the capabilities of their end device, be it a mobile phone or a desktop PC. A cloud repository provides the necessary scalability in terms of storage potential, while an extension to the MPEG-7 standard allows for intelligent content-based and semantic search, to allow users and providers to manage the database intuitively and efficiently. The platform architecture is built upon a highly modular philosophy of autonomous web services exchanging HTTP messages, which allows not only for highly structured development, but can ensure extensibility and interoperability with other systems.

In a world where mobile computing and communication devices become more and more ubiquitous, and vast amounts of usage data can be used to custom-tailor and personalize user experiences, the creation of a framework for organizing multimedia content—and especially for the emerging Web 3D technologies—its integration with the wide range of client devices currently available, and its adaptation with any user information available is necessary for any commercially applicable large-scale use case.

iPromotion is a three year research project in cooperation with private partners, currently in its second year of development. Most fundamental aspects of the system have already been implemented, including the cloud-based repository, the remote renderer, the extended MPEG-7 framework and its automatic extraction subsystem, as well as the basic versions of the HTML client. Other aspects of the platform are currently under development, such as the authoring tool, the content-based search services, and the MPEG-4 integration. Further aspects of the platform that have been designed but not yet implemented include the video streaming service instead of the current image streaming, interactivity with more complex devices, such as motion cameras, the fusion of local and remote rendering, and advanced functionalities such as sound integration in the VR scenes.

Work on the iPromotion project is still progressing, and we will continue to contribute software, standards and technologies to the research community, towards an adaptable framework that can distribute VR experiences to end-users throughout the internet, over a multitude of end devices, user contexts and environments.

Acknowledgments This work is granted by the European Union and the Hellenic General Secretary of Research and Technology under the "COOPERATION 2009 / 09SYN-72- 956" Framework.

References

1. IAB: Internet advertising revenue report, 2012 full year results, 2013. Last accessed 28 Aug 2013
2. Li, H., Daugherty, T., Biocca, F.: Impact of 3-D advertising on product knowledge, brand attitude, and purchase intention: the mediating role of presence. J. Advertising **31**(3), 43–57 (2002)
3. Mazursky, D., Vinitzky, G.: Modifying consumer search processes in enhanced on-line interfaces. J. Bus. Res. **58**(10), 1299–1309 (2005)
4. Suh, K.S., Lee, Y.E.: The effects of virtual reality on consumer learning: an empirical investigation. MIS Q. **29**(4), 673–697 (2005)
5. Baile, S., Debbabi, S., Daassi, M.: Effect of online 3D advertising on consumer responses: the mediating role of telepresence. J. Mark. Manage. **2**, 1472–1476 (2010)
6. Steven, M.: Edwards and Gan gadharbatla Harshavardhan. The novelty of 3D product presentations online. J. Interact. Advertising **2**(1), 10–18 (2001)
7. Çapin, T., Pulli, K., Akenine-Moller, T.: The state of the art in mobile graphics research. IEEE Comput. Graph. Appl. **28**(4), 74–84 (2008)
8. Lamberti, F., Zunino, C., Sanna, A., Fiume, A., Maniezzo, M.: An accelerated remote graphics architecture for PDAS. In: Bouville, C. (ed.) Proceeding of the Eighth International Conference on 3D Web Technology, Web3D 2003, 9–12 Mar 2003, pp. 55–62. Saint Malo, France, ACM (2003)
9. Lamberti, F., Sanna, A.: A streaming-based solution for remote visualization of 3D graphics on mobile devices. IEEE Trans. Vis. Comput. Graph. **13**(2), 247–260 (2007)
10. Tizon, N., Moreno, C., Cernea, M., Preda, M.: MPEG-4-based adaptive remote rendering for video games. In: Royan, J., Preda, M., Boubekeur, T., Polys, N.F. (ed.) 3D Technologies for the World Wide Web, Proceedings of the 16th International Conference on Web 3D Technology, Web3D 2011, Paris, France, 20–22 June 2011, pp. 45–50. ACM (2011)
11. Simoens, P., Praet, P., Vankeirsbilck, B., De Wachter, J., Deboosere, L., De Turck, F., Dhoedt, B., Demeester, P.: Design and implementation of a hybrid remote display protocol to optimize multimedia experience on thin client devices. In: Telecommunication Networks and Applications Conference, pp. 391–396. IEEE (2008)
12. Jurgelionis, A., Fechteler, P., Eisert, P., Bellotti, F., David, H., Laulajainen, J., Carmichael, R., Poulopoulos, V., Laikari, A., Perälä, P.H.J., De Gloria, A., Bouras, C.: Platform for distributed 3D gaming. Int. J. Comput. Games Technol. (2009)
13. Nijdam, N.A., Han, S., Kevelham, B., Magnenat-Thalmann, N.: A context-aware adaptive rendering system for user-centric pervasive computing environments. In: 15th IEEE Mediterranean Electrotechnical Conference, pp. 790–795. IEEE (2010)

14. Diepstraten, J., Görke, M., Ertl, T.: Remote line rendering for mobile devices. In: Proceedings of International Computer Graphics, pp. 454–461. IEEE Computer Society (2004)
15. Hanbury, A.: A survey of methods for image annotation. J. Vis. Lang. Comput. **19**(5), 617–627 (2008)
16. Dasiopoulou, S., Giannakidou, E., Litos, G., Malasioti, P., Kompatsiaris, Y.: A survey of semantic image and video annotation tools. In: Georgios Paliouras, Constantine D. Spyropoulos, and George Tsatsaronis, editors, Knowledge-Driven Multimedia Information Extraction and Ontology Evolution—Bridging the Semantic Gap, volume 6050 of Lecture Notes in Computer Science, pp. 196–239. Springer (2011)
17. Hu, W., Xie, N., Li, L., Zeng, X., Maybank, S.: A survey on visual content-based video indexing and retrieval. IEEE Trans. Syst. Man Cybern. Part C **41**(6), 797–819 (2011)
18. Salembier, P., Sikora, T., Manjunath, B.S.: Introduction to MPEG-7: Multimedia Content Description Interface. Wiley, New York (2002)
19. Iyer, N., Jayanti, S., Lou, K., Kalyanaraman, Y., Ramani, K.: Three-dimensional shape searching: state-of-the-art review and future trends. Comput. Aided Des. **37**(5), 509–530 (2005)
20. Tangelder, J.W.H., Veltkamp, R.C.: A survey of content based 3D shape retrieval methods. Multimedia Tools Appl. **39**(3), 441–471 (2008)
21. Bilasco, I.M., Gensel, J., Villanova-Oliver, M., Martin, H.: An MPEG-7 framework enhancing the reuse of 3D models. In: Gracanin, D. (ed.) Proceeding of the Eleventh International Conference on 3D Web Technology, Web3D 2006, Columbia, Maryland, USA, 18–21 Apr 2006, pp. 65–74. ACM (2006)
22. Chmielewski. J.: Building extensible 3D interaction metadata with interaction interface concept. In: Proceedings of the 1st International Conference on Knowledge Discovery and Information Retrieval KDIR, pp. 160–167. (2009)
23. Chmielewski, J.: Describing interactivity of 3D content. In: Cellary, W., Walczak, K. (eds.) Interactive 3D Multimedia Content, pp. 195–221. Springer, London (2012)
24. Halabala, p.: Semantic metadata creation. In: Proceedings of 7th Central European Seminar on Computer Graphics CESCG, pp. 15–25. (2003)
25. Koutsoudis, A., Stavroglou, K., Pavlidis, G., Chamzas, C.: 3DSSE—a 3D scene search engine: exploring 3D scenes using keywords. J. Cult. Heritage **13**(2), 187–194 (2012)
26. Le Bonhomme, B., Preda, M., Preteux, F., Extended MPEG-7 schema for 3d on demand. In: Workshop on Multimedia Semantics—The Role of Metadata, pp. 218–229. (2007)
27. Pittarello F., De Faveri, A.: Semantic description of 3D environments: a proposal based on web standards. In: Gracanin, D. (ed.) Proceeding of the Eleventh International Conference on 3D Web Technology, Web3D 2006, Columbia, Maryland, USA, 18–21 Apr 2006, pp. 85–95. ACM (2006)
28. Flotyński, J., Walczak, K.: Describing semantics of 3D Web content with RDFa. In: The First International Conference on Building and Exploring Web Based Environments, WEB 2013, pp. 63–68. IARIA (2013)
29. Dasiopoulou, S., Tzouvaras, V., Kompatsiaris, I., Strintzis, M.G.: Enquiring MPEG-7 based multimedia ontologies. Multimedia Tools Appl. **46**(2–3), 331–370 (2010)
30. Mell, P., Grance, T.: The NIST definition of cloud computing: Recommendations of the national institute of standards and technology, 2009. Version 15. Available online at http://csrc.nist.gov/groups/SNS/cloud-computing. Accessed 28 Aug 2013
31. Alhamad, M., Dillon, T.S., Chang, E.: SLA-based trust model for cloud computing. In: Enokido, T., Xhafa, F., Barolli, L., Takizawa, M., Uehara, M., Durresi, A. (ed.) The 13th International Conference on Network-Based Information Systems, NBiS 2010, Takayama, Gifu, Japan, 14–16 Sep 2010, pp. 321–324. IEEE Computer Society (2010)
32. Hu, W., Tian, C., Liu, X., Qi, H., Zha, L., Liao, H., Zhang, Y., Zhang, J.: Multiple-job optimization in mapreduce for heterogeneous workloads. In: Proceedings of Sixth International Conference on Semantics Knowledge and Grid (SKG), 2010, pp. 135–140 (2010)
33. Lam, T., Ding, J.J., Liu, J.C.: XML document parsing: operational and performance characteristics. IEEE Comput. **41**(9), 30–37 (2008)

34. Lombardo, S., Di Nitto, E., Ardagna, D.: Issues in handling complex data structures with noSQL databases. In: Proceedings of 14th International Symposium on Symbolic and Numeric Algorithms for Scientific Computing (SYNASC), 2012, pp. 443–448. IEEE Computer Society (2012)
35. The Xj3D Project website. http://www.xj3d.org/. Accessed 28 Aug 2013
36. Brutzman, D., Daly, L.: X3D: Extensible 3D Graphics for Web Authors. Elsevier Science, Amsterdam (2010). (The Morgan Kaufmann Series in Computer Graphics)
37. Behr, J., Eschler, P., Jung, Y., Zöllner, M.: X3DOM: a DOM-based HTML5/X3D integration model. In: Spencer, S.N., Fellner, D.W., Behr, J., Walczak, K. (ed.) Proceeding of the 14th International Conference on 3D Web Technology, Web3D 2009, Darmstadt, Germany, 16–17 June 2009, pp. 127–135. ACM (2009)
38. Burkard, R.E.: Assignment problems: Recent solution methods and applications. In: Prekopa, A., Szelezsaan, J., Strazicky, B. (eds.) System Modelling and Optimization, volume 84 of Lecture Notes in Control and Information Sciences, vol. 84, pp. 153–169. Springer, Berlin Heidelberg (1986)
39. Spala, P., Malamos, A.G., Doulamis, A., Mamakis, G.: Extending MPEG-7 for efficient annotation of complex web 3D scenes. Multimedia Tools Appl. 59(2), 463–504 (2012)
40. Zampoglou, M., Spala, P., Kontakis, K., Malamos, A.G., Ware, J.A.: Direct mapping of X3D scenes to MPEG-7 descriptions. In: Posada, J., Brutzman, D.P., Gracanin, D., Yoo, B., Oyarzun, D. (ed.) The 18th International Conference on Web3D Technology, Web3D '13, San Sebastian, Spain, 20–22 Jun 2013, pp. 57–65. ACM (2013)
41. Pereira, F., Ebrahimi, T.: The MPEG-4 Book. Prentice Hall Professional, New Jersey (2002)
42. Burnett, I.S., Pereira, F., van de Walle, R., Koenen, R.: The MPEG-21 Book. Wiley, Chichester (2006)
43. Hecht, R., Jablonski, S.: Nosql evaluation: a use case oriented survey. In: Proceedings of the 2011 International Conference on Cloud and Service Computing, CSC '11, pp. 336–341, Washington, DC, USA, 2011. IEEE Computer Society (2011)
44. Official webapage of Hadoop and related projects. http://hadoop.apache.org/

Printed in the United States
By Bookmasters